Josef Ballmann
Richard Eppler
Wolfgang Hackbusch (Eds.)

Panel Methods in Fluid Mechanics with Emphasis on Aerodynamics

Notes on Numerical Fluid Mechanics
Volume 21

Volume 1 Boundary Algorithms for Multidimensional Inviscid Hyperbolic Flows (K. Förster, Ed.)

Volume 2 Proceedings of the Third GAMM-Conference on Numerical Methods in Fluid Mechanics (E. H. Hirschel, Ed.) (out of print)

Volume 3 Numerical Methods for the Computation of Inviscid Transonic Flows with Shock Waves (A. Rizzi / H. Viviand, Eds.)

Volume 4 Shear Flow in Surface-Oriented Coordinates (E. H. Hirschel / W. Kordulla)

Volume 5 Proceedings of the Fourth GAMM-Conference on Numerical Methods in Fluid Mechanics (H. Viviand, Ed.) (out of print)

Volume 6 Numerical Methods in Laminar Flame Propagation (N. Peters / J. Warnatz, Eds.)

Volume 7 Proceedings of the Fifth GAMM-Conference on Numerical Methods in Fluid Mechanics (M. Pandolfi / R. Piva, Eds.)

Volume 8 Vectorization of Computer Programs with Applications to Computational Fluid Dynamics (W. Gentzsch)

Volume 9 Analysis of Laminar Flow over a Backward Facing Step (Ken Morgan / J. Periaux / F. Thomasset, Eds.)

Volume 10 Efficient Solutions of Elliptic Systems (W. Hackbusch, Ed.)

Volume 11 Advances in Multi-Grid Methods (D. Braess / W. Hackbusch / U. Trottenberg, Eds.)

Volume 12 The Efficient Use of Vector Computers with Emphasis on Computational Fluid Dynamics (W. Schönauer / W. Gentzsch, Eds.)

Volume 13 Proceedings of the Sixth GAMM-Conference on Numerical Methods in Fluid Mechanics (D. Rues / W. Kordulla, Eds.) (out of print)

Volume 14 Finite Approximations in Fluid Mechanics (E. H. Hirschel, Ed.)

Volume 15 Direct and Large Eddy Simulation of Turbulence (U. Schumann / R. Friedrich, Eds.)

Volume 16 Numerical Techniques in Continuum Mechanics (W. Hackbusch / K. Witsch, Eds.)

Volume 17 Research in Numerical Fluid Dynamics (P. Wesseling, Ed.)

Volume 18 Numerical Simulation of Compressible Navier-Stokes Flows (M. O. Bristeau / R. Glowinski / J. Periaux / H. Viviand, Eds.)

Volume 19 Three-Dimensional Turbulent Boundary Layers – Calculations and Experiments (B. van den Berg / D. A. Humphreys / E. Krause / J. P. F. Lindhout)

Volume 20 Proceedings of the Seventh GAMM-Conference on Numerical Methods in Fluid Mechanics (M. Deville, Ed.) (in preparation)

Volume 21 Panel Methods in Fluid Mechanics with Emphasis on Aerodynamics (J. Ballmann / R. Eppler / W. Hackbusch, Eds.)

Josef Ballmann · Richard Eppler
Wolfgang Hackbusch (Eds.)

Panel Methods in Fluid Mechanics with Emphasis on Aerodynamics

Proceedings of the Third GAMM-Seminar
Kiel, January 16 to 18, 1987

Springer Fachmedien Wiesbaden

CIP-Titelaufnahme der Deutschen Bibliothek

Panel methods in fluid mechanics with emphasis on aerodynamics: Kiel, January 16–18, 1987/
Josef Ballmann ... (ed.). – Braunschweig;
Wiesbaden: Vieweg, 1988
(Proceedings of the ... GAMM seminar; 3)
(Notes on numerical fluid mechanics; Vol. 21)

NE: Ballmann, Josef [Hrsg.]; Gesellschaft für Angewandte Mathematik und Mechanik: Proceedings of the ...; 2. GT

Manuscripts should have well over 100 pages. As they will be reproduced photomechanically they should be typed with utmost care on special stationary which will be supplied on request. In print, the size will be reduced linearly to approximately 75 %. Figures and diagramms should be lettered accordingly so as to produce letters not smaller than 2 mm in print. The same is valid for handwritten formulae. Manuscripts (in English) or proposals should be sent to the general editor Prof. Dr. E. H. Hirschel, Herzog-Heinrich-Weg 6, D-8011 Zorneding.

The addresses of the editors of the series are given on the last page.

Produced by W. Langelüddecke, Braunschweig

ISSN 0179-9614

ISBN 978-3-528-08095-2 ISBN 978-3-663-13997-3 (eBook)
DOI 10.1007/978-3-663-13997-3

Foreword

The GAMM Committee for Efficient Numerical Methods for Partial Differential Equations (GAMM Fachausschuß "Effiziente Numerische Verfahren für Partielle Differentialgleichungen") organizes conferences and seminars on subjects concerning the algorithmic treatment of partial differential equation problems.

The two first seminars "Efficient Solution of Elliptic Systems" (1985) and "Efficient Numerical Methods in Continuum Mechanics" (1986) were followed by a third one, co-organized together with the GAMM Committee for Discretizing Methods in Solid Mechanics and the special research project SFB 25 "Vortex Flows in Aeronautics" at the RWTH Aachen, sponsored by the Deutsche Forschungsgemeinschaft (DFG).

During the last decades Panel Methods and Boundary Element Methods became a very efficient tool to solve numerically the integral equations with surface or boundary singularity distributions in Aerodynamics and in Continuum Mechanics. First designed for elliptic problems with linear partial differential equations, the different variants of the methods now apply to nonlinearities introduced by free vortex sheets or compressible flows and to time-dependent problems as wave propagation and non-stationary flows.

The seminar was attended by 47 scientists from 10 countries. A greater part of the 22 papers presented at the seminar concerned the different approaches for stationary and non-stationary subsonic potential flows around aerodynamic configurations and propellers, including inverse methods for design problems. Other contributions dealt with hybrid methods for subsonic flows with embedded supersonic regions and weak shocks, panel methods for linearized supersonic flows and boundary element methods for three-dimensional viscous flows, for flows through porous media and transient loading by pressure waves.

An increase of the efficiency and accuracy of panel methods was achieved, e.g. by multi-grid methods with panel-clustering, by new error estimates in solving the integral equations and by realizing the applicability of the Fredholm-Radon method for domains with corners.

The editors and organizers of the seminar would like to thank the land Schleswig-Holstein and the DFG for their support.

December 1987

J. Ballmann
R. Eppler
W. Hackbusch

C O N T E N T S

Page

PANEL METHODS IN AERODYNAMICS; SOME HIGHLIGHTS

H.W.M. Hoeijmakers
National Aerospace Laboratory NLR
Anthony Fokkerweg 2, 1059 CM Amsterdam, The Netherlands

SUMMARY

A survey is presented of several aspects of the use and further development of panel methods in aerodynamics. Aspects discussed include possible types of boundary conditions, low versus higher-order formulations, simulation of subsonic and supersonic flow and modeling of wakes as well as of leading-edge vortex separation. Also discussed are computational aspects of panel methods and possible directions for new developments. The latter include the extension of the domain of applicability to compressible flow and the coupling with viscous flow methods, as well as ways to improve the efficiency of the panel method.

INTRODUCTION

The computation of the aerodynamic characteristics of aircraft configurations has been carried out by panel methods since the mid 1960's, following the pioneering work of Hess & Smith [1] and Rubbert & Saaris [2]. But even before the availability of large-scale digital computers work was done on surface singularity methods, notably in Germany by Prager [3] and Martensen [4]. Panel methods are presently the only computational aerodynamic tools that have been developed to an extent that they are routinely used in the aerospace industry for the analysis of the subsonic and supersonic flow about real-life, complex aircraft configurations. Panel methods have also been used for the analysis of the flow about propellers, automobiles, submarines, shiphulls, sails, etc. Panel methods have been used so heavily because of their ability to provide for complex configurations linear potential flow solutions of engineering accuracy at reasonable expense. The latter applies to the computer resources required for running the computer code as well as to the manhour cost involved in preparing the input.

The relatively easy input requirement of panel methods, very important from a user's point of view, is directly related to the circumstance that a discretization is required for the surface of the 3D configuration only. This is considered to be an order of magnitude simpler than the volume discretization of the space around the configuration generally needed for finite-difference and finite-element methods.

It may be noted that in aircraft development projects the application of panel methods is gradually shifting from the final design phase towards the preliminary design phase, primarily due to:

- the increase in computer power, decrease of its costs and improvement of turn-around times,
- modern data handling techniques,
- availability of graphic displays for visualizing geometry and flow solution.

Several panel methods, e.g. Ref. [5], [6], [7], have been developed and are in use in the aerospace industry that are variations on the approach described in Refs. [1] and [2]. Other investigators extended the panel method to linearized supersonic flow, e.g. Refs. [8], [9]. Because most of these "first-generation" panel methods do have some restrictions concerning their geometric and aerodynamic modeling capabilities and require improvement of their computational efficiency several efforts have been undertaken to develop a "second-generation" panel method, e.g. Refs. [10]-[18].

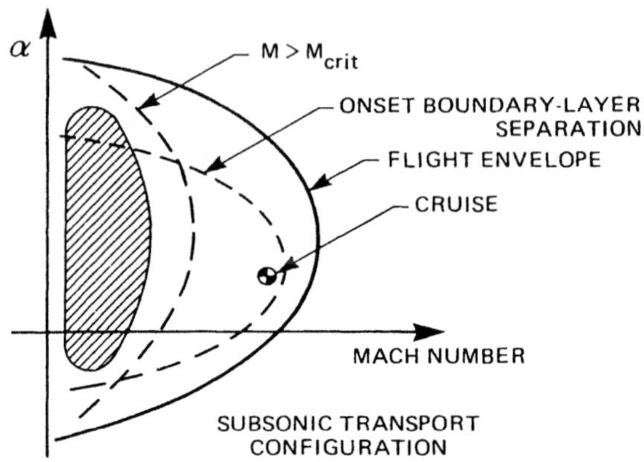

Fig. 1 Domain of applicability of panel method (adapted from [27])

The major drawback of the present panel methods is that their range of applicability is restricted to linear potential flow, i.e. nonlinear compressibility effects are not accounted for and the important case of local regions of supercritical flow and shock waves cannot be treated. The latter occur in the high speed region in a large portion of the flow field and possibly at low speeds (at the larger incidences used in start and landing) in smaller portions of the flow field. To account for such regions of supercritical flow would require the use of a transonic finite-difference code on a spatial grid covering the entire space around the configuration. However, up to the present time the spatial grid generation problem has not been solved satisfactorily for complex configurations as configurations with extended slats and flaps. In addition, the computer resources required for present-day finite difference codes are quite substantial.

A second drawback of current panel methods is that the computational effort, and cost, is proportional to N^2, or even N^3, where N is the number of panels. This implies that the method becomes rapidly impractical for N of the order of 2000-5000, panel numbers required for the resolution required for the coupling with boundary-layer methods. It is also an experience of the practise of applying panel methods that this number of panels is often quite easily "consumed", so that compromises have to be sought regarding resolution and accuracy.

In the present lecture an overview is given of several aspects related to the formulation and use of panel methods and the possibilities for extending the domain of applicability and improving the computational efficiency. The discussion deals primarily with methods for the steady flow about three-dimensional configurations. It is based on the literature on the subject and on past and current NLR research. The latter is aimed at the development of a higher-order panel method (AEROPAN) and of a panel method (PDAERO) to be used in preliminary design studies.

MATHEMATICAL MODEL

Background

In aerodynamics the Reynolds-averaged Navier-Stokes equations model essentially all flow details. However, turbulence and transition needs to be modeled in a manner appropriate for the flow considered. The computer resources required for numerically solving the equations on a mesh that sufficiently resolves the boundary layers, free shear layers (wakes), vortex cores, etc. are still quite excessive.

For most high-Reynolds-number flow of interest in aircraft aerodynamics viscous effects are confined to thin boundary layers, thin wakes and centers of vortex cores, i.e. the global flow features depend only weakly on Reynolds number. This implies that a model based on Euler's equations, which allow the occurrence of shock waves and convection of rotational flow, provides an appealing alternative. On a local scale, specifically at points where the flow leaves the surface (separates) and vorticity is shed into the flow field, some kind of model for viscous-flow dominated features will be required. Although the computer requirements of Euler codes can be met by the current generation of supercomputers, routine practical application of these codes (to simple configurations) is only just emerging.

If there are no strong shocks and if the rotational flow is confined to compact regions, the flow may be modeled as potential flow with embedded free vortex sheets and vortex cores. Now the rotational flow regions are "fitted" explicitly into the solution, rather than "captured" implicitly as part of the solution as is the case for above flow models. Although one has to decide a priori on the presence of vortex sheets and cores and generally the topology of the vortex system must be well-defined, "fitting" still requires that both the position and strength of the vortex sheets and cores have to be determined as part of the potential-flow solution. The treatment of vortex sheets and vortex cores, freely floating in a fixed spatial grid, still poses considerable problems for finite difference/volume methods solving the nonlinear full-potential equation. The computer requirements of full-potential codes are relatively modest, but application to general aircraft configurations is hampered by the grid-generation problem.

In case shockwaves are absent altogether and the perturbation on the free-stream due to the presence of the configuration is small, the potential-flow model is further simplified to the linear potential flow model, governed by the Prandtl-Glauert equation. (In incompressible flow the small-perturbation assumption is not required and Laplace's equation applies.) Now the flow and the position of the vortex sheets and cores can be solved for by employing a boundary-integral type of approach utilizing singularity distributions exclusively on the surface of the configuration and the vortex sheets, not requiring a spatial grid. This linear potential flow model is the model underlying the classical panel method. It is emphasized at this point that though the governing equation in linear, the problem is still nonlinear because the position of the vortex sheets appears nonlinearly in the boundary condition on the solid surface as well as in the boundary conditions on the vortex sheet itself. It can be argued that for most configurations in cruise condition the wakes remain fairly simple, i.e. do not roll up within one wing span downstream of the wing trailing edge. This leads to the conventional attached flow model in which the vortex sheet is chosen as some appropriate, user-specified rigid surface, rendering the resulting problem fully linear.

Domain of applicability of panel methods

The preceding discussion will have made clear that substantial assumptions had to be made to finally arrive at the framework in which the panel method may be applied soundly. The assumptions are summarized here as:

- large Reynolds number, essentially inviscid flow
- no flow separation other than at the trailing edges
- compact regions with vorticity, i.e. thin wakes
- incompressible flow; or small-perturbation compressible flow without shocks.

The restriction to small perturbation compressible flow implies that at the higher Mach numbers the linearized potential flow panel method applies to configurations with slender bodies and thin wings at low angles of attack and sideslip.

In general it can be stated that the panel method provides detailed but "simplified aerodynamics" for complex configurations. This is illustrated further in Fig. 1, which shows, for a subsonic transport configuration, the α-M_∞ plane. Curves indicate the occurrence of flow features as shockwaves and the onset of boundary-layer separation. The shaded area indicates the domain of applicability of the conventional panel method, possibly coupled with a boundary-layer method. It will be clear that extension of the panel method approach into regions with transonic flow, without sacrificing its ability to treat arbitrary configurations, will greatly enhance its value for the aircraft designer.

Another area where there is still a gain to be realized is to improve upon the treatment of the wakes. For configurations with extended flaps or for combat aircraft that operate at higher incidences the rigid-wake approach adopted in most "first-generation" panel methods is inadequate.

Governing equations

The full potential equation is with $\vec{u} = \vec{\nabla}\Phi$ written as

$$\frac{\partial}{\partial x}\left(\rho\frac{\partial\Phi}{\partial x}\right) + \frac{\partial}{\partial y}\left(\rho\frac{\partial\Phi}{\partial y}\right) + \frac{\partial}{\partial z}\left(\rho\frac{\partial\Phi}{\partial z}\right) = 0 \quad , \tag{1.a}$$

where $\rho = \rho_\infty\{1 + \frac{\gamma-1}{2} M_\infty^2(1 - \frac{|\vec{\nabla}\Phi|^2}{U_\infty^2})\}^{1/\gamma-1}$ (1.b)

is the density. The pressure coefficient follows from the isentropic formula:

$$C_p = \frac{p-p_\infty}{q_\infty} = \frac{2}{\gamma M_\infty^2}[\{1 + \frac{\gamma-1}{2} M_\infty^2(1 - \frac{|\vec{\nabla}\Phi|^2}{U_\infty^2})\}^{\gamma/\gamma-1} -1] \ . \tag{1.c}$$

In the case of incompressible flow $p \equiv p_\infty$, $M_\infty = 0$ and Eq. (1.a) reduces to Laplace's equation, i.e. with $\vec{u} = \vec{U}_\infty + \vec{\nabla}\varphi$, where φ is the perturbation velocity potential one finds

$$\frac{\partial^2\varphi}{\partial x^2} + \frac{\partial^2\varphi}{\partial y^2} + \frac{\partial^2\varphi}{\partial z^2} = 0 \ , \tag{2.a}$$

while the pressure coefficient follows from Bernoulli's equation

$$C_p = 1 - \frac{|\vec{U}_\infty + \vec{\nabla}\varphi|^2}{U_\infty^2} \tag{2.b}$$

In compressible flow Eq. (1.a) is linearized under the assumption that $\vec{\nabla}\varphi$ is $O(\varepsilon)$. In case the freestream is directed along the x-axis Eq. (1.a) reduces to the Prandtl-Glauert equation:

$$(1 - M_\infty^2)\frac{\partial^2\varphi}{\partial x^2} + \frac{\partial^2\varphi}{\partial y^2} + \frac{\partial^2\varphi}{\partial z^2} + O(\varepsilon^2) = 0 \ . \tag{3.a}$$

Note that this equation is elliptic for subsonic free-stream Mach numbers and hyperbolic for supersonic free-stream Mach numbers. Within the scope of the linearization it may be assumed that for small incidence and sideslip the compressibility axis remains the same, i.e. the x-axis. Note that for incompressible flow the Prandtl-Glauert equation reduces to Laplace's equation. Further note that expanding Eqs. (1.a-c) in the Rayleigh-Janzen expansion for small Mach number also results in Laplace's equation.
To the same order of approximation as used for Eq. (3.a), Eq. (1.b) reduces to

$$\rho = \rho_\infty\{1 - M_\infty^2(\vec{U}_\infty.\vec{\nabla}\varphi)/U_\infty^2 + O(\varepsilon^2)\} \ , \tag{3.b}$$

while the pressure coefficient follows as

$$C_p = -2(\vec{U}_\infty.\vec{\nabla}\varphi)/U_\infty^2 + O(\varepsilon^2) \ . \tag{3.c}$$

If in practise the perturbations are not small, Eq. (3.c) may attain unphysical values, i.e. lower than vacuum or exceeding stagnation. In most methods the computed value is limited to vacuum and stagnation values.

A natural way to extend the capability of linear potential flow methods is the apply Eqs. (3.a-c) everywhere in the flow field where the perturbation velocity is small and the full potential flow formulation, Eqs. (1.a-c), in the remaining small isolated regions.

Boundary conditions

(i) On the surface S_b of the configuration (Fig. 2) the condition is applied that the normal component of the velocity either vanishes (solid body) or is prescribed (boundary-layer transpiration concept, inlet fan face simulation, propeller slip-stream effects, jet entrainment, design option), i.e.

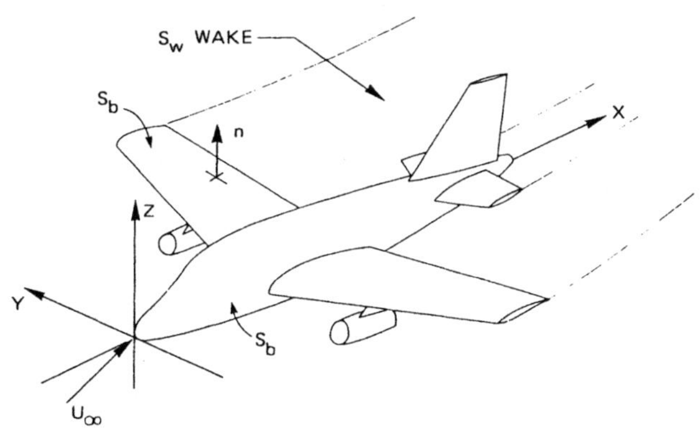

Fig. 2 Airplane configuration

$$(\vec{U}_\infty + \vec{\nabla}\varphi).\vec{n} = v_n \quad , \tag{4.a}$$

where $\vec{n}$ is the normal to the surface and v_n given.

(ii) On the vortex sheet S_w two conditions apply, the stream surface condition Eq. (4.a) with $v_n=0$ and the condition that the static pressure is continuous across the vortex sheet, i.e.

$$\Delta C_p = C_p(S_w^+) - C_p(S_w^-) = 0 \quad . \tag{4.b}$$

(iii) At the trailing edges of S_b the Kutta condition is applied that the flow leaves the surface "smoothly".

(iv) At infinity upstream the perturbation vanish. The free-stream velocity U_∞ may consist of (constant) components due to incidence and sideslip, but also ones due to (small) steady rates of pitch, yaw and roll. Sometimes a user-specified onset flow (and total-pressure increment) is added to model for instance propeller slipstream effects.

(v) No upstream influences in supersonic flow.

Only in case of the rigid wake assumption, valid for high-aspect ratio lightly loaded wings, not strongly interacting with other components of the configuration or with other vortex sheets, will the resulting boundary value problem be linear. In case of "relaxed wakes" the boundary conditions are mildly nonlinear in terms of φ but highly nonlinear in terms of the also to be solved for position of the vortex wakes.

Integral representation of the solution

The solution of the potential flow problem may be represented through Green's 3rd identity by singularity (source q and doublet μ, see Fig. 3 for definition) distributions over the surface S_b of the configuration and the vortex sheets S_w in the form (see Ref. [19] for incompressible and Ref. [20] for subsonic compressible flow):

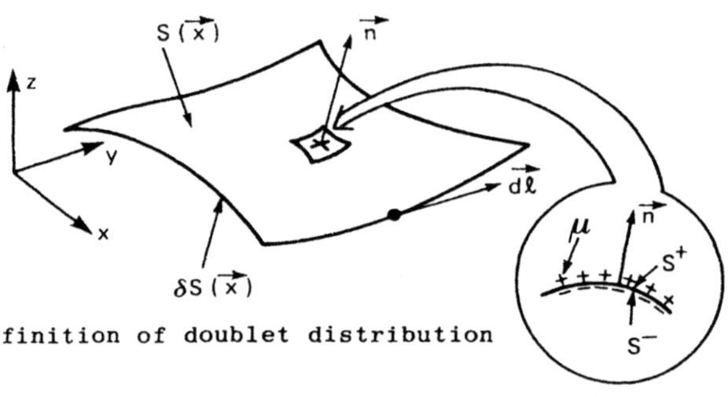

Fig. 3 Definition of doublet distribution

$$\varphi(x_o) = \varphi_q(\vec{x}_o) + \varphi_\mu(\vec{x}_o) \quad , \tag{5.a}$$

where

$$\varphi_q(\vec{x}_o) = \frac{-1}{4\pi} \iint_{S_b} q(\vec{x}) \frac{dS}{|\vec{R}|} \quad , \tag{5.b}$$

$$\varphi_\mu(\vec{x}_o) = -\frac{B^2}{4\pi}\iint_{S_{b+w}} \mu(\vec{x})\frac{[B^{-1}]\vec{n}.\vec{R}}{|\vec{R}|^3}\, dS \quad . \tag{5.c}$$

In Eq. (5) $\vec{R} = [B](\vec{x}_o - \vec{x})$, $\vec{n}$ is the outward normal and the matrix [B] is defined as

$$[B] = \begin{pmatrix} 1 & 0 & 0 \\ 0 & B & 0 \\ 0 & 0 & B \end{pmatrix} , \; B = (1 - M_\infty^2)^{\frac{1}{2}} , \; [B^{-1}] = \begin{pmatrix} 1 & 0 & 0 \\ 0 & 1/B & 0 \\ 0 & 0 & 1/B \end{pmatrix} . \tag{5.d}$$

The velocity potential as defined in Eq. (5) satisfies the Prandtl-Glauert equation Eq. (3.a) exactly. The integrals in Eq. (5) have a singular integrand, which results, for regular q and μ, in a situation where the potential is regular everywhere, except that it has a jump across S, i.e.

$$\varphi(\vec{x}_o \in S^{\pm}) = \varphi^P(\vec{x}_o \in S) \mp \tfrac{1}{2}\mu(\vec{x}_o) \; , \tag{5.e}$$

where the superscript P denotes the so-called Principal-Value of the integral.

The velocity field induced by the singularity distributions follows from Eq. (5) through differentiation, as:

$$\vec{u}(\vec{x}_o) = \vec{\nabla}\varphi(\vec{x}_o) = \vec{u}_q(\vec{x}_o) + \vec{u}_\mu(\vec{x}_o) \quad , \tag{6.a}$$

where

$$\vec{u}_q(\vec{x}_o) = \frac{[B]}{4\pi}\iint_{S_b} q(\vec{x}) \frac{\vec{R}}{|\vec{R}|^3}\, dS \quad , \tag{6.b}$$

$$\vec{u}_\mu(\vec{x}_o) = \vec{u}_\gamma(\vec{x}_o) \qquad\qquad + \vec{u}_v(\vec{x}_o) \tag{6.c}$$

$$= \frac{[B]}{4\pi}\iint_{S_{b+w}} ([B]\vec{\gamma}) \times \frac{\vec{R}}{|\vec{R}|^3}\, dS - \frac{[B]}{4\pi}\int_{\partial S_{b+w}} \mu(\vec{x})\frac{\vec{R}}{|\vec{R}|^3} \times ([B]d\vec{\ell}) \, . \tag{6.d}$$

Here we used the equivalence of the velocity induced by a doublet distribution μ with that induced by a surface vorticity distribution $\vec{\gamma}$ of strength $\vec{\gamma} = -\vec{n}\times\vec{\nabla}\mu$ plus that induced by a concentrated vortex of strength $\Gamma=\mu$ along the boundary of S. Note that in using the equivalence it is assumed that μ is continuous except at the boundary. The main advantage of using the doublet distribution rather than the surface vorticity distribution is that Kelvin's vortex laws (e.g. vortex lines are closed) are satisfied automatically. Across the surface the velocity is discontinuous, i.e.

$$\vec{u}(\vec{x}_o \in S^{\pm}) = \vec{u}^P(\vec{x}_o \in S) \pm \tfrac{1}{2}\{q(\vec{x}_o)\frac{\vec{n}}{B^2} + \vec{\gamma}\times\vec{m}\}/(\vec{n}.\vec{m}) \; , \tag{6.e}$$

where $\vec{m} = [B^{-1}]([B^{-1}]\vec{n})$. The vector $\vec{m}$ is parallel to the normal $\vec{n}$ in case of incompressible flow and in case $\vec{n}$ is normal to or along the x-axis (= "compressibility axis").
Considering above expressions we note that the jump in the velocity potential is solely due to the doublet distribution. The velocity due to the

source distribution has a jump in normal direction. The velocity induced by the doublet distribution has a jump in tangential and in normal direction, since $\vec{\gamma}\times\vec{m} = -(\vec{n}.\vec{m})\vec{\nabla}\mu + (\vec{m}.\vec{\nabla}\mu)\vec{n}$. In the case of incompressible flow the velocity across the doublet distribution experiences a jump in the tangential component only. This feature has a consequence that modeling vortex sheets (≡ jump in tangential velocity) in compressible flow requires a composite singularity distribution consisting of a doublet distribution of strength μ and a source distribution of strength $q = -B^2(\vec{m}.\vec{\nabla}\mu)$.

In the literature one also finds the so-called (linearized) mass-flux vector, defined as

$$\vec{W} = \rho(\vec{U}_\infty + \vec{u}) \tag{7.a}$$

$$= \rho_\infty(\vec{U}_\infty + \vec{u}) - \rho_\infty(\vec{U}_\infty.\vec{u}) \frac{M_\infty^2}{U_\infty^2} \vec{U}_\infty + O(\varepsilon^2) \ , \tag{7.b}$$

and also the (linearized) perturbation mass-flux vector $\vec{w} = \vec{W}-\rho_\infty\vec{U}_\infty$, which has like $\vec{u}$ order of magnitude ε. In the derivation of Eq. (7.b) we used Eq. (3.b). Employing the jump relation given in Eq. (6.e) it turns out that

$$\vec{w}(\vec{x}_o\epsilon S^{\pm}) = \vec{w}^p(\vec{x}_o\epsilon S) \pm \tfrac{1}{2}\rho_\infty\{q\vec{m}/(\vec{n}.\vec{m}) - \vec{\nabla}\mu\} \ , \tag{7.c}$$

showing that the doublet distribution results in a jump in the tangential component $\vec{w}$ only, while the source distribution causes a jump in both the normal and the tangential direction. We close this section by pointing out some relations that may be used to reduce the computational burden. The first one is the correspondence between the velocity field $\vec{u}_q(\vec{x}_o;q)$ induced by a source distribution q and the velocity $\vec{u}_\gamma(\vec{x}_o)$ induced by a vorticity distribution, namely one can write

$$\vec{u}_\gamma(\vec{x}_o) = [B] \sum_{k=1}^{3} \vec{e}_k\times\{[B^{-1}]\vec{u}_q(\vec{x}_o;q=[B]\vec{\gamma}.\vec{e}_k)\} \ . \tag{8}$$

In cases with port side/starboard side symmetry computing time can be saved by realizing that the singularity distribution on the port side will be identical to the one on the starboard side. If then the velocity induced at $\vec{x}_o$ by the starboard-side distribution is denoted by $\vec{u}_s(\vec{x}_o)$ one finds that the velocity induced at $\vec{x}_o$ by the complete configuration is

$$\vec{u}(\vec{x}_o) = \vec{u}_s(\vec{x}_o) + [S]\vec{u}_s([S]\vec{x}_o) \ , \tag{9.a}$$

where [S] is the so-called symmetry matrix defined as

$$[S] = \begin{pmatrix} 1 & 0 & 0 \\ 0 & -1 & 0 \\ 0 & 0 & 1 \end{pmatrix} \ . \tag{9.b}$$

A final point is that in case the geometry is symmetric, but the flow is not symmetric because of the boundary conditions (e.g. side-slipping configuration) the problem can be reduced by almost a factor of 2 by formulating the problem in terms of the average and the difference of the singularity distributions on port and starboard side.

In supersonic flow the solution of the Prandtl-Glauert equation can also be written in terms of a source and a doublet distribution on the surface of the configuration, see Ref. [20]. However, here the hyperbolic character of the equation is to be accounted for by restricting the surface of integration in Eqs. (5.b-c) and (6.b-d) to the area within the forward Mach cone from the point $\vec{x}_o$ and changing $1/4\pi$ into $1/2\pi$.

IMPLEMENTATION OF BOUNDARY CONDITIONS

Neumann condition on S_b

On the solid surface S_b of the configuration the stream-surface conditions, Eq. (4.a) is applied, with v_n known. The most obvious manner to impose this condition is to substitute the integral representation Eq. (6) directly into Eq. (4.a). However, first one has to fix a degree of freedom still in the formulation. This relates to the circumstance that the flow inside the volume enclosed by S_b is arbitrary. It will be fixed by specifying some relation between the two singularity distributions or by choosing one of them. In the "first-generation" panel methods the following possibilities have been implemented:

	q	μ
BODIES	Unknown	0 (Lift-Carry-Over)
WINGS	Unknown	"Mode function"

where the "Mode function" (given shape, unknown amplitude) doublet distribution is situated on the wing surface itself or on some auxiliary surface (often the camber surface) inside the wing and is continued onto S_w in all cases. One of the problems encountered over and over again in the application of these methods is that the doublet distribution (vortex system) of the wing has to be continued into or onto fuselages, tiptanks, etc., this to avoid, or to position in a physically correct manner, the concentrated vortex associated with the second term in Eq. (6.d).
The formulation of the problem is then the following Fredholm integral equation of the 2nd kind:

$$\frac{1}{2B^2(\vec{n}.\vec{m})}\, q(\vec{x}_o) + \vec{n}(\vec{x}_o).\{\iint\limits_{S_b} q(\vec{x})\vec{K}_q(\vec{x}_o,\vec{x})dS(\vec{x})+\vec{u}_{\text{mode f.}}\}=v_n-\vec{U}_\infty.\vec{n}(\vec{x}_o) \qquad (10)$$

for all $\vec{x}_o \in S_b$, supplemented with as many "Kutta conditions" as there are "mode functions". In Eq. (10) the kernel $\vec{K}_q$ follows from Eq. (6.b). This type of integral equation provides a sound basis for the discretization. Once the solution of Eq. (10) is obtained the tangential velocity is obtained from the evaluation of the integral representation, Eq. (16).

Dirichlet condition on S_b

Most of the "second generation" panel methods offer an attractive alternative to the direct application of the Neumann condition. With the jump conditions the Neumann condition for the flow external to the volume V_b enclosed by the surface S_b can be converted into a Dirichlet condition for the flow inside V_b. An example, see Fig. 4, is the formulation in which the perturbation velocity φ is set equal to zero for all points $\vec{x}_o \in S_b^-$. This implies that $\varphi=0$ everywhere inside V_b and also that $\vec{n}.\vec{\nabla}\varphi=0$ for $\vec{x}_o \in S_b^-$. It then follows from Eq. (6.e) and Eq. (4.a) that

$$\vec{u}^P(\vec{x}_o)\cdot\vec{n} \qquad - \tfrac{1}{2}\{q/B^2 + (\vec{m}\cdot\vec{\nabla}\mu)\}/(\vec{n}\cdot\vec{m}) = 0 \ , \ \text{for } \vec{x}_o \in S_b^- \tag{11.a}$$

and

$$\{\vec{U}_\infty + \vec{u}^P(\vec{x}_o)\}\cdot\vec{n} + \tfrac{1}{2}\{q/B^2 + (\vec{m}\cdot\vec{\nabla}\mu)\}/(\vec{n}\cdot\vec{m}) = v_n, \ \text{for } \vec{x}_o \in S_b^+ \tag{11.b}$$

respectively. Subtraction then yields the following relation between q and μ:

$$q = B^2(\vec{n}\cdot\vec{m})(v_n - \vec{U}_\infty\cdot\vec{n}) - B^2(\vec{m}\cdot\vec{\nabla}\mu) \quad . \tag{11.c}$$

This formulation leads again to a Fredholm integral equation of the 2nd kind, now for μ:

$$-\tfrac{1}{2}\mu(\vec{x}_o) - \frac{1}{4\pi}\iint\limits_{S_{b+w}} \mu(\vec{x})K_\mu(\vec{x}_o,\vec{x})dS = \frac{1}{4\pi}\iint\limits_{S_b} q(\vec{x})K_q(\vec{x}_o,\vec{x})dS \ , \tag{12}$$

where K_μ and K_q follow from Eq. (5.c) and (5.b), respectively. This formulation does not require fictitious internal Lift-Carry-Over surfaces. However, note that at the intersection of the wake of a wing with a body the doublet distribution on the body has a jump (equal to the wake doublet strength) (Fig. 5).

Once Eq. (12) is solved for the doublet distribution the tangential velocity on S_b^+ can be obtained as follows. The Dirichlet boundary condition on S_b^- implies the $\varphi^P(x_o) = -\tfrac{1}{2}\mu$, see Eq. (5.e). From this same equation it follows also that

$$\varphi(\vec{x}_o \in S_b^+) = -\mu(\vec{x}_o) \tag{13.a}$$

and

$$\vec{u}(\vec{x}_o \in S_b^+) = \vec{U}_\infty + \{q/B^2 + (\vec{m}\cdot\vec{\nabla}\mu)\}\vec{n}/(\vec{n}\cdot\vec{m}) - \vec{\nabla}\mu \quad . \tag{13.b}$$

This last expression does not involve any evaluation of an integral representation, just the gradient of the doublet distribution.

<u>Lifting surface approximation</u>

In the derivation of the Prandtl-Glauert equation it was implicitly assumed that wings are relatively thin. Within the framework of linear theory the stream-surface condition can be simplified to a boundary condition on a reference surface, e.g. the camber surface or any other surface

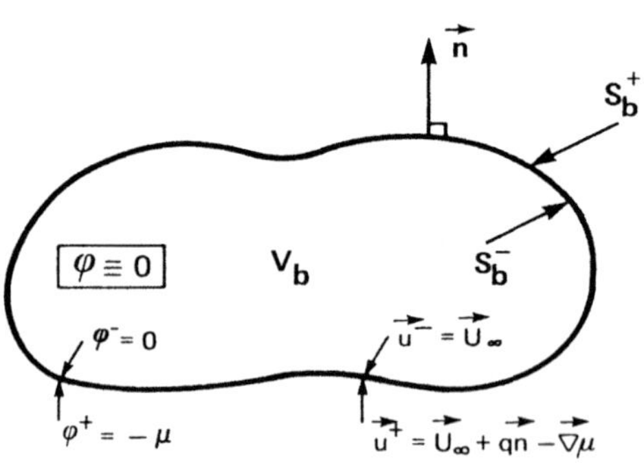

Fig. 4 Dirichlet condition

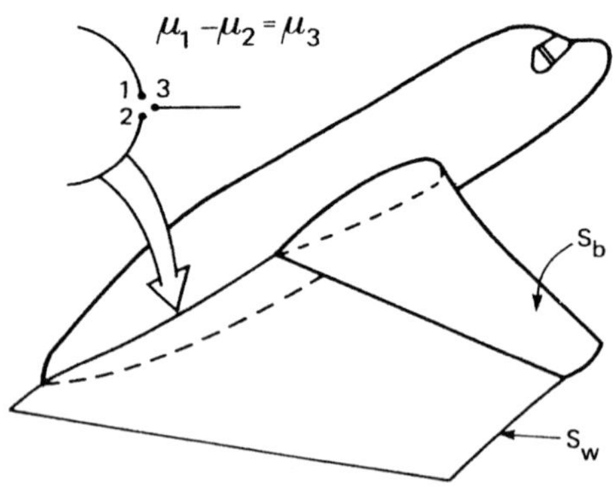

Fig. 5 Lift-Carry-Over in case of Dirichlet condition

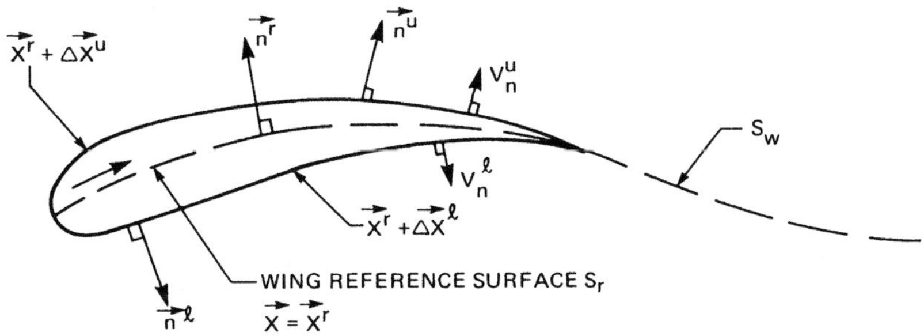

Fig. 6 Lifting surface approximation

near the true upper and lower wing surface (Fig. 6). Now points on the wing surface are defined as

$$\vec{x} = \vec{x}^r + \Delta\vec{x}^{u,\ell} \quad , \tag{14.a}$$

where r, u and ℓ denote the reference ("lifting") surface, upper and lower wing surface respectively. Under the assumptions that $\Delta\vec{x}^{u,\ell}$ and its first derivatives are small of order ε one may write

$$\vec{n}^{u,\ell} = \pm \{\vec{n}^r - \vec{T}^{u,\ell}\times\vec{n}^r\} + O(\varepsilon^2) \quad , \tag{14.b}$$

$$\vec{u}(\vec{x}^r + \Delta\vec{x}^{u,\ell}) = \vec{U}_\infty + \vec{u}(\vec{x}^{r\pm}) + O(\varepsilon^2) \quad , \tag{14.c}$$

where $\vec{n}$ denotes the normal, while the + and the - refer to the upper and lower side respectively. $\vec{T}^{u,\ell}$, of order ε, is tangential to the reference surface and contains a linear combination of first derivatives of $\Delta\vec{x}^{u,\ell}$. The normal velocity condition on the true upper and lower wing surface are expanded in terms of ε. Adding the two resulting expressions yields

$$\tfrac{1}{2}\{\vec{u}(\vec{x}^{r+}) - \vec{u}(\vec{x}^{r-})\}.\vec{n}^r = \tfrac{1}{2}(v_n^u + v_n^\ell) + \vec{U}_\infty.\tfrac{1}{2}(\vec{T}^u-\vec{T}^\ell)\times\vec{n}^r \quad , \tag{15.a}$$

where the second term on the right-hand side can be recognized as the slope of the thickness distribution of the wing, which is zero in case $\Delta\vec{x}^u=\Delta\vec{x}^\ell$, i.e. a wing of infinitesimal thickness). Substitution of the jump relation, Eq. (6.e), then yields the following relation for q:

$$q = 2B^2(\vec{n}.\vec{m})\{\tfrac{1}{2}(v_n^u + v_n^\ell) + \vec{U}_\infty.\tfrac{1}{2}(\vec{T}^u-\vec{T}^\ell)\times\vec{n}^r\} - B^2(\vec{m}.\vec{\nabla}\mu) \quad , \tag{15.b}$$

where the last term on the right-hand side is zero in case $\vec{m}$ is parallel to the normal $\vec{n}$.

Subtracting the two expressions yields:

$$\{\vec{U}_\infty + \vec{u}^P(\vec{x}_o)\}.\vec{n}^r = \tfrac{1}{2}(v_n^u-v_n^\ell) + \vec{U}_\infty.\tfrac{1}{2}(\vec{T}^u+\vec{T}^\ell)\times\vec{n}^r \quad . \tag{15.c}$$

The second term on the right-hand side corresponds to the slope of the "added" camber distribution, which will be zero in case $\Delta\vec{x}^\ell=-\Delta\vec{x}^u$, i.e. in case of a symmetric airfoil section with its camber surface chosen as reference surface.
It follows from Eq. (15.c) upon substitution of Eq. (6.d) that the lifting-surface approximation leads to the following integral equation for μ:

$$\frac{\vec{n}^r}{4\pi}\cdot\Big[\iint_{S_{r+w}} \{[B](\vec{n}^r\times\vec{\nabla}\mu)\}\times \vec{K}_\mu dS - \int_{\partial S_{r+w}} \mu\vec{K}_v\times d\vec{\ell}\Big] = \tfrac{1}{2}(v_n^u - v_n^\ell) + \vec{U}_\infty\cdot\tfrac{1}{2}(\vec{T}^u+\vec{T}^\ell)\times\vec{n}^r$$

$$- \vec{U}_\infty\cdot\vec{n}^r - \frac{\vec{n}^r}{4\pi}\cdot\iint_{S_r} q\vec{K}_q dS \ , \qquad (15.b)$$

where $\vec{K}_\mu$ and $\vec{K}_v$ follow from Eq. (6.d). Eq. (15.b) probably has to be classified as Fredholm integral equation of the first kind. It has proven that this type of integral equation is not so amenable to numerical solution techniques. However, at present no alternative formulation is known.

Boundary conditions on S_w

On the wake surface two boundary conditions apply: the stream-surface condition Eq. (4.a) and the zero-pressure-jump condition Eq. (4.b). Applying the stream-surface condition at both sides of the wake S_w gives, using Eq. (6.e):

$$q = - B^2(\vec{m}.\vec{\nabla}\mu) \ , \qquad (16.a)$$

$$(\vec{U}_\infty + \vec{u}^P).\vec{n} = 0 \ . \qquad (16.b)$$

It follows from the linear pressure formula Eq. (3.c), again using Eq. (6.e), that with Eq. (16.a):

$$\Delta C_p = 2(\vec{U}_\infty.\vec{\nabla}\mu)/U_\infty^2 + O(\varepsilon^2) = 0 \ . \qquad (16.c)$$

This then leads to the classical rigid-wake approximation in which the wake is directed along $\vec{U}_\infty$ (also the compressibility axis) with μ=constant (equal to its value at the trailing edge) in streamwise direction. Since now $\vec{m}.\vec{\nabla}\mu=0$ this wake does not carry a source distribution.
The rigid-wake approximation with straight vortex lines renders the problem linear, and note that the average normal-velocity condition, Eq. 16.b, is not satisfied.

A frequently used slight variation on the classical planar wake is to let the user specify the curves along which the doublet distribution is taken as constant, i.e. the vortex lines (Fig. 7). The vortex lines on this

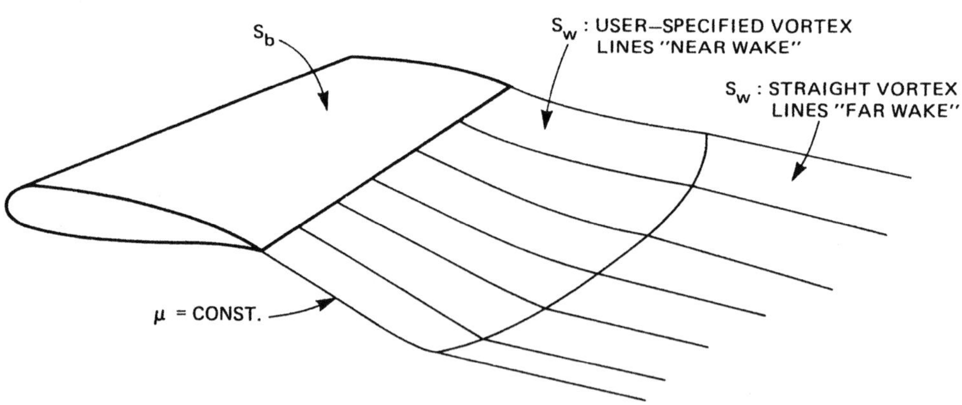

Fig. 7 Wake surface

"near wake" are continued as straight lines onto the "far wake". Although the user-specified near-wake option does improve the modeling capabilities of the method, there are cases where the interaction between the wake and the flow about a nearby component of the configuration is so strong that the full nonlinear boundary conditions have to be satisfied. The stream-surface condition leads to the two relations given in Eqs. (16.a and b). For the zero-pressure-jump condition the quadratic approximation of the isentropic pressure formula, Eq. (1.c), results in

$$(\vec{U}_\infty + \vec{u}^P).\vec{\nabla}\mu + O(\varepsilon^3) = 0 \tag{16.d}$$

which, combined with Eq. (16.b), implies that the vortex line is parallel to the local velocity.

To solve for both the position of the wake and the singularity distribution simultaneously is a difficult task, so frequently one relies on a hierarchical "wake relaxation" procedure in which one first obtains the singularity distribution for a rigid wake (as in Fig. 7) and subsequently aligns the vortex lines with the local velocity which defines an improved location of the wake surface, etc. However, in cases where the wake interacts very strongly with the flow about a nearby component of the configuration (e.g. slender wing with leading-edge vortex sheets) such simple hierarchical iteration scheme frequently diverges and the wake boundary conditions Eqs. (16.b and d) have to be solved simultaneously. The two resulting integral equations for μ and $S_w(\vec{x})$ are highly nonlinear in $\vec{x}$, while in terms of μ Eq. (16.b) is a Fredholm integral equation of the first kind and Eq. (16.d) is nonlinear.

Super-inclined surface

In compressible subsonic or supersonic flow the perturbations are assumed to be small. This usually implies that the true surface is only slightly inclined with respect to the free-stream direction and generally will be sub-inclined with respect to the Mach angle in supersonic flow. However, for instance at inlet faces where an inflow is prescribed the surface will be super-inclined, but the perturbation may still be small. These super-inclined portions of the surface require special boundary conditions, not given here, or the specification of an artificial sub-inclined cap covering the inlet.

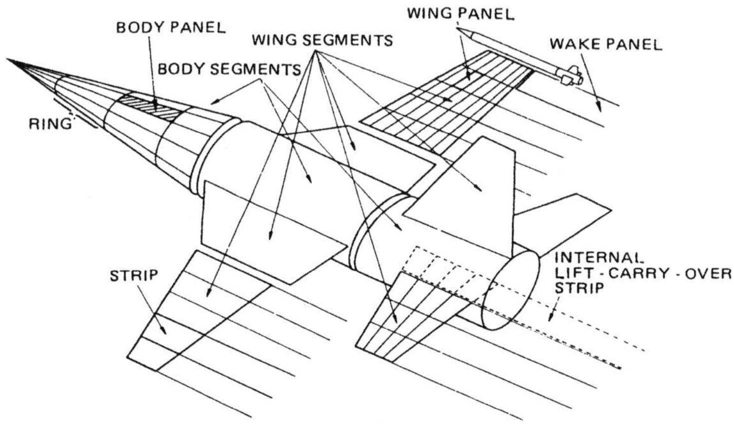

Fig. 8 Example of configuration paneling

APPROXIMATION AND DISCRETIZATION

General approach

In the approach taken in panel methods the following tasks can be distinguished:

1. Subdivide the surface S_b of the configuration and its wake S_w into (small) quadrilateral elements, the panels. This is mostly accomplished in a hierarchical manner (Fig. 8) in which the configuration is subdivided into parts, each part into segments ("networks") and each segment into a number of strips (rings) of the same number of panels. Subsequently the geometry (centroid, normal vector, curvature, etc.) of each panel is computed.

2. Approximate, to the required accuracy, the integrals over the panel surface corresponding to the contribution of the singularity distributions on the panel in the potential or velocity at the (collocation) points where the boundary conditions are to be imposed.

3. Choose on each panel sufficiently accurate local representations for the singularity distributions μ and q, and in case the position of the vortex sheets is simultaneously solved for also for x describing S_w.

The computational heavy ($\doteq N^2$) task, constituted by step 2 and 3, yields the so-called aerodynamic influence coefficients (AIC's).

4. Impose the boundary conditions at the collocation points. Solve the resulting non-sparse system of linear (non-linear in case of non-rigid wakes) equations for the unknown parameters in the local representations for the singularity distributions (and geometry). Solution of the system of linear equations requires order N^3 operations in case a direct solver is used and order N^2 operations in case an iterative solver is used.

5. Find to the required accuracy the velocity distribution on the surface S_b of the configuration.

6. Compute the pressure, integrated forces and moments, induced drag, (lifting surface) edge-suction forces, surface streamlines, isobars, velocity and pressure at off-body points, stability derivatives, trimmed-flight conditions, boundary-layer quantities, updated wake position, etc., etc.

The user of the method will interface with step 1., where the geometric input to the program is digested and with step 6. where the results of the flow simulation are presented. These two steps will determine the capability and user-friendliness of the program. The design of the remaining steps will determine the accuracy of and computer resources required for the application. As far as the accuracy and cost is concerned, the aim is to obtain the surface-velocity distribution to certain accuracy, i.e.

$$\vec{u}^h(\vec{x}_o \in S_b) = \vec{u}(\vec{x}_o \in S_b) + O(h^n), \quad \text{for } h \to 0 \tag{17}$$

for lowest costs. Here n is the "order" of the panel method. Most of the "first-generation" panel methods are first-order methods, most of the "second-generation" methods are second-order methods.
In the following we consider some aspects of developing a method of consistent order of approximation. The discussion will be restricted to first and second-order methods.

Small-curvature expansion for velocity

Consider the expression for the velocity induced by a source distribution, Eq. (6.b), which in order to simplify the discussion somewhat is taken for the limiting case of incompressible flow, i.e.

$$\vec{u}_q(\vec{x}_o) = \sum_i \frac{1}{4\pi} \iint_{S_i} q(\vec{x}) \frac{\vec{r}}{|\vec{r}|^3} dS(\vec{x}) \quad , \tag{18}$$

where $\vec{r}=\vec{x}_o-\vec{x}(s,t)$, S_i denotes the surface of the i-th panel and (s,t) is some surface coordinate system (Fig. 9). The integrand in Eq. (18) is singular for $\vec{x}_o$ coinciding with the point $\vec{x}(s,t)$ on S_i. Therefore numerical quadrature is not appropriate. Hess [21] proposed an expansion in which the characteristic behaviour of the integrand is maintained, resulting in a

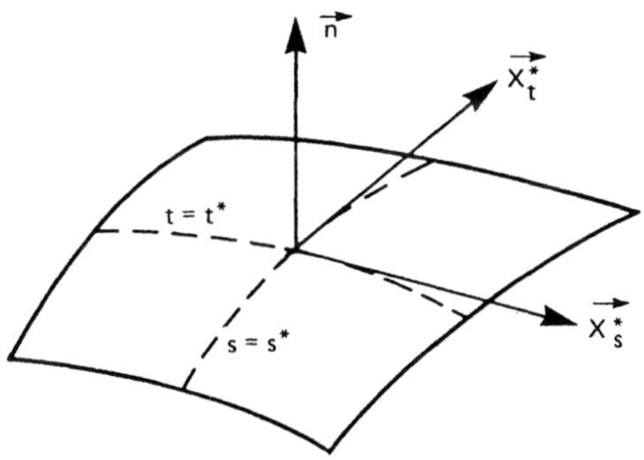

Fig. 9 Panel coordinate system

consistent approximation of Eq. (18) that possesses the correct behaviour as $\vec{x}_o$ crosses the surface S_b. In the expansion the nominator is split into a linear part and a part containing the curvature and twist of the panel, i.e.

$$\vec{r} = \vec{x}_o - [\{\vec{x}^* + \Delta s\vec{x}^*_s + \Delta t\vec{x}^*_t\} + \vec{r}_2\} + \ldots] \tag{19.a}$$

$$= \vec{D} - \vec{r}_2 + O(\delta^3) \quad , \tag{19.b}$$

with $\Delta s=s-s^*$ and $\Delta t=t-t^*$.

In Eq. (19.a) the term inside the curly brackets corresponds to the plane tangent to the panel at $\vec{x}(s^*,t^*)$ (the so-called expansion point). The panel curvature/twist term $\vec{r}_2 = \frac{1}{2}\Delta s^2\vec{x}^*_{ss} + \Delta s\Delta t\vec{x}^*_{st} + \frac{1}{2}\Delta t^2\vec{x}^*_{tt}$ is of order δ^2, where δ is a linear measure for the panel size.
The term $\vec{D} = \vec{x}_o - \{\vec{x}^* + \Delta s\vec{x}^*_s + \Delta t\vec{x}^*_t\}$ is order δ in the "near field" of the panel and order 1 in the "far field" of the panel. In terms of the surface coordinate system the surface element $dS = |\vec{x}_s \times \vec{x}_t| dsdt$ is expanded as:

$$dS(\vec{x}) = |\vec{x}^*_s \times \vec{x}^*_t| [1 + \vec{n}^*.\{\Delta s\vec{T}_1 + \Delta t\vec{T}_2\} + O(\delta^2)] \quad , \tag{19.c}$$

where $\vec{T}_1$ and $\vec{T}_2$ are expressions containing the curvature and twist terms. The singularity distributions q and μ are expanded in a similar fashion, i.e.

$$q = q^* + \Delta s q_s^* + \Delta t q_t^* + O(\delta^2) , \tag{19.d}$$

$$\mu = \mu^* + \Delta s \mu_s^* + \Delta t \mu_t^* + \tfrac{1}{2}\Delta s^2 \mu_{ss}^* + \Delta s \Delta t \mu_{st}^* + \tfrac{1}{2}\Delta t^2 \mu_{tt}^* + O(\delta^3) . \tag{19.e}$$

Substitution of Eqs. (19.a-d) into Eq. (18) and expansion yields

$$\begin{aligned}
\vec{u}_q(\vec{x}_o) = \frac{1}{4\pi} \sum_i |\vec{x}_s^* \times \vec{x}_t^*| \, [q^* \iint_{S_i} \frac{\vec{D}}{|\vec{D}|^3} \, dsdt & \qquad + O(\frac{\delta^2}{D^2}) \\
+ q_s^* \iint_{S_i} \Delta s \frac{\vec{D}}{|\vec{D}|^3} dsdt + q_t^* \iint_{S_i} \Delta t \frac{\vec{D}}{|\vec{D}|^3} dsdt & \qquad + O(\delta\frac{\delta^2}{D^2}) \\
+ q^* \{ \iint_{S_i} (- \frac{\vec{r}_2}{|\vec{D}|^3} + 3(\vec{r}_2 . \vec{D}) \frac{\vec{D}}{|\vec{D}|^5}) dsdt & \qquad + O(K\delta\frac{\delta^3}{D^3}) \\
+ \vec{n}^* . \iint_{S_i} (\Delta s \vec{T}_1 + \Delta t \vec{T}_2) \frac{\vec{D}}{|\vec{D}|^3} \, dsdt \} & \qquad + O(K\delta\frac{\delta^2}{D^2}) \\
+ O(\delta^2\frac{\delta^2}{D^2} , K\delta^2\frac{\delta^2}{D^2} , \ldots)] \quad , &
\end{aligned} \tag{20}$$

where K is a measure of the panel curvature and twist. For all the remaining integrals closed-form expressions can be derived.

It follows from Eq. (20) that for a first-order panel method a panel-wise constant representation for q on a flat-panel approximation suffices, while a second-order panel method requires a panel-wise linear representation for q on a curved-panel approximation. However, note that if the panel curvature and twist are such that $K\delta_{local} < \bar{\delta}^2$, where $\bar{\delta}$ is the average panel size, the terms in Eq. (20) due to the curvature (the most complex ones) are small of higher order and may be omitted. Reduction of δ_{local} by sub-paneling, in which the number of sub-panels increases with increasing K but also with decreasing $\bar{\delta}$, is another possibility to get rid of the most complex terms in Eq. (20). Applying the small-curvature expansion to Eq. (6.d), the velocity induced by the doublet distribution, shows that a first-order method requires a panelwise linear representation for μ on a flat-panel approximation, while a second-order method requires a panelwise quadratic representation for μ on a curved-panel approximation.

All the surface integrals in Eq. (20) contain the same type of trancendental functions (logarithms and inverse tangent) implying that if carefully designed and programmed a higher-order formulation does not need to be very much more expensive than a lower-order formulation. However, it will be clear that a third-, or even higher-order method, which amongst others will involve derivatives of the curvature, is beyond practical limits.

The small-curvature expansion warrants that the induced velocity has the proper behaviour as the point $\vec{x}_o$ crosses the surface at $\vec{x}=\vec{x}(s^*,t^*)$, and the panel expansion point is usually chosen as the collocation point on the panel, irrespective of the location of $\vec{x}_o$.

Far-field expansion for velocity

Although the small-curvature expansion is uniformly valid, it is computationally expensive and therefore only applied in the region where it is really needed, i.e. in the "near field" of the panel. In the "far-field" of the panel the nominator of Eq. (18) is expressed as

$$\vec{r} = \vec{r}_o - \vec{r}_1 - \vec{r}_2 + O(\delta^3) \quad , \tag{21}$$

where $\vec{r}_o = \vec{x}_o - \vec{x}^*$ is $O(1)$ and $\vec{r}_1 = \Delta s \vec{x}^*_s + \Delta t \vec{x}^*_t$ is $O(\delta)$. Substitution in Eq. (18) then yields:

$$\begin{aligned}\vec{u}_q(\vec{x}_o) = \frac{1}{4\pi} \sum_i |\vec{x}^*_s \times \vec{x}^*_t| \; [q^* \frac{\vec{r}_o}{|\vec{r}_o|^3} \iint_{S_i} dsdt \quad & \leftarrow O(\delta^2) \\ + (\ldots) \iint_{S_i} (\Delta s \text{ or } \Delta t) dsdt \quad & \leftarrow 0 \\ + O(\delta^4)] \quad . &\end{aligned} \tag{22}$$

The second term in Eq. (22) is zero if the expansion point (s*,t*) coincides with the panel midpoint. The truncation error in Eq. (22) is sufficiently small to obtain a composite error of $O(\delta^2)$. Eq. (22) indicates that for the far-field expansion both a first-order and a second-order method require a panelwise constant representation for q on a flat-panel approximation. For the doublet distribution one requires a linear representation on a flat-panel approximation.
Often the region where the near-field expansion is to be applied is reduced in size by defining an "intermediate" field in which some form of numerical (e.g. Gauss) quadrature is applied.

Small-curvature expansion for potential

The expression for the velocity potential induced by a source and a doublet distribution can be expanded in a similar fashion as illustrated above for the velocity. However, note that in order to evaluate the velocity to $O(\delta^2)$ the velocity potential has to be evaluated to $O(\delta^3)$.
For instance, the small-curvature expansion for the velocity potential induced by a source distribution, Eq. (5.b), becomes in incompressible flow:

$$\begin{aligned}\varphi_q(\vec{x}_o) = \frac{1}{4\pi} \sum_i |\vec{x}^*_s \times \vec{x}^*_t| \quad [q^* \iint_{S_i} \frac{dsdt}{|\vec{D}|} \quad & \leftarrow O(\delta \tfrac{\delta}{D}) \\ +\{q^*_s \iint_{S_i} \frac{\Delta s}{|\vec{D}|} dsdt + q^*_t \iint_{S_i} \frac{\Delta t}{|\vec{D}|} dsdt\} \quad & \leftarrow O(\delta^2 \tfrac{\delta}{D}) \\ -q^* \{\iint_{S_i} \frac{\vec{r}_2 . \vec{D}}{|\vec{D}|^3} dsdt \quad & \leftarrow O(K\delta^2 \tfrac{\delta^2}{D^2}) \\ +\vec{n}^* . \iint_{S_i} (\Delta s \vec{T}_1 + \Delta t \vec{T}_2) \frac{dsdt}{|\vec{D}|} \} \quad & \leftarrow O(K\delta^2 \tfrac{\delta}{D}) \\ +O(\delta^3 \tfrac{\delta}{D} \, , \, K\delta^3 \tfrac{\delta}{D} \, , \, \ldots)] \quad . &\end{aligned} \tag{23}$$

It follows from Eq. (23) and from a similar expression for the potential induced by a doublet distribution that for a first or second-order method with the Dirichlet condition the same type of panel-wise representations are required as for the corresponding method with the Neumann condition. The condition under which the last two terms in Eq. (23) may be neglected is $K\delta^2_{local} < \delta^3$. Comparison of Eqs. (20) and (23) indicates that the same type of integrals appear in both expressions, so that also the same trancendental functions have to be computed. This implies that computationally the formulation in terms of Dirichlet conditions is not cheaper than the formulation in terms of Neumann conditions, but of course the former formulation require less storage.

Further aspects of the computational model

In most panel methods the panel midpoint or centroid is chosen as the collocation point. This leads to the simplest expressions in the far-field expansion, see Eq. (22). With this choice problems are also avoided with (nearly) cancelling weakly singular contributions associated with discontinuities across panel edges of the geometry and of the singularity distributions.

In the foregoing the contribution of the singularity distributions on a panel was expressed in terms of quantities at the panel expansion point. The next choice to be made concerns the numerical scheme to express these quantities, i.e. q^*, q^*_s, q^*_t, μ^*, μ^*_s, μ^*_t, μ^*_{ss}, μ^*_{st}, μ^*_{tt} and in case of free vortex sheets also $\vec{x}^*$, $\vec{x}^*_s$, $\vec{x}^*_t$, $\vec{x}^*_{ss}$, $\vec{x}^*_{st}$ and $\vec{x}^*_{tt}$, in terms of the parameter values to be solved for. For the specific choice of the panel collocation point the numerical scheme must be chosen such that the resulting method is stable. Proving stability of candidate numerical schemes for solving the integral equations, given in Eqs. (10), (12), (15.b) or (15.b) + (16.d) for the Neumann, Dirichlet, lifting-surface or wake boundary condition, is a difficult task. Most investigators base their choice on arguments from what is known about stability of interpolatory splines (e.g. quadratic spline: provide function value or second derivative at interval midpoints, or first derivative at the nodes), supplemented with numerical experimentation.

For thick configurations with the Neumann boundary condition and with the source distribution as unknown, which resulted in the integral equation of Eq. (10), a stable scheme is the central scheme based on the function value at panel midpoints as unknown parameters. For a first-order method this is simply a constant source distribution on a flat panel approximation. For methods employing a higher-order representation the precise form of the numerical scheme can take many forms. The scheme may have been derived from finite-difference type of expressions on a segmentwise defined rectangular computational domain, or from a least-squares fit of the parameters at immediate neighbouring midpoints, etc. The efficiency of a higher-order panel method will depend strongly on the compactness of the numerical scheme used.

For thick configurations with the Dirichlet boundary condition and with the doublet distribution as unknown, which resulted in the integral equation of Eq. (12), a stable scheme is also the central scheme based on mid-point function values.

For lifting surfaces, for which we derived integral equation Eq. (15.b) the choice of the numerical scheme is less trivial. It turns out that in subsonic flow the leading term of the equation imposed at $\vec{x}(s^*_i,t^*_j)$ is proportional to the second derivatives of μ at (s^*_i,t^*_j), so that for a second- order formulation a central scheme based on mid-point function values will provide stability. Stability for a method with a linear representation for the doublet distribution turns out to require a shift of the collocation point to the panel edges, with its own problems.

Finding a stable numerical scheme for the two coupled wake boundary conditions is even more difficult. The first condition is the lifting-surface condition above, the second one, Eq. (16.d) is quadratic in μ and involves the first derivative, while both conditions are highly nonlinear in terms of the unknown geometric parameters. Usually some kind of mixed numerical scheme, found by intuition and experimentation is arrived at.

In the case of subsonic flow a disturbance decreases in magnitude with increasing distance from its source, specifically a discontinuity in (in function value or derivatives) the numerical model and usually does not cause serious problems.
In supersonic flow disturbances, due to actual discontinuities but also due the ones introduced by the discretization, propagate to large distances downstream of where the discontinuity and associated Mach cone originated. This quite seriously hampers the application of low-order methods to complex multi-component aircraft configurations, and a higher-order method is almost obligatory for such configurations. Another problem constitutes the waves that propagate into the interior of the configuration and may give rise to a sequence of spurious internal reflections that eventually destroy the accuracy of the solution. This is illustrated in Fig. 10 (Ref. [9]) which shows the effect of a refined (in axial direction) paneling on the aft-cone of a cone-cylinder-cone configuration at zero incidence. In the interior of the aft cone the reflecting Mach waves cause the source distribution to oscillate severely, the "best" answer is obtained for the coarsest paneling. In Ref. [22] some of these problems could be alleviated by devicing a special composite source-doublet (triplet) singularity distribution with the property that it has a (partially) cancelling interior velocity field. However, for multiple-component configurations spurious reflections are still present. It appears that a higher-order formulation employing the Dirichlet boundary condition is superior in avoiding much of the undesired spurious reflections discussed above.
For a configuration computing time is generally less in supersonic flow than in subsonic flow. This is achieved by considering the region of dependence of the point where the influence is computed. If the complete part, segment (or strip) is outside the upstream Mach cone from the point its influence is zero and the individual panel influences do not need to be considered.

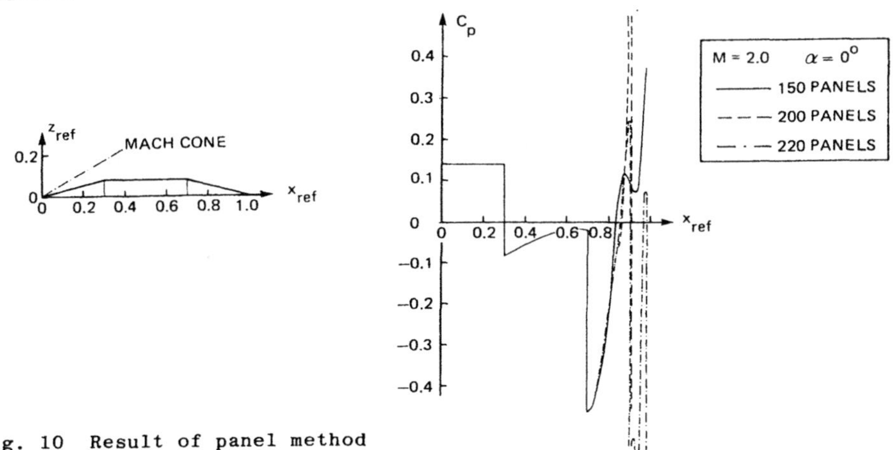

Fig. 10 Result of panel method with Neumann b.c. for supersonic flow [9]

Table 1: Partial list of methods presently in use

	Panel	Source	Doublet	B.C.	Remarks
DOUGLAS-NEUMANN [1], 1962/1972	Flat	Const.	Mode function (external)	Neumann	$M_\infty<1$
BOEING-TEA230 [2], 1967	Flat	Const.	Mode function (internal)	Neumann	$M_\infty<1$ Göthert rule
NLR [6], 1969	Flat	Const.	Mode function (internal)	Neumann	$M_\infty<1$ NLR comp. rule
MBB [6], 1970	Flat	Const.	Mode function (internal)	Neumann	$M_\infty<1$
HUNT-SEMPLE [7], 1976	Flat	Const.	Mode function (internal,opt.)	Neumann	$M_\infty<1$
USSAERO [8], 1973	Flat	TB:Const. LS:Linear	- Non-polynomial	Neumann Linear-ized	$M_\infty<1,>1$
NLRAERO [9], 1980	Flat	TB:Const. LS:Linear	- Quadratic	Neumann Linear-ized	$M_\infty<1,>1$
PAN AIR [10], 1975	Flat Sub-panels	Linear	Quadratic	Neumann/ Dirichlet	$M_\infty<1,>1$
DOUGLAS H.O. [11], 1980	Quadratic	Linear	Mode function (external,quad.)	Neumann	$M_\infty<1$
ROBERTS [12], 1975	Cubic	Cubic	Mode function (internal,Birnb.)	Neumann	$M_\infty=0$
MCAERO [13], 1980	Flat Sub-panels	Linear	Quadratic	Dirichlet	$M_\infty=0$?
SAAB [14], 1984	Quadratic	Linear	Quadratic	Dirichlet	$M_\infty=0$?
HISSS [15], 1984	Flat Sub-panels	Linear	Quadratic	Dirichlet	$M_\infty<1,>1$
VSAERO [16], 1981	Flat	Const.	Const.	Dirichlet	$M_\infty=0$(?) Wake relax.
QUADPAN [17], 1983	Flat	Const.	Const.	Dirichlet	$M_\infty<1$

BC: Boundary Condition
TB: Thick Bodies
LS: Lifting Surface

Description

At present there are many panel methods in use capable of computing the linearized potential flow about 3D configurations (see Table 1). Three categories can be distinguished:

- first-generation methods, Refs. [1]-[2], [5]-[9], all with the direct Neumann boundary condition and the flat-panel approximation. Lift is generated through a mode-function doublet distribution in the interior or on the surface of wings or by lifting surfaces. All these methods are first-order methods.
- second-generation methods, Refs. [10]-[15], all accounting for panel curvature and employing higher-order representations for the singularity distributions. Some methods still use the mode-function formulation, solving for q through the Neumann boundary condition, other employ the Dirichlet boundary condition and solve for μ. A number of such methods are under development, e.g. [18] and at NLR AEROPAN and PDAERO.
- advanced low-order methods, Refs. [16]-[17]. These apparently quite successful methods employ the Dirichlet boundary condition on the flat-panel approximation with a constant source as well as a constant doublet distribution, claiming higher-order accuracy.

Fig. 11 shows a comparison (Ref. [23]) of results of two first-generation methods (Refs. [6] and [7]) with the higher-order (3rd order!) method of Roberts [21]. The case considered is the incompressible flow about the "RAE wing" at 5 deg incidence for a panel scheme of 12 strips of 60 panels each. The left-hand side of the figure shows the comparison of the x- and the y-component of the velocity for a wing thickness of 5%. The x-component of the velocity does not differ much for the three methods considered. However, the y-component of the velocity computed by the method of Ref. [6], which employs as mode function an internal piecewise constant doublet distribution (equivalent to a vortex lattice, see Eq. (6.d)), of

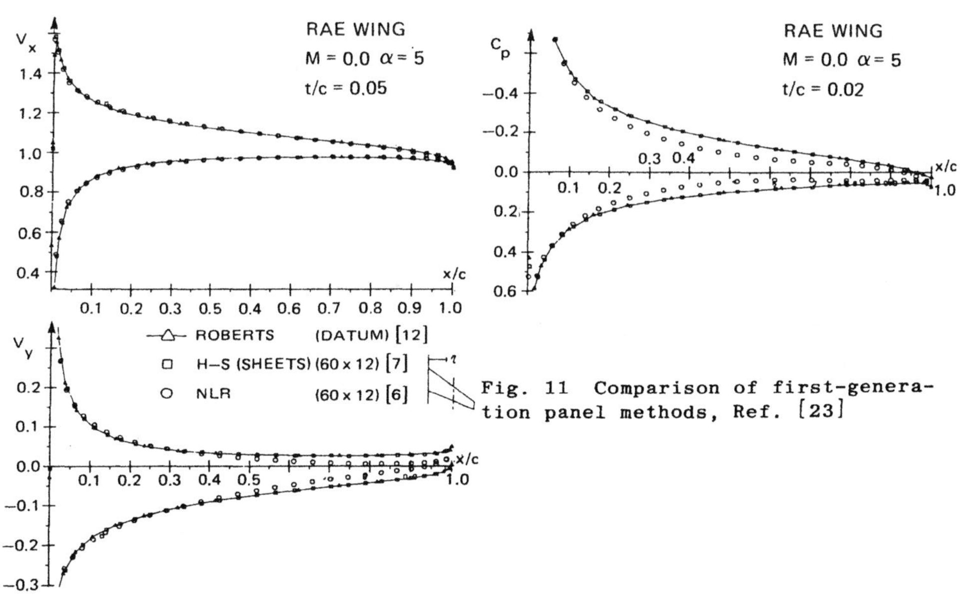

Fig. 11 Comparison of first-generation panel methods, Ref. [23]

user-specified shape, shows large deviations from the "datum" solution of Roberts and from the solution of the method of Ref. [7] that employs a more continuous "optimized" internal mode function. For even thinner wings the necessity of the more advanced internal mode function becomes even more apparent, as is demonstrated in the right-hand side of Fig. 11.
The advanced lower-order methods [16]-[17] employ the Dirichlet boundary condition, Eq. (12) with the apparent inconsistent discretisation of a panelwise constant source and constant (instead of linear) doublet distribution on the flat-panel approximation. In order to find the surface velocity Eq. (13.b) is to be evaluated. Using at least first-order accurate numerical differentiation to obtain $\vec{\nabla}\mu$, as is required for a fully consistent first order method, it is suggested that one still finds the surface velocity with O(h) accuracy. For the 2D case Oskam [24] investigated the accuracy of a method with the internal Dirichlet condition $\Phi=0$ (which leads

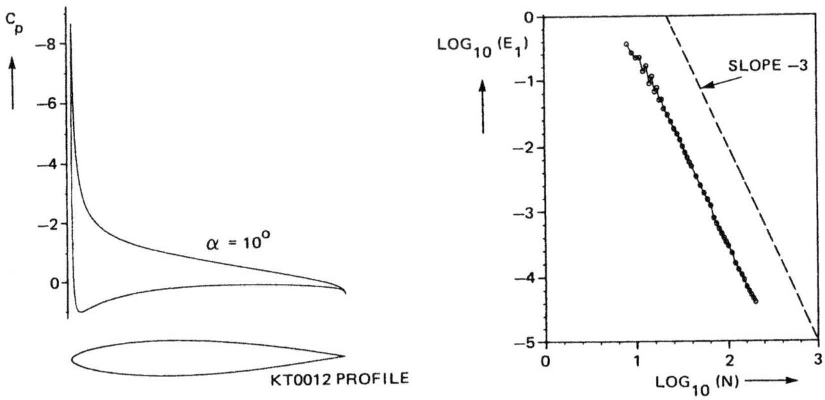

Fig. 12 Accuracy of solution for 2D flow using Dirichlet boundary condition and second-order formulation [24]

to $q\equiv0$ if $v_n\equiv0$). In Eq. (12) a quadratic representation for the doublet distribution and a curved-panel approximation was used, while in Eq. (13.b) a fourth-order accurate numerical differentiation was employed to find $\vec{\nabla}\mu$. The investigation showed that the resulting method instead of 2nd- is 3rd-order accurate (Fig. 12), as one might have hoped from the results of Refs. [16]-[17]. This is a puzzling situation and one may wonder what causes this anomaly, is the analysis based on the small-curvature expansion perhaps too conservative, or is there some hidden annihilation of error terms in the process of obtaining μ from Eq. (12) and the surface velocity from Eq. (13.b)?!

Neumann vs Dirichlet boundary condition

In this section we consider some of the strengths and weaknesses of methods that employ the Neumann boundary condition and methods that employ the Dirichlet boundary condition. The discussion is based on the literature and on experience gained at NLR.

The methods using the Neumann boundary condition solve for the source distribution on the surface of the configuration. Lift is generated through a (mode function) doublet distribution on an artificial surface in the interior of the configuration or on the actual surface of the configuration. Alternatively lifting components of the configuration are treated as lifting surfaces. Some positive (+) and negative (-) assets are:

+ The formulation appears to be forgiving for irregular paneling (M_∞<1).
- Lift-carry-over through user-specified or automatically generated artificial surfaces is subject to some arbitrariness.
- Thin wings may cause problems (Fig. 11).
- Internal flows cannot be modeled because of "leakage".
- Spurious Mach-wave reflections in the interior of the configuration.

The methods using the Dirichlet boundary condition solve for the doublet distribution on the surface of the configuration, while the source distribution follows from an algebraic relation. For this methods the following applies:

+ Lift-carry-over is partly implicitly accounted for. However, for a wing/-body part the intersection of the wake of wing with the body has to be identified as the edge of a segment across which the doublet distribution is discontinuous (Fig. 5).
+ More accurate for the same computing cost, or for the same accuracy less computing cost?
+ Better behaved in supersonic flow.
+ Less storage required.
- More sensitive to irregular paneling and gaps in the geometry.

In view of above points it is quite evident that the methods with the Dirichlet boundary condition are considered to be a definite improvement over the (older) methods with the more direct Neumann boundary condition. However, much expertise in applying the latter methods to practical situations, which often violate the underlying assumptions locally, has been built up, which does not carry over to the newer methods.

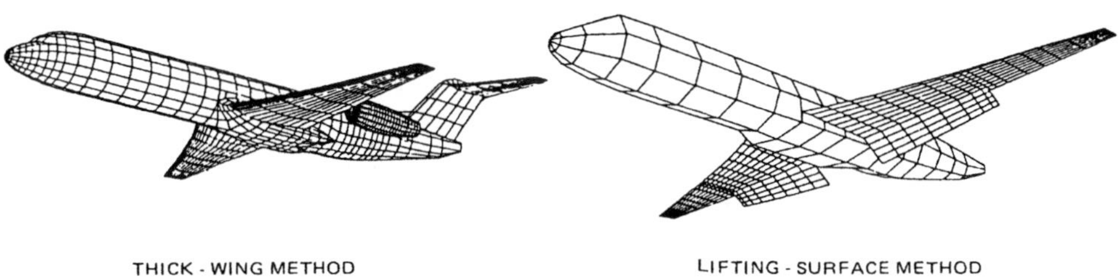

Fig. 13 Type of panel method (Courtesy Fokker Aircraft B.V.)

Lower-order vs higher-order methods

Regarding the matter if choosing (for the development and/or application) of a lower-order versus a higher-order method several arguments, pro or contra, can be put forward. For a low-order method it can be remarked that it:

++ is less complex to design, program and maintain. Less information is required to define the geometry and less AIC integrals have to be worked out.
+ has more flexibility because no higher-order continuity is pre-assumed or required.
- can introduce non-physical features in the flow field such as discrete vortices which may give rise to spurious, numerical effects.
- is not suited for supersonic flow unless "triplets" [22] or some kind of averaging (see Ref. [25]) is introduced.

Similarly, the higher-order methods are:

+ possibly more accurate for the panel size tending to zero. This does not imply that for a certain (coarse) paneling the solution of the higher-order method is better than the one obtained with the lower-order method.
+ more economic when a fine paneling is required for a subsequent computation of the effect of the boundary layer.
+ required for $M_\infty > 1$, and for wake relaxation.
- less flexible because more ordering is required in the specification of the geometry.
- computationally more expensive in case continuity of geometry and singularity distributions across segment boundaries is to be enforced.
- less attractive to develop because a thorough analysis is required to minimize the computational effort.
- difficult to code and maintain.

It should be noted here that in many practical situations the panel scheme chosen, because of restrictions in computing budget or computer core memory, is just fine enough to resolve the relevant flow features, so that in these cases there is no advantage in using the higher-order method. However, with computing cost decreasing and core sizes increasing the user will tend to increase the number of panels and the higher-order method will become more economic. It should also be realized that at locations where the solution is (nearly) singularly behaved, as frequently occurs at the edges of lifting surfaces, but also at sharp trailing edges, etc., higher-order accuracy is formally only attained if the singular parts in the solution are extracted and treated explicitly (e.g. Ref. [26]). This may be pursued in 2D but is far too complicated to be extended to the general 3D framework.

What type of panel method to use

The type of panel method to be used depends strongly on the purpose of the application (Fig. 13). If the method is to be used during the preliminary design of an aircraft project in which many possible candidate configurations are studied, a method that provides the six component forces and moments, stability derivatives and spanwise load distributions will suffice. For this type of application a relatively coarse paneling is

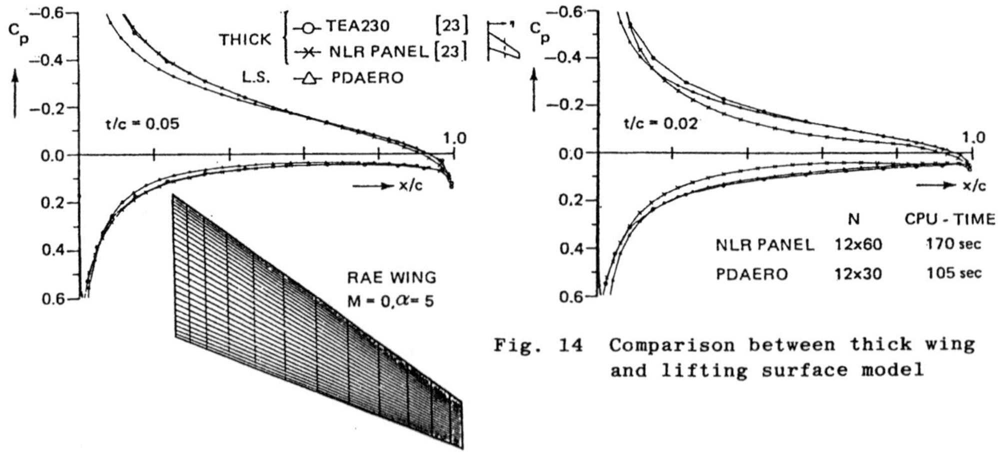

Fig. 14 Comparison between thick wing and lifting surface model

allowed. Also allowed will be the lifting surface approximation which reduces the computer cost (the number of panels on the wing is halved, see Fig. 14) even further. This type of consideration may lead to the for this application desired situation where the "turn-around-time" is such that the method can be used interactively. In this environment the designer can investigate rapidly the effect on the aerodynamic characteristics due to changing the position and type of the propulsion system, changes in wing-tail lay-out, flap settings, roll angles (for missiles) etc. Fig. 14 provides an insight in the accuracy of the predicted pressure distribution. It shows that for both wing thicknesses the thick-wing and the lifting-surface model give comparable results, except near the blunt leading edge where the lifting-surface approximation is invalidated.

More detailed investigations, at a later phase in an aircraft project, requiring accurate pressure distributions, will ask for a finer and also for the actual wing surfaces to be modeled, or maybe even for wake relaxation. Clearly this requires an accurate, reliable and computationally efficient method.

In cases where the configuration operates at subsonic and supersonic speeds (combat aircraft, missiles) it will be advantageous to use a method that applies to both subsonic and supersonic flow. The main benefit here will be the saving in manhours to prepare the input for a subsonic flow method and a separate one for a supersonic flow method.

Above discussion tends to lead to the conclusion that a general purpose panel method, probably second-order, with several aerodynamic modeling options as thick wings, lifting surfaces, near wakes, inflow and outflow segments, subsonic or supersonic flow, etc., as well as automatic (re)-paneling options is the aerodynamic tool that is needed. A prerequisite for such a "building-block" system is that, in spite of the many options, it remains extendable, maintainable, economic and above all "user-friendly".

Regarding the latter it must be kept in mind that in the application of panel methods the costs involved in the manhours required for preparing the input and analyzing the results of the computation generally far exceed the bare computing costs.

PANEL METHOD ENVIRONMENT

As sketched in Fig. 15 the panel method is embedded between pre- and post-processing. The main task of the pre-processing is the generation of the input for the method, which includes:

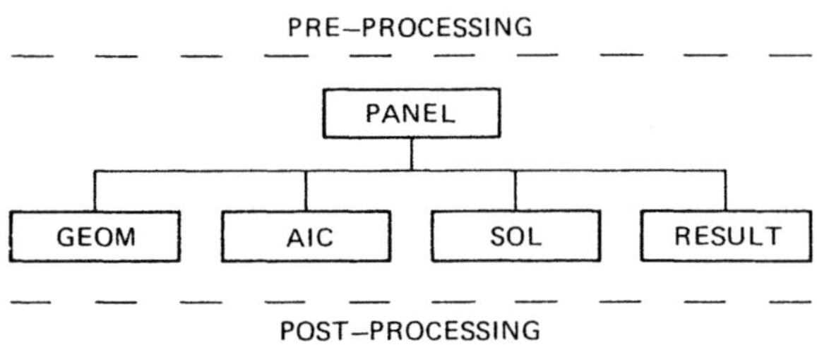

Fig. 15 Panel method environment

- definition of the geometry, subdivision into parts and segments, determination of intersections between wings and bodies, etc.
- paneling of the individual segments, or in case the method features automatic paneling features, specification of the panel scheme parameters
- specification of near-wake surfaces, inlet faces, lift-carry-over segments, internal surfaces carrying the mode-function doublet distribution, etc.
- inspection of the paneled configuration (Fig. 16)
- specification of the normal velocity component v_n that simulates viscous effects, inlet flow, etc.
- specification of slipstream data as velocity and total pressure increment, other user-specified onset flows, etc.
- specification of free-stream direction, Mach number, steady rotation rates, etc.

The first few points of above list require that the user has access to a geometry package for geometry manipulation. Inspection will involve the visualization of the paneled geometry as a wire frame (Fig. 13), with/without "hidden lines", as a "solid model", as a "pin cushion" to check on the direction of the normals, etc. The last couple of points on above list may require the interfacing with other methods like a boundary-layer calculation method, a method for wake relaxation, a method for isolated propeller aerodynamics, etc.

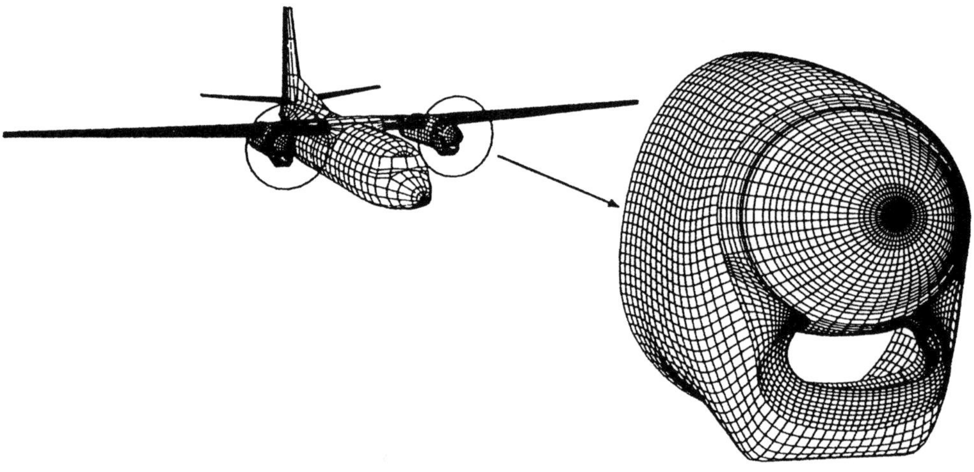

Fig. 16 Geometry modeling (Courtesy Fokker Aircraft B.V.)

The main task of the post-processing is the digestion of the output of the panel method. It may involve:
- generation and visualization of pressure plots, isobars, surface (Fig. 17) and free streamlines
- comparison with other data from theory or experiment
- (weak-interaction) boundary-layer computation
- wake relaxation
- archiving of aerodynamic data in a data-base system.

In the practical use of panel methods the rapid and user-friendly visualization of geometry and of flow results on advanced graphical (color) workstations is essential.

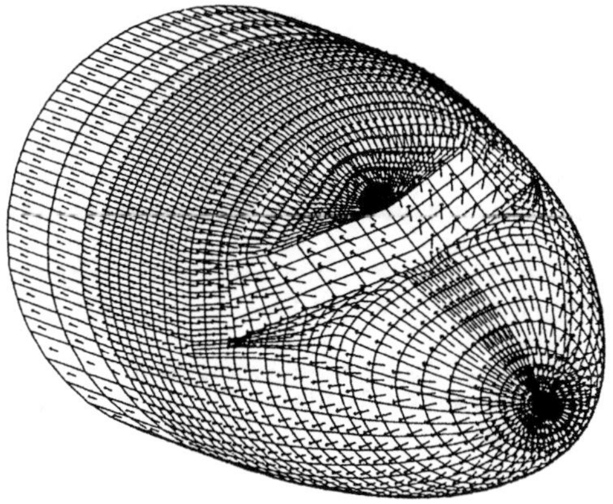

Fig. 17 Surface velocity vectors (Courtesy Fokker Aircraft B.V.)

OPPORTUNITIES FOR IMPROVEMENT

There are several areas where (existing) panel methods may be improved (see also [27]). Referring to Fig. 15, where the various parts of a panel method are indicated, the following items are considered.

GEOM: This part of the program handles the geometric input, in "stand-alone" panel codes it acts as preprocessor. In this part of the method an automatic procedure for generating a (curvature- or even solution-)adaptive panelling would result in an increased accuracy.

AIC: In this part of the program the influence integrals are evaluated, the operational count is $O(N^2)$. In present-day panel methods, using an iterative solver for the system of equations, this part of the program accounts for up to 80% of the total CPU time. Here attention has to be paid to the vectorization of the code, such that it runs efficiently on super-computers. As an example of the speed-up that is obtained on super-computers the table below gives the CPU time required for running the (scalar) NLR PDAERO panel method (Neumann b.c. on bodies/lifting surface b.c. on wings) for a configuration with 1500 unknown and 800 known singularity parameters.

	CYBER 180		CRAY XMP
	NOS	NOSVE	48+SSD
AIC	600	?	76
SOL(LU)	3000	?	44
TOTAL	3650	1550	123

PDAERO CPU-SEC.

	CORE MEMORY:
CYBER NOS	60 KWORDS
CYBER NOSVE	1 MWORDS
CRAY	1 MWORDS

It shows that even for the unmodified code a substantial speed-up is realized. Note that because of differences in core memory used, part of the reduction of the CPU time is due to the smaller amount of I/O activity required. It is expected that the CPU-time required for AIC can be reduced further by rearranging the computation such that a greater part of the code vectorizes.
Another area of interest is reduction of the $O(N^2)$ operational count, without sacrificing the accuracy of the solution. Although some studies have been initiated in this area, e.g. [28], progress has been slow.

SOL: In this part of the program the system of equations is solved, either using an iterative procedure or a direct (LU) procedure. The direct solution require $O(N^3)$ operations, but can be vectorized to a large extent, see above table. The iterative solution procedure, mostly block-Jacobi or Gauss-Seidel, requires $O(it*N^2)$ operations with it the number of iterations. The main problem with the iterative procedure is that for complex configurations, depending on the paneling, the number of iterations may be excessive or it may occur that the procedure fails to convergence. Also for lifting surfaces in subsonic flow the convergence is rather slow [9]. More robust iteration procedures, as multi-grid, are to be investigated.

RESULT: In this part of the program the velocity and pressure distribution are computed and are used to compute loads, etc. Also a file is prepared for use during post-processing.

RECENT AREAS OF INTEREST AND APPLICATION

There are numerous areas where the panel method technique is successfully applied. Some areas that have attracted some attention recently are:

Propulsion installation effects. The renewed interest in propeller propulsion of transport aircraft has brought along the need to predict the effects of the slipstream on the configuration aerodynamics. During preliminary design studies mostly a simple model is used in which the flow field due to the propeller in isolation is superimposed on the free stream onset flow. However, it appears that this simple model is not always adequate and an improved modeling is required.

Viscous effects. The pressure distribution computed by the panel method can be used to calculate the boundary layer on the surface of the configuration. Under cruise conditions the flow will be attached and a weak interaction may be assumed. However, for take-off and landing configurations a strong inviscid-viscous flow coupling is to be taken into account, while for wing-flap configurations also the viscous wake modeling needs to be considered. An important item here is that the arbitrary-geometry capability is typically further developed for panel methods than for boundary-layer methods.

Wake relaxation. The classical rigid-wake approximation of straight trailing vortex lines is totally inadequate for configurations with closely-coupled components as wing-flap-tail configurations (take-off and landing) and canard-wing configurations (combat aircraft and missiles at $M_\infty<1$, >1). The rigid-wake approximation with a user-specified "near wake" (Fig. 7) will improve the modeling only if the vortex lines (μ=const.) on this part of the wake are sufficiently aligned with the streamlines. A possible partial alleviation of this problem is to fix the shape of the near wake and its paneling but to "relax" the doublet distribution (vortex lines) on the near wake by imposing Eq. (16.d). This later condition is quadratic in μ, so that an iterative procedure is required, which however does not involve a (partial) recomputation of the AIC's. The shape of the near-wake, still to be specified by the user, might be obtained from a method that solves the non-linear problem in an approximate framework. An example of this is the method which within the framework of the 2D time-dependent analogy computes, employing a 2D second-order panel method, the roll-up of general vortex wakes [29]. In Fig. 18a a result of this method is presented, which also demonstrates the complexity of the wake of configurations with deflected flaps.

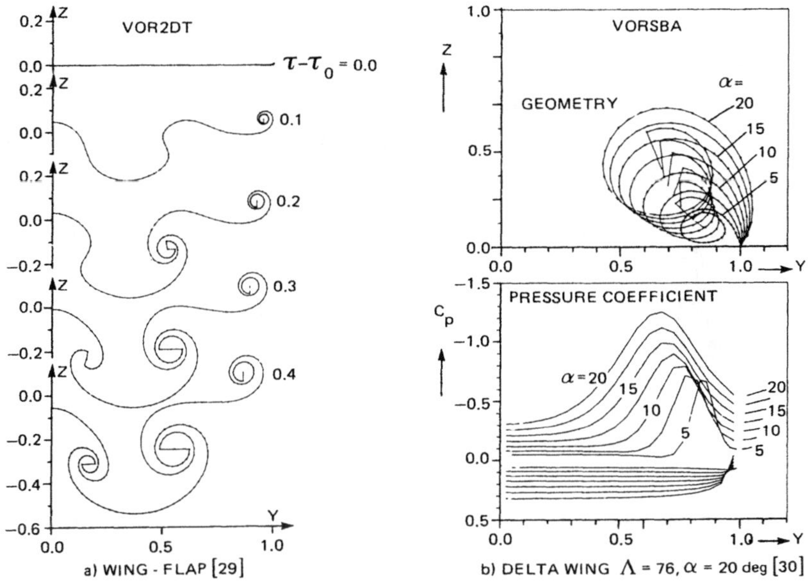

Fig. 18 Example of results of panel methods for configurations with free vortex sheets; approximate frame works, a) time-dependent analogy, b) slender-body approximation

In cases where the interaction of the wake and the solid geometry is stronger, e.g. separation from flap side edges, wing tips or for slender wings with leading-edge vortex sheets the two wake boundary conditions (Eqs. (16.b) and (16.d)) have to be solved simultaneously. The fully non-linear 3D-wake relaxation problem is a tough problem. Here also methods formulated in an approximate framework, as slender-body theory, are used for preliminary studies or as preprocessor that constructs the initial guess for the method for fully 3D flow. Fig. 18b shows the result of such a non-linear second-order panel method [30], formulated in the slender-body-approximation, for the flow about a thin delta wing of unit aspect ratio at several incidences. Subsequently such a solution is used to construct initial guess for the method for fully 3D flow, see Fig. 19 for a typical result.

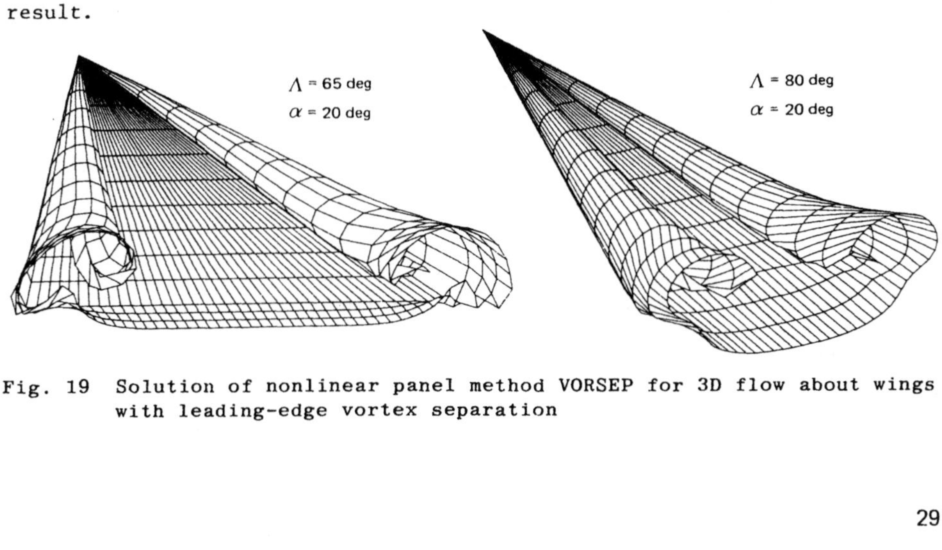

Fig. 19 Solution of nonlinear panel method VORSEP for 3D flow about wings with leading-edge vortex separation

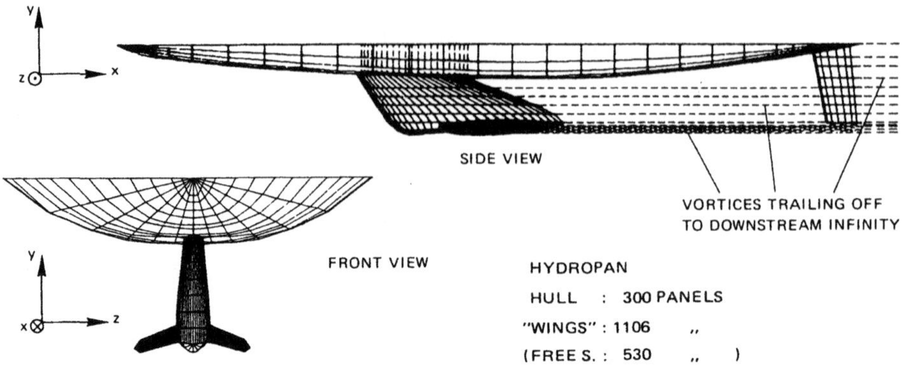

Fig. 20 Example of panelling of hydrodynamical application [31]

Free-surface effects. Panel methods are also applied in hydrodynamics. For the flow about submerged or partly submerged objects the effect due to the free surface may be substantial. At NLR the NLR panel method [5] has been extended to hydrodynamical problems by including the free surface [31]. On the paneled free surface, which is approximated as a rigid surface, the linearized free-surface boundary conditions are applied. From the computed solution the wave resistance is deduced. Fig. 20 shows the paneling for one of the recent applications.

EXTENSION OF DOMAIN OF APPLICABILITY

The domain of applicability of the panel method for linearized potential flow is limited to sub-critical flow. However, extension of the capability of the panel method approach to flows with regions in which non-linear compressibility effects cannot be neglected is possible.
In one approach one includes in the integral representation, Eq. (5), for the solution of the Prandtl-Glauert equation the contribution due to a source distribution σ in the field, i.e. the solution of Eq. (3.a) is expressed as

$$\varphi(\vec{x}_o) = \varphi_q(\vec{x}_o) + \varphi_\mu(\vec{x}_o) + \varphi_\sigma(\vec{x}_o) \quad , \tag{24.a}$$

where φ_q and φ_μ are defined in Eqs. (5.b-c) and

$$\varphi_\sigma(\vec{x}_o) = \frac{-1}{4\pi}\iiint_V \sigma(\vec{x}) \frac{dV(\vec{x})}{|\vec{R}|} \quad . \tag{24.b}$$

In Eq. (24.b) $V(\vec{x})$ denotes the region(s) with nonlinear compressibility effects and $\sigma(\vec{x})$ the spatial (field-)source distribution. In the "field-panel" method the spatial source distribution is found by satisfying the full-potential equation Eq. (1.a) at the points within $V(\vec{x})$. It has been shown in Ref. [32] that for the 2D (Transonic Small Perturbation) case modeling of super-critical flow with shock waves is possible. In Ref. [33] the 2D field-panel approach was extended to the full-potential equation (1.a), using established techniques of contemporary finite-volume methods.

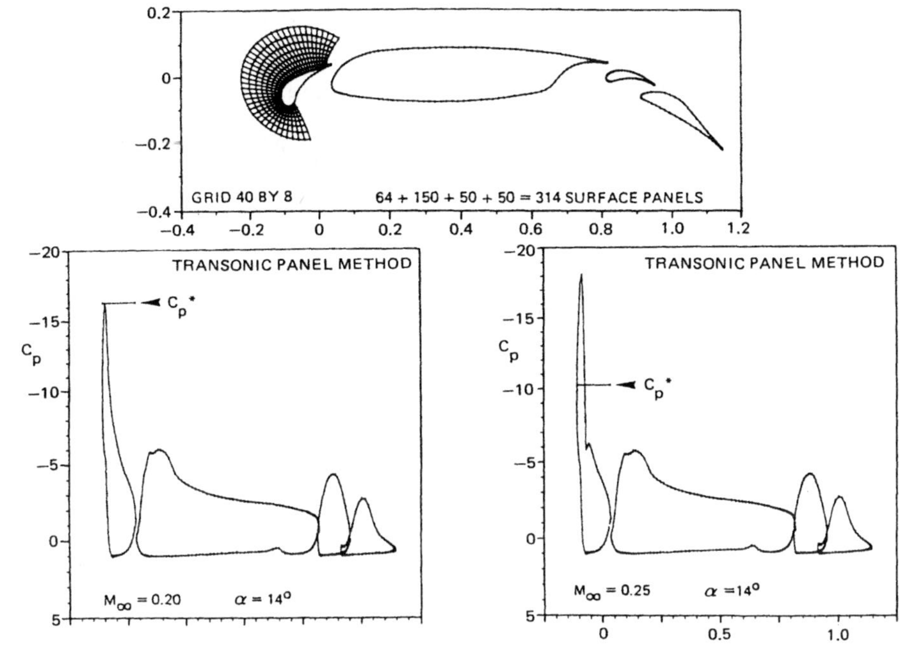

Fig. 21 Application of field panel method to 4 component airfoil section [33]

Fig. 21 shows the result of the application of this method to the flow about a 4-component airfoil section at 14 deg incidence at M_∞=0.2 and 0.25. It is observed that a small region of super-critical flow terminated by a shock developes on the highly loaded slat as M_∞ increases from 0.2 to 0.25. This example indicates that for configurations with separate compact regions of nonlinear compressibility effects the panel method can be applied with success, without sacrificing the capability of linear panel methods to treat complex geometries. However, note that the number of field panels increases very rapidly as the extent of the transonic flow region(s) becomes larger.

If the field-panel method is to be applied successfully to 3D configurations the computational costs for evaluating the influence integrals will have to be reduced considerably, by vectorization or preferably by lowering the $O(N^2)$ operational count. Another point worth noting is that ideally the regions with nonlinear flow would be detected automatically by the program in some kind of iterative procedure. Ref. [34] describes an application of the field-panel concept to the flow about delta wings with leading-edge-vortex separation, using the vortex-lattice method to simulate the linear potential flow.

An alternative to the approach using a field source distribution is the zonal (hybrid) type of approach. Here the full-potential equation Eq. (1.a) is solved in the regions where nonlinear compressibility effects are non-negligible and the Prandtl-Glauert equation elsewhere. The two zones are coupled iteratively through the the boundary conditions on the interface between the zones.

At Boeing [35] the PAN AIR code is being extended to transonic flow by superimposing onto the arbitrary surface-paneled configuration a spatial rectangular grid. The volume integrals on this rectangular grid, that is not body-conforming, are evaluated using Fast Fourier Transforms.

Other investigators [25] suggest that there are prospects that the non-linear compressible flow problem can be formulated in terms of surface integrals only, though the latter have to be re-evaluated in an iterative procedure.

CONCLUDING REMARKS

- Panel methods are important aerodynamic tools with powerful and flexible modeling capabilities, which are used at an increasing rate in aircraft design projects.
- Panel methods still form a fruitful area of research, in particular with reference to the extension into the transonic flow regime and improved handling of wakes of closely-coupled components, the results having a direct implication on the applicability of the method.

There exist possibilities for improving the computational efficiency of the panel method through:

. vectorization on supercomputers
. reduction of operational count for the evaluation of influence integrals
. new formulations and improved numerics
. better, more robust (and faster) iterative procedures for solving large, non-sparse systems of equations

which will lead to a further utilization of panel methods.

- Pre- and post-processing, are an essential part of the "panel-method environment".
- At all times it should be realized that panel methods are modeling real flow under a great number of assumptions.

REFERENCES

[1] Hess, J.L., Smith, A.M.O.: "Calculation of non-lifting potential flow about arbitrary three-dimensional bodies", Douglas Aircraft Report No. E.S. 40622 (1962), J. of Ship Res. 8, No. 2 (1964), pp. 22-44.

[2] Rubbert, P.E., Saaris, G.R.: "A general three-dimensional potential flow method applied to V/STOL aerodynamics", SAE Paper 68004 (1968).

[3] Prager, W.: "Die Druckverteilung an Körper in einer Potentialströmung", Physik. Zeitschr. (1928), pp. 865-869.

[4] Martensen, E.: "Die Berechnung der Druckverteilung an dicken Gitterprofilen mit Hilfe von Fredholmschen Integralgleichungen zweiter Art", Mitt. Max-Planck-Inst. Strömungs Forschung No. 23 (1959).

[5] Labrujère, Th. E., Loeve, W., Slooff, J.W.: "An approximate method for the calculation of the pressure distribution on wing-body combinations", AGARD CP-71 (1970).

[6] Kraus, W.: "Das MBB-UFE Unterschall Panelverfahren", Report MBB-UFE 633-70 (1970).

[7] Hunt, B., Semple, W.G.: "The BAC(MAD) program to solve the 3-D lifting subsonic Neumann problem using the plane panel method", Report ARG 97 BAC(MAD) (1976).

[8] Woodward, F.A.: "An improved method for the aerodynamic analysis of wing-body-tail configurations in subsonic or supersonic flow", NASA CR-2228 (1973).

[9] Hoeijmakers, H.W.M.: "A panel method for the determination of the aerodynamic characteristics of complex configurations in linearized subsonic or supersonic flow", Report NLR TR 80124 (1980).

[10] Carmichael, R.L., Erickson, L.L.: "PAN AIR-A higher order panel method for predicting subsonic or supersonic linear potential flows about arbitrary configurations", AIAA Paper 81-1255 (1981).

[11] Hess, J.L.: "A higher-order panel method for three-dimensional potential flow", Report MDC J8519 (1979).

[12] Roberts, A., Rundle, K.: "Computation of incompressible flow about bodies and thick wings using the spline-mode system", BAC(CAD) Rep. Aero Ma 19 (1972).

[13] Bristow, D.R.: "Development of panel methods for subsonic analysis and design", NASA CR 3234 (1980).

[14] Lötstedt, P.: "A three-dimensional higher-order panel method for subsonic flow problems - Description and applications", SAAB-SCANIA Rep. L-0-1 R100 (1984).

[15] Fornasier, L.: "HISSS - A higher-order subsonic/supersonic singularity method for calculating linearized potential flow", AIAA Paper 84-1646 (1984).

[16] Maskew, B.: "Prediction of subsonic aerodynamic characteristics: A case for low-order panel methods", AIAA Journal, Vol. 19, No. 2 (1982), pp. 157-163.

[17] Youngren, H.H., Bouchard, E.E., Coopersmith, R.M., Miranda, L.R.: "Comparison of panel method formulations and its influence on the development of QUADPAN, an advanced low-order method", AIAA Paper 83-1827 (1983).

[18] Lê, T.H., Morchoisne, Y., Ryan, J.: "Techniques numériques nouvelles dans les methodes de singularités pour l'application à des configuration tri-dimensionelles complexes", Paper 6, AGARD-CP-412 (1986).

[19] Kellogg, O.D.: "Foundations of potential theory", Dover (1954).

[20] Ward, G.N.: "Linearized theory of steady high-speed flow", Cambridge University Press (1955).

[21] Hess, J.L.: "Consistent velocity and potential expansions for higher-order surface singularity methods", Report MDC J691L (1975).

[22] Woodward, F.A., Landrum, E.J.: "The supersonic triplet - A new aerodynamic panel singularity with directional properties", AIAA Journal, Vol. 18, No. 2 (1980), pp. 138-142.

[23] Sytsma, H.A., Hewitt, B.L., Rubbert, P.E.: "A comparison of panel methods for subsonic flow computation", AGARDograph No. 241 (1979).

[24] Oskam, B.: "Asymptotic convergence of higher-order accurate panel methods", J. of Aircraft, Vol. 23, No.2 (1986), pp. 126-130.

[25] Hunt, B., Hewitt, B.L.: "The indirect boundary-integral formulation for elliptic, hyperbolic and non-linear fluid flows", Ch. 8 of Development in Boundary Element Methods, Vol. 4, Elsevier Applied Science Publishers (1986).

[26] Hoeijmakers, H.W.M.: "Aspects of second- and third-order panel methods demonstrated for the two-dimensional flat plate problem", NLR TR 81074 U, (1981).

[27] Slooff, J.W.: "Requirements and developments shaping a next generation of integral methods", Paper IMA Conf. on Num. Meth. Aeron. Fl. Dyn., Reading, (1981). Also NLR MP 81007 U.

[28] Schippers, H.: "On the evaluation of aerodynamic influence coefficients", Paper presented at this seminar.

[29] Hoeijmakers, H.W.M.: "An approximate method for computing inviscid vortex wake roll-up", NLR TR 85149 U (1985).

[30] Hoeijmakers, H.W.M.: "An approximate method for computing the flow about slender configurations with vortex-flow separation", NLR TR 86011 U (1986).

[31] van Beek, C.M., Piers, W.J., Slooff, J.W.: "Boundary integral method for the computation of potential flow about ship configurations with lift and free surface effects", NLR TR 85142 U (1985).

[32] Piers, W.J., Slooff, J.W.: "Calculation of transonic flow by means of a shock-capturing field panel method", AIAA Paper 79-1459 (1979).

[33] Oskam, B.: "Transonic panel method for the full potential equation applied to multicomponent airfoils", AIAA Journal, Vol. 23, No. 9 (1985), pp. 1327-1334.

[34] Kandil, O.A., Yates, E.C.: "Transonic vortex flow past delta wings: Integrals equation approach", AIAA Journal, Vol. 24, No. 11 (1986), pp. 1729-1736.

[35] Ericksson, L.L., Strande, S.M.: "A theoretical basis for extending surface-paneling methods to transonic flow", AIAA Journal, Vol. 23, No. 12 (1985), pp. 1860-1867. See also AIAA Paper 87-0034 (1987).

TIME DOMAIN BOUNDARY ELEMENT SOLUTIONS OF HYPERBOLIC EQUATIONS FOR 2-D TRANSIENT WAVE PROPAGATION

H. Antes
Ruhr-Universität Bochum, Institut für Mechanik
Universitätsstr. 150, D-4630 Bochum, Germany

SUMMARY

Transient wave propagation in two-dimensional elastic, isotropic media and in compressible, inviscid fluids is treated via time-dependent boundary integro-differential equations. The derivation is based on Graffi's reciprocal theorem [1] in the time domain. The results of numerical calculations from elasto- and fluid-dynamics, e.g. pressure wave propagation in a semi-infinite fluid domain, indicate that time-stepping boundary element techniques are avery efficient way of dealing with transient problems.

INTRODUCTION

Most of the research carried out so far on boundary element methods is concerned with the solution of elliptic and parabolic-type differential equations, and the large number of investigations that have been performed in this area show that the boundary element method is an efficient technique for treating these types of problems. Considerably less effort has been directed towards the solution of hyperbolic differential equations, although this type of equations is of paramount importance in structural dynamics. For example, the wave propagation in elastic media as well as in compressible fluids is governed by hyperbolic equations.

The most important advantage of the boundary element method (BEM) in handling dynamic problems is that it can easily be applied to problems involving infinite domains. Due to the incorporation of the proper fundamental solutions, it automatically takes into account the effect of radiation damping at infinity.

Time domain boundary element analyses of hyperbolic equations, i.e. BEM solutions coupled with a time-stepping technique have so far been performed only for special problems. Among these is the work of Karabalis and Beskos [2], Spyrakos et al. [3,4], Mansur [5] and Antes [6] in elastodynamics, and of Mansur and Brebbia [7,8] in fluid-dynamics. Antes et al. [9,10] also treated the dynamic interaction between compressible fluids and elastic soil-structure systems.

This paper presents essential aspects of time-dependent boundary integral equations and of the time-stepping boundary element procedure for the vectorial as well as the scalar wave equation. Its aim is to demonstrate ways of handling hyperbolic equations by the boundary element method.

BASIC DIFFERENTIAL, INTEGRAL AND INTEGRO-DIFFERENTIAL EQUATIONS

The present two-dimensional study is based on the following assumptions:
a) the material is linearly elastic, isotropic and homogeneous
b) the fluid is compressible but has no internal viscosity
c) the motion of the fluid particles as well as that of the elastic material is limited to small amplitudes.

Under these premises, the equations of motion for elastic material can be written in the following vectorial form

$$L(\mathbf{u}) = (c_1^2 - c_2^2)\nabla\nabla\cdot\mathbf{u} + c_2^2\nabla\cdot\nabla\mathbf{u} - \ddot{\mathbf{u}} = -\mathbf{b}/\rho \ , \tag{1}$$

whereas for the compressible fluid the scalar equation

$$L(\Phi) = c^2 \ \nabla\cdot\nabla\Phi - \ddot{\Phi} = -\gamma \tag{2}$$

holds. In (1) and (2), ∇ denotes the vector of spatial derivatives and overdots indicate derivatives with respect to time. The velocities of elastic dilatational and distortional waves are denoted by c_1 and c_2, whereas c stands for the pressure wavespeed in fluids. Also, ρ stands for the mass density and b_i for the body forces of the elastic medium, while γ and ρ_f give the source distribution and the density of the fluid.

Along the boundary segment Γ_1 either displacements $\bar{\mathbf{u}}(\mathbf{x},t)$ or the velocity potential $\bar{\Phi}(\mathbf{x},t)$ is prescribed, whereas on Γ_2 the tractions $\bar{\mathbf{T}}(\mathbf{x},t)$ or the flux $\bar{\phi}(\mathbf{x},t)$ are prescribed. The traction components and the flux can also be expressed as functions of the displacement and potential derivatives. This leads to

$$\mathbf{T} = \rho[(c_1^2-c_2^2)\ \mathbf{u}\cdot\nabla\mathbf{n} + c_2^2(\mathbf{u}\nabla\cdot\mathbf{n} + \mathbf{u}\cdot\mathbf{n}\nabla)] , \tag{3}$$

$$\phi = c^2\ \mathbf{n}\cdot\nabla\ \Phi \quad . \tag{4}$$

Moreover, at t = 0 the solutions have to satisfy appropriate initial conditions, that is, prescribed displacements and velocities $\mathbf{u}(\mathbf{x},0) = \mathbf{u}_o(\mathbf{x})$ and $\dot{\mathbf{u}}(\mathbf{x},0) = \mathbf{v}_o(\mathbf{x})$, or prescribed potential and pressure values $\Phi(\mathbf{x},0) = \Phi_o(\mathbf{x})$ and $\dot{\Phi}(\mathbf{x},0)=p(\mathbf{x},0)/\rho_f = p_o(\mathbf{x})/\rho_f$.

Obvi usly, the initial-boundary value problem of wave propagation in compressible fluids is a special case of the more general wave propagation problem in elastic media. This is due to the fact that one obtains equation (2), together with the appropriate boundary and initial conditions, from equation (1) and its respective conditions by formally setting $c_1 = c_2 = c$, and considering scalar instead of vector-valued states. This holds for all analytical and numerical solution steps. Thus, for lack of space and convenience of derivation, we may restrict ourselves in considerations to the simpler scalar case. Details of the procedure for the more general case of elastic media may be found in reference [6].

The basic idea of the proposed method is to describe the initial-boundary value problem by integral equations. In order to derive these using Graffi's elastodynamic reciprocal theorem [1], it is necessary to consider two distinct states which satisfy equations (2), whereby one is subjected to the actual source distribution γ and the other to a unit impulsive source at time τ and point ξ. Thus, one obtains [9,10]

$$\begin{aligned} d\ \Phi(\xi,t) = \int_0^{t^+}\{ \oint_\Gamma [\overset{*}{\Phi}(\mathbf{x},\xi,t-\tau)\phi(\mathbf{x},\tau) - \overset{*}{\phi}(\mathbf{x},\xi,t-\tau)\Phi(\mathbf{x},\tau)]d\Gamma_{\mathbf{x}} \\ + \int_\Omega \overset{*}{\Phi}(\mathbf{x},\xi,t-\tau)\gamma(\mathbf{x},\tau)d\Omega_{\mathbf{x}}\}d\tau \\ + \frac{1}{c^2}\int_\Omega [\overset{*}{\Phi}(\mathbf{x},t)p_o(\mathbf{x})/\rho_f - \frac{\partial}{\partial t}\overset{*}{\Phi}(\mathbf{x},t)\ \Phi_o(\mathbf{x})]d\Omega_{\mathbf{x}} \quad . \end{aligned} \tag{5}$$

On a smooth boundary ($\xi\epsilon\Gamma$) we have d=0.5 while for interior points ($\xi\epsilon\Omega$) the value d=1 holds. The upper limit of integration t^+ represents $t+\varepsilon$, ε being arbitrarily small. The necessary fundamental solution of equation (2) is given by [5,6]

$$\overset{*}{\Phi}(\mathbf{x},\xi,t-\tau) = \frac{1}{2\pi c}\frac{1}{R}\ H[c(t-\tau) - r] \ , \tag{6}$$

where

$$R = (c^2t'^2-r^2)^{1/2} \quad ; \quad t' = t-\tau \quad ; \quad r = |\,\mathbf{x}-\boldsymbol{\xi}\,| \tag{7}$$

and H is the Heaviside step function guaranteeing the causality of the propagating waves.

The corresponding singular flux $\overset{*}{\psi}(\mathbf{x},t)$ can be determined by use of relation (4), but it should be noted that it contains derivatives of the Heaviside function, i.e. the Dirac function, in combination with a 1/R singularity as well as the Heaviside function together with $1/R^3$. These terms are not amenable to numerical solution procedures. In order to eliminat them, some integrations by parts with respect to time need to be carried out. Assuming zero source density $\gamma(\mathbf{x},t)$ and initial conditions, the result of such regularization is the following boundary integro-differential equation for wave propagation in compressible fluids:

$$d\,\Phi(\xi,t) = \int_0^{t^+}\oint_\Gamma \{\, \overset{*}{\Phi}(\mathbf{x},\xi,t')\psi(\mathbf{x},\tau) + \frac{c}{2\pi R}\, H(ct'-r)r_{,n}[\frac{ct'-r}{R^2}\;\Phi(\mathbf{x},\tau) + \frac{1}{c}\dot{\Phi}(\mathbf{x},\tau)]\,\}d\Gamma d\tau. \tag{8}$$

By similar transformations an appropriate system of integro-differential equations describing the wave propagation in elastic media can be derived [6].

The only singularities in these integro-differential equations are those that occur when r and t' tend to zero simultaneously. Thus, these time-dependent equations contain kernels which only have a 1/R singularity, so that they can be integrated numerically without difficulty, provided that appropriate shape-functions are used.

NUMERICAL IMPLEMENTATION AND BOUNDARY ELEMENT EQUATIONS

Equation (8) can now be used to determine the time-dependent potential in the interior of Ω as well as along the boundary Γ. However, one must first find values for the unkown potential Φ and the flux ψ along the boundary. The procedure thus consists of

(i) a discretization of the boundary in which the potential and the flux are assumed to be constant over each boundary element Γ_1.

(ii) a step-by-step integration in time whereby the potential as well as the flux are taken to be linear and constant over each time interval Δt.

In order to arrive at systems of algebraic equations, collocation is used at every node ξ_λ and at all time steps $t_m = m\Delta t$. Then, integrations over each time interval and over each boundary element have to be carried out, according to (i) and (ii), e.g.

$$\mathbf{H}^{nm} := (\, H^{nm}_{\lambda 1}\,) = \int_{(n-1)\Delta t}^{n\Delta t} \int_{\Gamma_1} \overset{*}{\Phi}(\mathbf{x},\xi_\lambda,m\Delta t-\tau)d\Gamma_{\mathbf{x}}d\tau\,. \tag{9}$$

The integrations with respect to time have been performed analytically:

$$H^{nm}_{\lambda 1} = \frac{1}{2\pi c^2}\int_{\Gamma_1} \ln(\frac{n-m+1\;+\;D_{nm}(r,1)}{n-m\;\;+D_{nm}(r,0)})\;\; d\Gamma_{\mathbf{x}}\,, \quad (r=|\mathbf{x}-\xi_\lambda|), \tag{10}$$

$$G^{nm}_{\lambda l} = -\frac{1}{2\pi c^2}\int_{\Gamma_l}\frac{\mathbf{n}\cdot(\nabla r)}{r}[D_{nm}(r,1)-2D_{nm}(r,0)+D_{nm}(r,-1)]d\Gamma_{\mathbf{x}} \;, \tag{11}$$

where

$$D_{nm}(r,q) = \begin{cases}\sqrt{(n-m+q)^2 - (r/c\Delta t)^2} & \text{if } (n-m+q) > r/c\Delta t\;, \\ 0 & \text{otherwise}\;.\end{cases} \tag{12}$$

Moreover, to avoid numerical instabilities due to quadrature inaccuracies, the integrations along the "singular" boundary elements, i.e. along elements containing the singular point ξ_λ, have also been carried out analytically for the case of straight boundary segments. If this singular point ξ_λ is the midpoint of the element Γ_l, the values are found to be

$$H^{nn}_{ll} = \begin{cases}[d_l \ln\left(\frac{1 + D_{nn}(0.5d_l,1)}{D_{nn}(0.5d_l,0)}\right) + 2c\Delta t \arcsin(\frac{d_l}{2c\Delta t})]/2\pi c^2 & \text{if } d_l < 2c\Delta t \\ \Delta t/2c & \text{if } d_l > 2c\Delta t,\end{cases} \tag{13}$$

where d_l denotes the length of the element Γ_l, and

$$G^{nn}_{ll} = 0 \;, \tag{14}$$

because $\mathbf{n}\cdot(\nabla r) = 0$ holds along straight boundaries.

The integrals over the "non-singular" boundary segments have been evaluated approximately using eight point Gaussian quadrature. Finally we obtain the discrete analogues of equation (8). It can be cast in the following form [6,12]

$$\hat{G}\cdot\Phi = (0.5I + G)\cdot\Phi := \sum_n\sum_m (0.5I^{nm} + G^{nm})\cdot\Phi^m = \sum_n\sum_m H^{nm}\cdot\phi^m =: H\cdot\phi. \tag{15}$$

According to the causality condition the matrices G and H are lower triangular, whereby, if all time-steps Δt have the same duration, all blocks along the diagonals are equal (see Figure 1), because they depend only on the time difference between the impulse and the observation. Thus, for each time step t_m, it is necessary to determine only one extra blockmatrix G^m and H^m.

$$\begin{bmatrix} G^{11} & & & & \\ G^{21} & G^{22} & & \mathbf{0} & \\ G^{31} & G^{32} & G^{33} & & \\ \vdots & & & & \\ G^{n1} & G^{n2} & \cdots & & G^{nn}\end{bmatrix} = \begin{bmatrix} G^{1} & & & & \\ G^{2} & G^{1} & & \mathbf{0} & \\ G^{3} & G^{2} & G^{1} & & \\ \vdots & & & & \\ G^{n} & G^{n-1} & \cdots & & G^{1}\end{bmatrix}$$

Figure 1: Structure of the G-matrix

The equivalent numerical implementation procedure for the equations describing the wave propagation behavior of elastic media cannot be given here for lack of space, but they may be found in the papers of Antes et al. [6,11]. Only the final result, the system of algebraic equations for the determination of the unknown boundary displacements and tractions is presented:

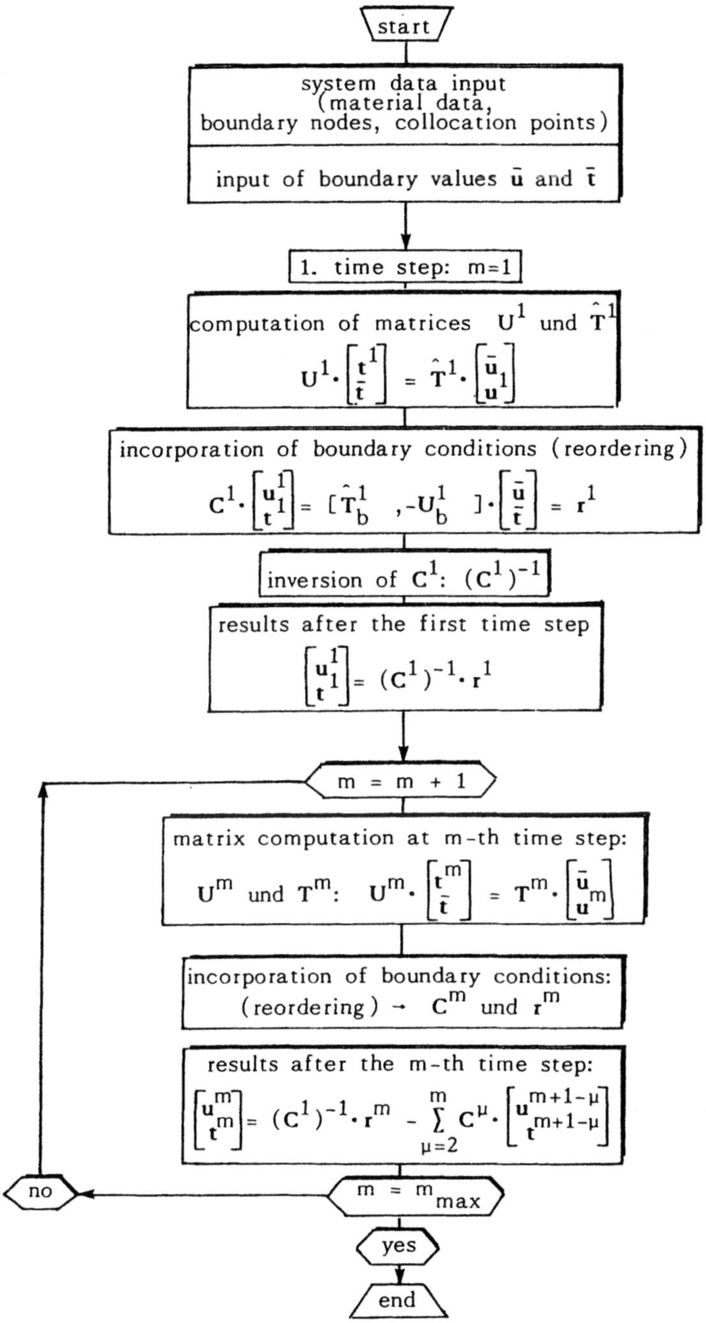

Figure 2: Program flow chart

$$\hat{\mathbf{T}}\cdot\mathbf{u} = (0.5\mathbf{I} + \mathbf{T})\cdot\mathbf{u} := \sum_n \sum_m (0.5\mathbf{I}^{nm} + \mathbf{T}^{nm})\cdot\mathbf{u}^m = \sum_n \sum_m \mathbf{U}^{nm}\cdot\mathbf{t}^m =: \mathbf{U}\cdot\mathbf{t}. \quad (16)$$

For these equations, which correspond to (15), a program flow chart has been given in Fig. 2 which demonstrates the sequence of the solution procedure steps.

PRESSURE WAVES IN A SEMI-INFINITE FLUID DOMAIN

In this section an application of the above procedures is presented which shows the correctness of the solution, and which also demonstrates their advantages in the case of semi-infinite domains.

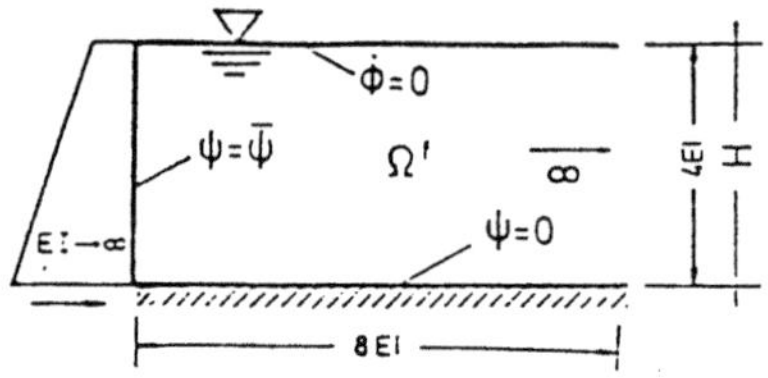

Figure 3: Geometry and discretization of the infinite reservoir

Consider a rigid dam of height H=100 m storing an "infinite" reservoir of the same depth which rests on a rigid bed (see Figure 3). The wavespeed c in water is taken as 1438 m/s. Let this dam now be subjected to a horizontal ground motion acting as a unit impulse of duration Δt=0.0346 secs.

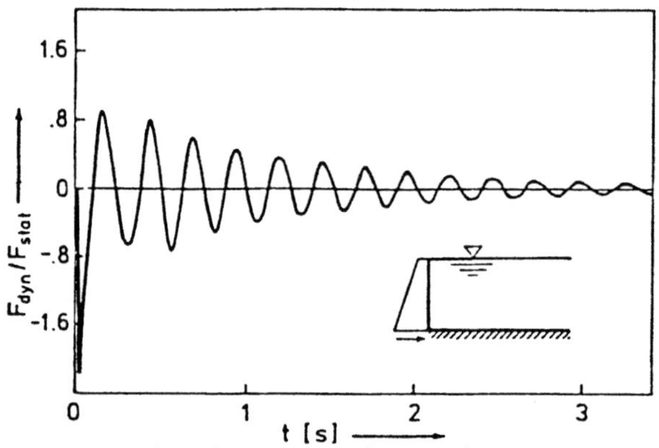

Figure 4: Time history of the total hydrodynamic force on the dam

Figure 4 presents the time history of the total hydrodynamic force acting on the dam. It obviously is damped very quickly by radiation of energy to infinity.

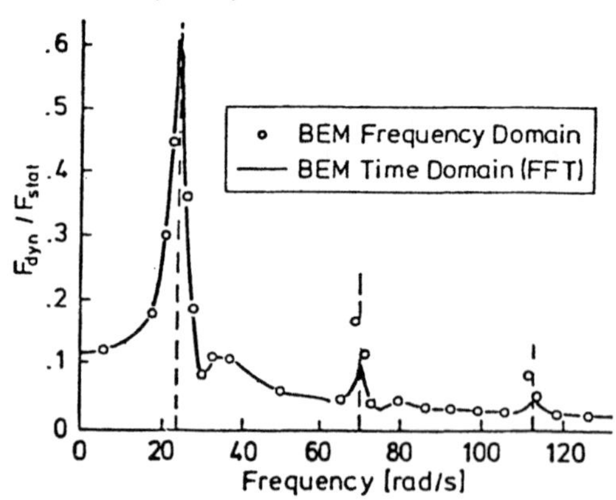

Figure 5: Total hydrodynamic force on dam versus frequency

In order to check this time domain solution we have also performed a Fast-Fourier transformation into the frequency domain. The result, the total hydrodynamic force on the dam face versus frequency is then compared to the values which were determined directly in the frequency domain for distinct harmonic excitation frequencies (see Fig. 5). The general agreement is excellent and, in particular, the first three eigen-frequencies (ω_1=23.8, ω_2=69.8, ω_3=112.9) agree almost exactly.

Next, Figure 6 shows, again in the case of a horizontal impulse along the left rigid wall, how the pressure waves propagate into the interior of the fluid domain: first, when the reservoir extends to infinity in the horizontal direction, and second, when it has finite dimensions. Before the pressure wave has reached the right-hand side, there is no difference between the solutions for a reservoir which is infinite or finite, i.e. open or closed. If we compare the results after 10 time steps, however, we see that in the first case the wave radiates to infinity, while in the second case the wave is reflected along the rigid right boundary.

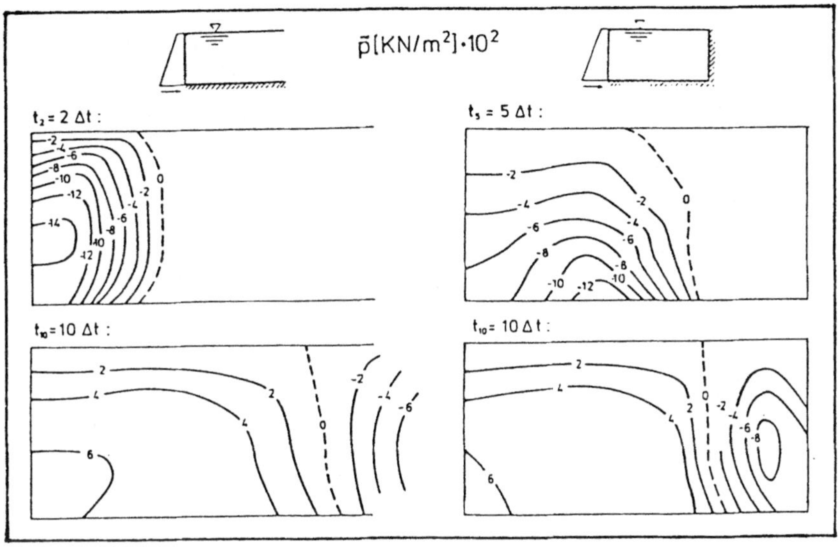

Figure 6: Pressure wave propagation into the interior of the reservoir

Results for the more general problem of wave propagation in elastic media, i.e. applications of the corresponding time-domain-boundary-element procedure for elastic media may be found in [12-15].

REMARK

A characteristic phenomenon of transient wave propagation, that needs special attention, is a particular type of causality constraint, namely the appearance of so-called shadow-zones [16].

CONCLUSIONS

A time-stepping boundary element procedure has been presented for the efficient solution of the hyperbolic equations describing wave propagation problems. This method properly accounts for the effect of radiation of energy to infinity. Results have been obtained for a semi-infinite fluid region, and comparisons between known frequency and transformed time domain calculations show excellent agreement.

ACKNOWLEDGEMENT

The author would like to acknowledge the financial support of collaborative research NATO grant RG. 85/0640 which allowed him to constructively cooperate with Prof. C. Spyrakos, Faculty of Civil Engineering, West Virginia University.

REFERENCES

[1] Graffi, D.: "Sul Teorema di Reciprocita nella Dinamica dei Corpi Elastici", Mem. Accad. Sci. Bologna 18 (1947), pp. 103-109.

[2] Karabalis, D.L. and Beskos, D.E.: "Dynamic Response of 3-D Rigid Surface Foundations by Time Domain Boundary Element Method", Earthquake Engng. Struct. Dyn. **12** (1984), pp. 73-93.

[3] Spyrakos, C.C. and Beskos, D.E.: "Dynamic Response of Rigid Strip-Foundations by Time Domain Boundary Element Method", Int. J. Num. Meth. Engng. **23** (1985) pp. 1547-1565.

[4] Spyrakos, C.C. and Beskos, D.E.: "Dynamic Analysis of Embedded Rigid Strip Footings by Time Domain Boundary Element Method", Proc. 7th. Int. Conf. on BEM, Como, Springer Verlag (Eds.: Brebbia C.A. and Maier G.) vol. I (1985) pp. 6-71 to 6-78.

[5] Mansur, W.J.: "A Time-Stepping Technique to Solve Wave Propagation Problems Using the Boundary Element Method", Ph. D. Thesis, Southampton University 1983.

[6] Antes, H.: "A Boundary Element Procedure for Transient Wave Propagation in Two-Dimensional Isotropic Elastic Media", Finite Elements in Analysis and Design **1** (1985), pp. 313-322.

[7] Mansur, W.J. and Brebbia, C.A.: "Formulation of the Boundary Element Method for Transient Problems Governed by the Scalar Wave Equation", Appl. Math. Modelling **6** (1982) pp. 307-311.

[8] Mansur, W.J. and Brebbia, C.A.: "Numerical Implementation of the Boundary Element Method for Two-Dimensional Transient Scalar Wave Propagation Problems", Appl. Math. Modelling **6** (1982) pp. 299-306.

[9] Antes, H. and Estorff, O. von: "Dynamic Soil-Fluid Interaction Analysis by the Boundary Element Method, in BETECH 86 (Eds.: Brebbia, C.A. and Connor, J.J.), Proc. 2nd. Bound. Elem. Techn. Conf., M.I.T., Boston, CML Publ. (1986) pp. 687-698.

[10] Antes, H. and Estorff, O. von: "Analysis of Absorption Effects on the Dynamic Response of Reservoir Systems by Boundary Element Methods", Earthqu. Engng. Struct. Dyn., (1987), (to be published).

[11] Spyrakos, C.C. and Antes, H.: "Time Domain Boundary Element Method Approaches in Elastodynamics: A Comparative Study", Computers & Structures, **24** (1986) pp. 529-535.

[12] Estorff, O. von: "Analysis of Dynamic Interaction Between Structures and Surrounding Media by Time-Dependent BEM (in German), Report TWM 86-10, Inst. f. Konstr. Ingenieurbau, Ruhr-Universität Bochum, Germany.

[13] Antes, H. and Estorff, O. von: "Dynamic Soil-Structure Interaction by BEM in the Time and Frequency Domain", Vol. 2 (1986) pp. 5.5-33 to 5.5-40, Proc. 8th. Europ. Conf. Earthqu. Engng., Lab. Nac. Eng. Civil, Lisbon, Portugal.

[14] Antes, H. and Estorff, O. von: "Dynamic Response Analysis of Rigid Foundations and Elastic Structures by Boundary Element Procedures", Soil Dyn. Earthqu. Engng. (submitted for publication).

[15] Antes, H. and Estorff, O. von: "Transient Behavior of Strip Foundations Resting on Different Soil Profiles by a Time Domain BEM", Proc. Conf. on Soil Dynamics Earthqu. Engng., Princeton, June 1987 (Ed.: Cakmak, A.S.), (to be published).

[16] Antes, H. and Estorff O. von: "On Causality in Dynamic Response Analysis by Time-Dependent Boundary Element Methods", Earthqu. Engng. Struct. Dyn. (1987), (to be published).

COMPUTATION OF THE TRANSIENT FLOW IN ZONED ANISOTROPIC POROUS MEDIA BY THE BOUNDARY ELEMENT METHOD

E. BRUCH and S. GRILLI
L.H.C.H., University of Liège
Laboratories of Hydrodynamic, Applied Hydraulic and Hydraulic Constructions
Quai Banning, 6, B - 4000 Liège, Belgium

SUMMARY

In this paper, we present new developments in the application of the BEM to the transient two-dimensional flows in zoned anisotropic porous media, including the iterative determination of the free surface seepage position. We justly introduce the classical BEM equations, for the transient porous media potential flows, using a finite difference in time. We then discretize them by linear, quadratic or cubic elements and use special singular numerical quadrature rules. We finally improve the method by the addition of a sub-region division, and develop special compatibility equations on their multiple intersection nodes. All these techniques improve the result accuracy and allow to avoid the typical oscillation problems described in the literature. Examples are presented, where we study the discretization effect on the convergence of the iterative solving.

INTRODUCTION

Our main purposes are briefly presented in the previous summary, but what we mean by a 2D transient flow has still to be pointed out. We consider a perfect fluid flow with small velocities. So, it satisfies to the Darcy law, both for the steady and transient flows, in anisotropic porous media (with k_x and k_y principal permeabilities). Furthermore, we get a moving boundary due to the unknown free seepage position.

Thus, we obtain an elliptic, Dirichlet-Neuman type, problem with a local stepping, due to the non-linear geometric boundary condition, and a time-stepping. The classical example is the seepage through a dam. It is drawn on the figure 1.

A second important question is to explain why we are using the BEM [1] for this problem. It is, indeed found more efficient than the other methods, for the several following reasons :

- we are only interested in the results along the boundaries. Thus, the BEM which eliminates any interior discretization, highly reduces, the meshing problems (especially in 3D), and the number of unknowns.
- one boundary is moving (the phreatic surface). Thus, updating of the discretization is easier than with the domain decomposition method,
- the introduction of water sources or local pumping inside the domain, which leads to high singularities, is straightforward.

At last, it is interesting to point out the similarities or the differences between the so-called Panel method (PM) and the Boundary Element Method (BEM). Basically, these are the same methods, based on integral equations, but some differences appear in their numerical treatment. in fact, we can consider the BEM, as an improvement or a generalization of the PM by :

- the "wedding" of some Finite Element Method techniques (FEM), i.e. the geometry and field discretization, by isoparametric elements and the use of numerical quadratures instead of an analytical one.

With the Boundary Integral Equations Method, i.e. the 3d Green identity : leading to singular integral equations, collocated at boundary nodes,

- formulating integral principles, starting from a weighting residual method (Petrov-Galerkin),
- obtaining softwares, independent of the elements degree and shape, and performing regular or singular integrals on the same way, by implementing time-efficient quadrature rules,
- using a sub-region division, providing larger but sparser system matrices, solved by a direct inversion (well-suited to vector computers).

The background of this study is a quite old one. Indeed, for almost 20 years, many authors have carried out the numerical resolution of porous media flows. Before 1970, only the Finite Difference Method is used. In 1973 [2], a significant improvement is brought out by using the FEM. Since 1977 [3], the solutions are regularly improved in accuracy and in time efficiency, by using the BEM. For a complete review of steady state and transient problems by BEM, see [4] and [5]. However many authors have pointed out typical oscillation problems, of the free surface, in space and time, which sometimes yield the divergence of the iterative procedure. Recent papers present new techniques attempting to avoid this problem by either a smoothing [6] or an increase in accuracy [7]. In this paper, we present new developments of the original method of [3] which allow to avoid the solution oscillations.

2. EQUATIONS AND BOUNDARY CONDITIONS OF A POROUS MEDIA FLOW

2.1. In two dimensions the continuity equation gives :

$$\nabla^2 u = 0, \qquad u : \text{the velocity potential of the flow}$$

$$\nabla^2 = k_x \frac{\partial}{\partial x^2} + k_y \frac{\partial}{\partial y^2}, \tag{1}$$

k_x, k_y : the principal permeabilities of the anisotropic medium (axis x,y).

Indeed, (1) is also an expression of the Darcy Law, for the velocity v_j :

$$v_j + (k_\ell \frac{\partial u}{\partial x_j})\,\delta_{j\ell} = -\frac{1}{g}\frac{\partial V_j}{\partial t} - \frac{V_i}{\varepsilon g}\frac{\partial V_j}{\partial x_i}\ ; \qquad i,\ j,\ k = 1,2\ ,$$

$$u = \frac{p}{\rho} + Y, \qquad p : \text{the pressure}, \quad Y : \text{the vertical coordinate}, \quad g : \text{the gravity}, \quad \varepsilon : \text{the porosity}, \tag{2}$$

Therefore, if V_j is small (2) gives the classical law.

$$V_j = -(k_\ell \frac{\partial u}{\partial x_j})\,\delta_{j\ell} \quad \text{or} \quad V_j = -k\frac{\partial u}{\partial x_j}, \quad k = \text{cst}. \tag{3}$$

2.2. The boundary conditions are the following (see figure 1), by (2) :

$$\left[\begin{array}{lll} AB : \; u = H(t), & BC : \left[\begin{array}{l} u = Y(t), \\ \dfrac{\partial u}{\partial t} = - \dfrac{\cos\gamma}{\varepsilon \cos\alpha} \dfrac{\partial u}{\partial n} \end{array}\right. & \text{see [3], [8]}, \\ CD : \; u = Y(t), & DE : \; u = k(t), \quad EA : \; \dfrac{\partial u}{\partial n} = 0 . & \end{array}\right. \tag{4}$$

3. APPLICATION OF THE BEM TO THE 2d POTENTIAL PROBLEMS

3.1. By [1], we can write, by considering the figure 2 :

$$\left[\begin{array}{ll} c_i \, u_i = \int_\Gamma u_i^* \, q d\Gamma - \int_\Gamma q_i^* \, u \, d\Gamma , & i = \text{collocation node} \\ \Gamma = \Gamma u \cup \Gamma q : \text{the boundary of the } \Omega \text{ domain} & \\ \text{with } u = \bar{u} \text{ on } \Gamma u & \text{(Dirichlet condition),} \\ \qquad q = \bar{q} \text{ on } \Gamma q & \text{(Neuman condition).} \end{array}\right. \tag{5}$$

Where (with ∇^2, the operator of (1)) :

$$\nabla^2 u_i^* + \delta_i = 0, \qquad \delta_i : \text{the Dirac function,} \tag{6}$$

gives the following fundamental solutions :

$$\left[\begin{array}{l} u_i^* = - \dfrac{1}{2\pi\sqrt{k_x k_y}} \ln \left[\dfrac{(x-x_i)^2}{k_x} + \dfrac{(y-y_i)^2}{k_y} \right]^{1/2} , \\ \\ q_i^* = - \dfrac{1}{2\pi\sqrt{k_x k_y}} \dfrac{n_x(x-x_i) + n_y(y-y_i)}{\dfrac{(x-x_i)^2}{k_x} \quad \dfrac{(y-y_i)^2}{k_y}} = \dfrac{\partial u_i^*}{\partial n} , \\ \\ \underline{n} = (n_x, n_y) : \text{the outwards normal unitary vector,} \\ q = k_x \dfrac{\partial u}{\partial x} n_x + k_y \dfrac{\partial u}{\partial y} n_y = \dfrac{\partial u}{\partial n} : \text{the normal potential gradient.} \end{array}\right. \tag{7}$$

3.2. The coefficient c_i depends on the domain geometry :

$$c_i = 1 + \lim_{u \to 0} \int_{\Gamma u} q_i^* \, d\Gamma , \quad \lambda = \sqrt{\frac{k_y}{k_x}} ,$$

$$c_i = 1 - \frac{1}{2\pi} \left[\operatorname{arctg} \left(\frac{1}{\lambda} \operatorname{tg} \gamma \right) \right]_{\gamma_a}^{\gamma_b} . \tag{8}$$

Thus : (see figure 3) :

$$\left[\begin{array}{ll} c_i = \dfrac{\alpha}{2\pi} , \quad \alpha = \gamma_b - \gamma_a & : \text{isotropic case ,} \\ \\ c_i = f(\lambda, \gamma_a, \gamma_b) & : \text{anisotropic case .} \end{array}\right. \tag{9}$$

4. DISCRETIZATION, NUMERICAL INTEGRATION

We use linear, quadratic or cubic isoparametric elements, to discretize the potential and the potential gradient. Therefore we need singular integration methods, to evaluate the terms of (5), when i belongs to the considered element. Thus :

$$\left[\begin{array}{ll} u = N_j\, U_j, & N_j \text{ : shape function related to j, polynomial of degree 1, 2 or 3 (depending on the number of all nodes),} \\ q = N_j\, Q_j, & U_j,\ Q_j \text{ : nodal values of u and q.} \end{array}\right. \tag{10}$$

The discretized terms of (5) are the following :

$$K^e u_{ij} = \int_{\Gamma ue} N_j\, u_i^* d\Gamma = \int_{-1}^{+1} N_j(\xi) u_i^* \, [x(\xi), y(\xi)] \frac{\partial s}{\partial \xi} d\xi . \tag{11}$$

If i does not belong to the element e (part Γue of Γu), (11) is regular and we use the classical Gauss quadrature rule (case e" of the figure 2).
If i belongs to e (case e' of the figure 2), we can evaluate the singular integral (12) by using a kernel transformation and by applying it to the parts having a logarithmic singularity :

$$\int_0^1 f(x) \ln \frac{1}{x} dx = \sum_{s=1}^{m} w_s\, f(x_s) ,$$

$$(w_s, x_s) \text{ : special weights and points (see [7]),} \tag{12}$$

b. $$K^e_{q_{ij}} = \int_{\Gamma qe} N_j\, q_i^* \, d\Gamma = \int_{-1}^{+1} N_j(\xi) q_i^* \, [x(\xi), y(\xi)] \frac{\partial s}{\partial \xi} d\xi . \tag{13}$$

On the same way, for the case e" of the figure 2 we again apply the Gauss method. But for the other case e', we have no longer a singularity, due to the scalar product in (7). However, it is more accurate to use a special method, based on an integration by parts which gives (see [7]) :

$$K^e_{q_{ij}} = \begin{cases} -\frac{\alpha}{2\pi} [\operatorname{arctg} v(-1) + I_j] & \text{if } j = 1 \\ -\frac{\alpha}{2\pi} I_j & \text{if } 1 < j < n \\ +\frac{\alpha}{2\pi} [\operatorname{arctg} v(+1) - I_j] & \text{if } j = n , \end{cases} \tag{14}$$

$$I_j = \int_{-1}^{1} \frac{dN_j(\xi)}{d\xi} \operatorname{arctg} v(\xi)\, d\xi , \tag{15}$$

$$\alpha = 1, \quad v(\xi) = \frac{1}{\lambda} \frac{y(\xi) - y_i}{x(\xi) - x_i} = v_1 \qquad \text{if } 0 \leqslant |\beta| \leqslant \frac{\pi}{4} ,$$

$$\alpha = -1, \quad v(\xi) = \frac{1}{v_1} \qquad \text{if } \frac{\pi}{4} \leqslant |\beta| \leqslant \frac{\pi}{2} . \tag{16}$$

For the regular integral I_j, we again use the Gauss method.

For example, the evaluation of (13) on a semi-elliptic element, gives the following comparison (see figure 5) :

I : analytic solut., I_s : singular integ., I_g : Gauss integ.,

$E = |(I_g - I) / (I_s - I)|$, 6 integrations points.

Table 1

$K^e_{q_{ij}}$	I	I_s	I_g	E
11	- 0,0031638213	- 0,0031603911	- 0,0031721240	2,4
12	0,119057620	0,119044831	0,119101717	3,4
13	0,095116636	0,09512600	0,09503730	8,5
21	0,0355254237	0,0354255070	0,0357608528	2,4
22	0,281365535	0,281565368	0,279011244	11,8

We see, the relative error is up to 12 times higher with the Gauss integration, than with the singular one.

c. The matric system corresponding to (5) is the following linear system of N by N equations :

$$c_i\, U_i = [\sum_{e=1}^{Ne} K^e_{u_{ij}}]\, Q_j - [\sum_{e=1}^{Ne} K^e_{q_{ij}}]\, U_j \qquad i,j = 1, N$$

or

$$c_i\, U_i = Ku_{ij}\, Q_j - Kq_{ij}\, U_j \qquad i,j = 1, N \,, \qquad (17)$$

Ne : Number of elements, N : number of nodes.

5. SUB-REGION DIVISION

This technique is used in two cases. The first one is the case of a zoned domain, i.e. composed of several regions of different permeabilities. The second one is for purely numerical purposes.
Indeed, if the domain geometry is quite complicated, or narrowed in some parts, it will be more accurate to use several sub-regions. Besides, we will obtain in each case, a system matrix containing null blocks which will be taken into account, in order to optimize the solving process.

However, due to the multiple intersection nodes, between the regions (see figure 6, P) and between their interfaces and the exterior boundary (A, B and C), special problems occur. In fact, near these points, the potential values are found in good concordance, but the normal gradients (velocities) are not so well fitted. The reason is that the normal and tangential velocities are not independent, but locked by special compatibility equations which must be developed and implemented.
Doing that, we highly improve the results accuracy (see examples in [7]).

6. ITERATIVE DETERMINATION OF THE FREE SURFACE POSITION

6.1. The phreatic surface (BC in figure 1), is one of the problem unknowns but we have 2 conditions, in (4), valid on this boundary. The second one, on the gradient, will directly be introduced in the system (17) (see below), and the first one : will be used to interactively determine the position of BC, at a given time t+Δt, like :

$$\left[\begin{array}{l} Y_j^{t+\Delta t} = \omega u_j^{t+\Delta t} + (1-\omega)\ Y^* , \\ Y^* = Y^t \ :\ j = 1; \quad Y^* = Y_{j-1}^{t+\Delta t} \ :\ j > 1 , \end{array}\right. \tag{18}$$

ω : coefficient generally taken equal to 0.5.

(18) is checked for each node of BC, until the convergence is reached, according to the following criterion :

$$\mathrm{Max}_j(\varepsilon) = \mathrm{Max}_j \ \left| (Y_j^{t+\Delta t} - Y^*) \ / \ Y_j^{t+\Delta t} \right| \ \leqslant \varepsilon_o \ . \tag{19}$$

6.2. When the solution (i.e. potential and BC position) is known at one time t, by the joint resolution of (17) to (19), the unknowns (i.e., potential or gradient) are found at the time t+Δt, by using a finite difference in time, updating (17): By (4) (see figure 1) :

$$\frac{u_j^{t+\Delta t} - u_j^t}{\Delta t} = -\frac{1}{\varepsilon}\ \frac{\Theta \cos\gamma_j^{t+\Delta t}\ q_j^{t+\Delta t} + (1-\Theta)\cos\gamma_j^t\ q_j^t}{\Theta' \cos\alpha_j^{t+\Delta t} + (1-\Theta')\cos\alpha^t} \ . \tag{20}$$

(20) is simplified by considering small Δt intervals, thus :

- the number of iterations is reduced and $\Theta' \simeq 0$,
- $\cos\gamma_j^{t+\Delta t} \simeq \cos\gamma_j^t \simeq \cos\gamma$,
- $\Theta' = 0 \rightarrow \left[\begin{array}{ll} \cos\alpha \simeq \cos\alpha^t & \text{for } j = 1 \\ \cos\alpha \simeq \cos\alpha_{j-1}^{t+\Delta t} & \text{for } j > 1 . \end{array}\right.$

Therefore, (20) becomes :

$$q_j^{t+\Delta t} = (u_j^t - u_j^{t+\Delta t})\ \frac{\varepsilon \cos\alpha}{\Theta\,\Delta t \cos\gamma} - \frac{1-\Theta}{\Theta}\, q_j^t \ . \tag{21}$$

As mentioned, for several reasons, the value of $u_j^{t+\Delta t}$ is the unknow chosen on BC. So, the value of $q_j^{t+\Delta t}$ is found from (21) and introduced as a Neuman boundary condition in (17). After solving this system, (18) is checked, and the procedure is started again, until the convergence (19) is reached. Doing that, we got a better convergence than mentioned so far in the literature, where most of the authors takes q_j as the unknown on BC.

At last, we pointed out that the choice of Θ , in (21) is of prime importance on the solution stability. This problem has been studied in [4], for a simple domain (see figure 7). The optimal value of Θ, avoiding any divergence was found equal to :

$$\left[\begin{array}{l} \Theta_{op} = -\left(\frac{1}{\rho} + \frac{1}{e^{-\rho}-1}\right) , \\ \rho = \frac{\pi\sqrt{k_x k_y}}{\varepsilon} \frac{\Delta t}{\Delta x} \tanh\left(\frac{\pi H}{\Delta X}\sqrt{\frac{k_x}{k_y}}\right) . \end{array}\right. \tag{22}$$

$$\text{If} \left[\begin{array}{l} \rho \to 0 : \quad \Theta_{op} \to 0,5 , \\ \rho \to \infty : \quad \Theta_{op} \to 1 . \end{array}\right.$$

But the practical range is $\Theta_{op} \in [0.5, 0.7]$.

7. COMPUTATIONAL EXAMPLES

7.1. Steady state flow in a real zone dam.

We computed the case of the figure 8, for several decreasing kernel permeabilities k_2 (k_1 = .001 m/s, $\rho = k_2/k_1$, ε_o = .001).

The figures shows the free surface discretization (2 nodes elements) and position. If ρ becomes quite low, a seeping surface occurs inside the dam, like in the downstream part. This possibility had of course been implemented in the software. In the table 2, we see the discharge Q is decreasing a lot with ρ, and in the same time, the left and right phreatic surfaces are going up and down. The number of nodes NDE and number of iterations NIT show we had to adapt the discretization, in order to get the convergence.

Table 2

ρ	NDE	NIT	H_A	H_{B1}	H_{B2}	H_C	$Q(10^{-3})$
1	91	9	8,79	7,70	7,70	4,61	3,31
0,5	85	12	9,12	7,44	7,44	4,18	2,67
0,2	100	11	9,49	7,73	6,50	3,53	1,70
0,1	100	10	9,71	7,90	5,48	3,11	1,05
0,05	97	10	9,84	8,02	4,54	3	0.60
0,01	102	12	9,97	8,14	3,14	3	0.13
0	-	-	10	8,14	3	3	0

7.2. Transient flow in a homogeneous dam

The figure 9 shows the free surfaces corresponding to the transient flow in the previous dam, for ρ = 1 (ε_o = .005, Θ_{op} = .7, NDE = 55, ω = .5, $k_1 = 10^{-2}$ m/min). At the time t < 0, the downstream level is the same than the upstream one, and we suppose, it has been put down instantaneously to the drawn level at t=0. The time step Δt is equal to 24 min. and $n = t/\Delta t$. For the first 5 steps, each resolution at a given time took 4 to 5 iterations (NIT), but this number quickly became 3, then 2 or 1. The total computation time, up to n = 70 is of 37.8 s (CRAY-XMP). Since we have an homogeneous dam, the last free surface (n=70) is very close to the steady state one described in the table 2, for ρ = 1.

CONCLUSION

These two computations showed the good behaviour of our solutions in space and time and, particularly, that we avoid the typical oscillation problems encountered by many authors. Furthermore, we pointed out that the choice of the discretization (i.e, number of nodes, degree of the elements) is of prime importance on the procedure convergence.

9. REFERENCES

[1] BREBBIA, C.A. and WALKER, M.A. (1980). Boundary Element Techniques in Engineering. Newnes-Butterworths London-Boston.

[2] CHENG, R.T. and LI, C (1973). On the Solution of Transient Free-Surface Flow Problems by the Finite Element Method. J. Hydrol., 20, 49-63.

[3] LIGGETT, J.A. (1977). Location of Free Surface in Porous Media. J. Hyd. Div., ASCE 103 (HY4),353-65.

[4] LIGGETT, J.A. and LIU, P.L.F. (1983). The Boundary Integral Equation Method for Porous Media Flow. George Allen and Unwin. London.

[5] LU, Z.K., BREBBIA, C.A; and ADEY, R.A. (1985). Calculation of Free Surface Seepage through Zoned Anisotropic Dams. Proceedings of the 7th Boundary Element Conference, Villa Olmo, Springer-Verlag. Berlin and New-York.

[6] CHANG, C.S. (1986). Application of Boundary Element Method in transient Flow Problems with Moving Boundaries. Proceeding of the 8th Boundary Element Conference, Tokyo. Springer Verlag, 849-58.

[7] BRUCH, E., GRILLI, S.,, LEJEUNE, A. (1986). Computation of the Fluid Flow in Zoned Anisotropic Porous Media and Determination of the Free Surface Seepage. Proceedings of the 8th Boundary Element Conference, Tokyo. Springer Verlag, 889-903.

[8] CAVOR, R., CVJETKOVIC, N. (1986). Unsteady Flow between Parallel Drains Solved by Boundary Element Method. Proceeding of the 6th International Conference on Finite Elements in Water Resources, Lisboa. Springer-Verlag. 167-176.

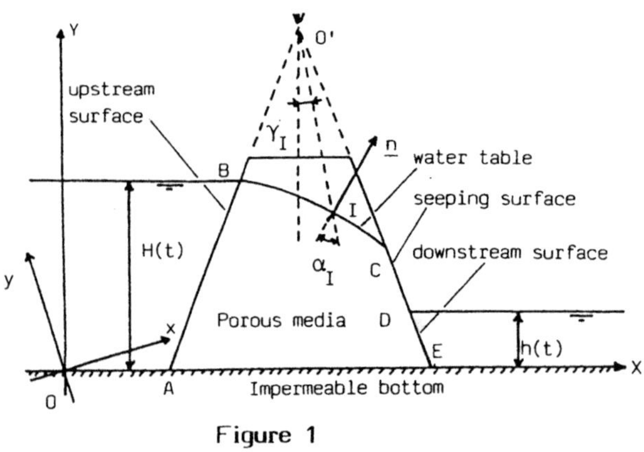

Figure 1

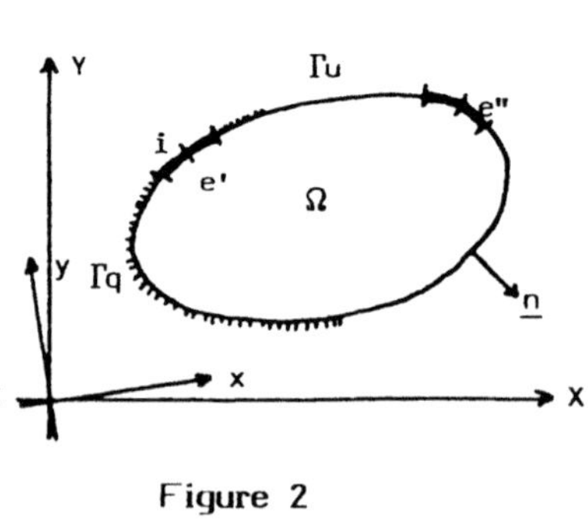

Figure 2

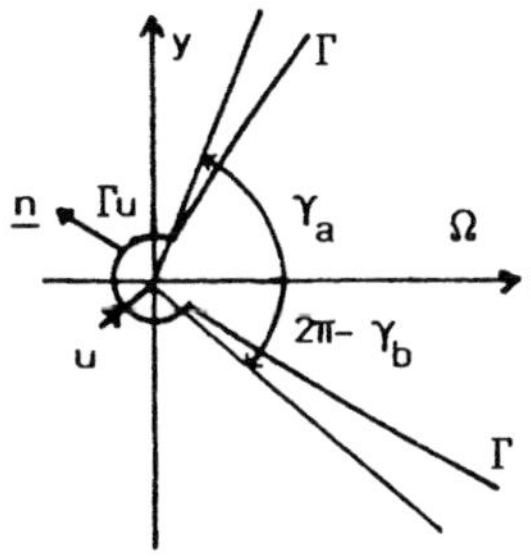

Figure 3

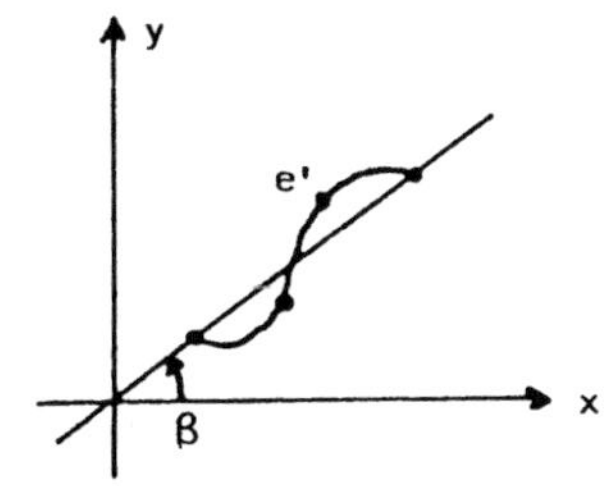

Figure 4

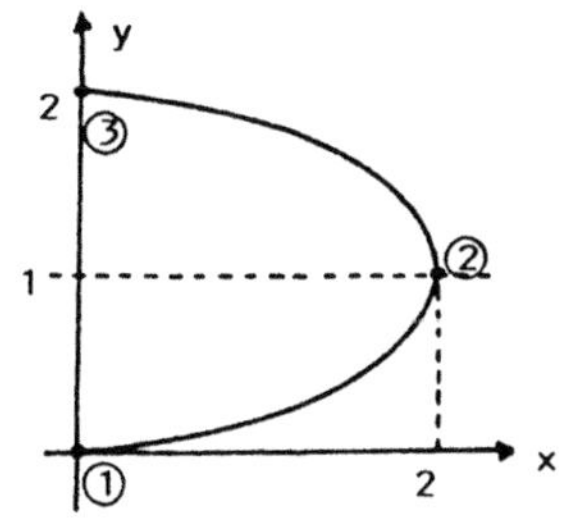

Figure 5

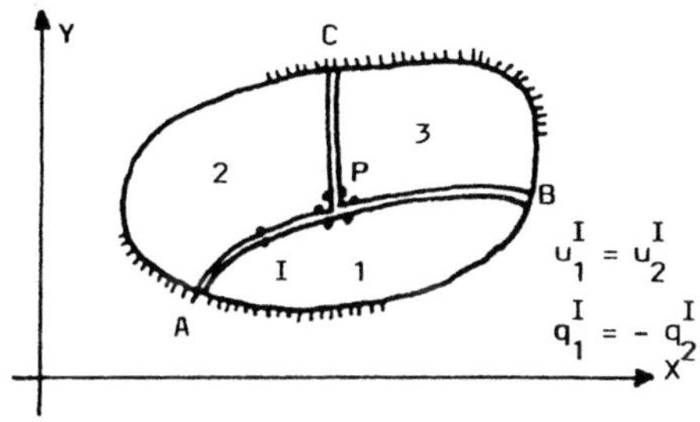

$u_1^I = u_2^I$

$q_1^I = - q_2^I$

Figure 6

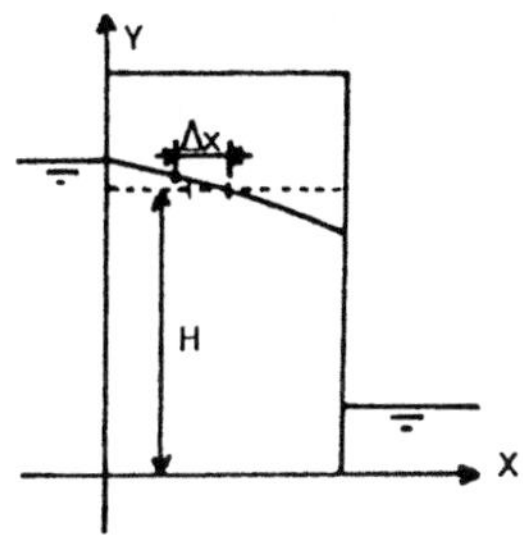

Figure 7

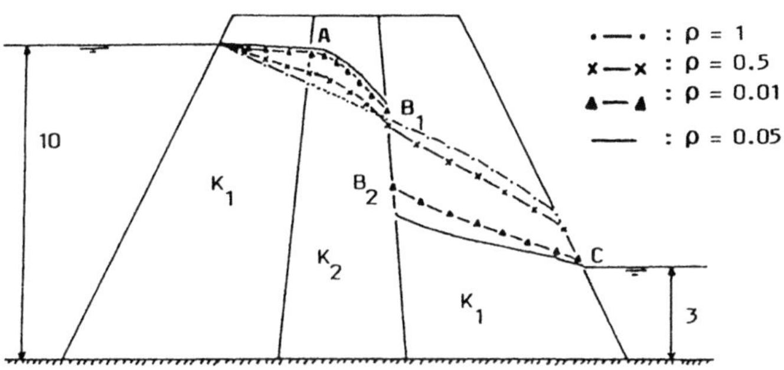

Figure 8

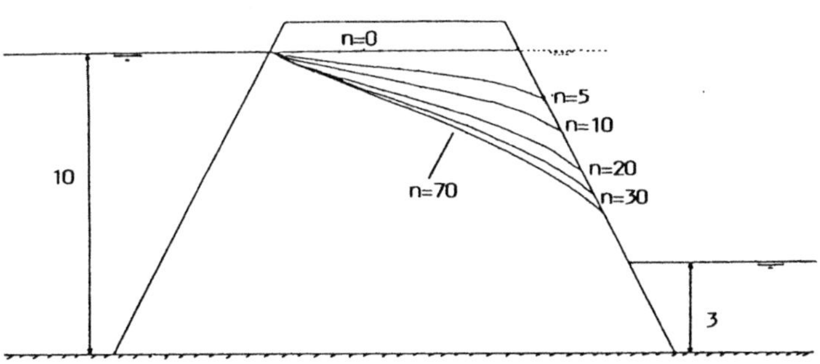

Figure 9

HISSS - A Higher-order Panel Method for Subsonic and Supersonic Attached Flow about Arbitrary Configurations.

L. Fornasier

Theoretical Aerodynamic Department LKE 122
Helicopter and Military Aircraft Division
Messerschmitt-Boelkow-Blohm GmbH
Postfach 80 11 60
D-8000 Muenchen 80 - FRG

SUMMARY

Extension to supersonic flow of the surface singularity techniques used to solve the Prandtl-Glauert equation at subsonic speeds has often been afflicted by numerical stability problems, resulting in physically inconsistent surface pressure fluctuations. The origin of these instabilities is discussed and a series of conditions necessary to obtain stable results are indicated. A panel method derived in compliance with the supersonic requirements is described. The method uses higher-order surface singularity distributions and features a very general boundary value specification capability which allows to solve either subsonic or supersonic flow problems around geometrically complex three-dimensional configurations. A survey of testcases is presented to demonstrate the effectiveness of the present approach in the analysis of supersonic configurations.

1. INTRODUCTION

In the last years a numerical method based on the use of surface singularity distributions has been developed for solving the linearized potential equation at subsonic and supersonic speeds about three-dimensional configurations. Aiming at overcoming limitations demonstrated by available panel codes in the analysis of complex aircraft configurations at supersonic speeds [1], the present code - called HISSS - follows the mathematical formulation pioneered by the PANAIR method [2],[3].

Very recently extensive calculations have been carried out for an advanced fighter-type airplane wind tunnel model [4]. The goals of this investigation were manifold. The main aim was the assessment of the range of problems for which HISSS yields adequate solutions for such a configuration. To this purpose numerical results have been compared with experimental data and, to a limited extent, with calculations from the EUFLEX Euler code [5]. In the same time, the code has been faced up with a realistic benchmark, giving the opportunity to exercise and check out a large number of modelling options. Interface procedures for pre-processing of the geometrical data and post-processing of the computed results have been established, paving the way for future applications to similar configurations.

This paper is focused more on the method itself than on its applications. The background theory and the origin of the numerical problems encountered when the surface singularity technique is applied to supersonic flow are addressed in Section 2. Implementation of the theory into the code is discussed in Section 3. The characteristics of the code are given in Section 4. Two examples of application are presented in Section 5. Final considerations conclude the paper.

2. THE BACKGROUND THEORY.

2.1 The governing equation.

The method relies upon solution of the second-order, partial differential equation

$$(1-M_\infty^2)\phi_{xx} + \phi_{yy} + \phi_{zz} = 0, \tag{1}$$

called Prandtl-Glauert equation where M_∞ is the freestream Mach number and ϕ is the potential function of the perturbation velocity. This equation holds for both subsonic and supersonic flows: in particular, it becomes the Laplace equation when $M_\infty = 0$ and it is then possible to apply the 3rd Green's theorem to transform it into the equivalent integral form

$$\phi(x,y,z) = \frac{1}{4\pi}\iint\left\{-\frac{\sigma}{|r|} + \mu\frac{\partial}{\partial n}\left(\frac{1}{r}\right)\right\} dS, \tag{2}$$

where σ and μ are respectively source and doublet singularities distributed over the boundary surface S enclosing the configuration of interest, r is the distance from the field point P(x,y,z) to the point Q lying on S and $\partial/\partial n$ is the normal derivative of S at the point Q.

Uniqueness of ϕ is guaranteed inside the domain whenever either the function ϕ itself (Dirichlet's problem) or its normal derivative $\partial\phi/\partial n$ (Neumann's problem) are known over the surface S. Perturbation velocity components are computed by differentiating ϕ with respect to the coordinates of P, i.e.

$$\mathbf{v}(x,y,z) = \frac{1}{4\pi}\iint\left\{-\sigma\,\mathrm{grad}\frac{1}{|r|} + \mu\,\mathrm{grad}\frac{\partial}{\partial n}\left(\frac{1}{r}\right)\right\} dS. \tag{3}$$

Although in general equations (2) and (3) must be solved numerically, analytical solutions can be derived, provided that the surface S and the singularity distributions σ and μ be expressed in polynomial form. In the panel method approach, the solution is achieved by dividing the boundary surface S into elements (panels) carrying singularity distributions of parameterized shape which approximate the actual σ and μ distributions over S. The values of the singularity parameters are then determined by enforcement of the boundary conditions at a set of surface points (control points), whose number must be equal to the number of unknown singularities. Once the σ and μ distributions are determined, eq. (3) can be used to evaluate the induced flowfield by summing up the contributions of all panels. The energy equation can then be used to derive pressure from velocity. Finally forces and moments acting on the configuration are derived from integration of pressure and of momentum.

A formal identity with (2) exists at supersonic speed for the case $M_\infty = \sqrt{2}$, provided that the geometrical distance $\mathbf{r}$ and the normal at the surface $\mathbf{n}$ be replaced by the hyperbolic distance $\mathbf{r}_h$ and the conormal $\tilde{\mathbf{n}}$ defined as follows:

$$\mathbf{r} = (r_x, r_y, r_z), \quad \mathbf{r}_h = (-r_x, r_y, r_z) \tag{4}$$

and

$$\mathbf{n} = (n_x, n_y, n_z), \quad \tilde{\mathbf{n}} = (-n_x, n_y, n_z). \tag{5}$$

The solution for any other intermediate Mach number is obtained from the relevant fundamental solution by applying Goethert's similarity rule.

Since the numerical solution can not be more accurate than the governing equation, assumptions under which equation (1) is derived restrict the range of validity of the method. Existence of a velocity potential implies inviscid, irrotational flow. Moreover, equation (1) refers to steady state. Further assumptions necessary to linearize the compressible potential equation along the x-axis postulate that the perturbation in this direction be small in comparison to the freestream velocity and that the freestream Mach number be neither too large nor close to one. The first assumption is locally violated in stagnation regions while the latter excludes applicability of this method to the transonic and hypersonic regimes. It is to note, however, that in the incompressible case (i.e. $M_\infty = 0$) the small perturbation assumption is not applied so that - for attached flow - singularity method and Euler equation solutions must coincide.

2.2 Features of linearized supersonic flow.

Although equation (2) holds for both subsonic and supersonic flow and a formal analogy exists for the two fundamental solutions, some special care must be paid when the surface singularity technique is used for solving numerically the wave equation. Essentially, two problems arise, both of them being originated by the hyperbolic character assumed by equation (1). Due to the mechanism peculiar to the wave propagation, numerical influences just like physical disturbances do not decay with distance from the generating source but propagate unattenuated along the characteristic lines throughout the whole domain of dependence. Therefore, the effect of the singular term induced into the perturbation velocity by a finite doublet strength at panel edges is transported into the computational domain by the Mach waves originated at the edges. Unless the doublet distributions are made continuous over the whole boundary surface - in which case the contributions from two adjacent panels cancel exactly each other - a false perturbation is felt by control points lying close to these Mach lines. As a result, severe local numerical instabilities can appear, which make the results extremely sensitive to random combination of panel spacings, control point locations and freestream Mach numbers. The other problem is related to the property of singularity distributions to propagate disturbances not only into the flowfield but also into virtual domains, like interiors of closed volumes. In this way, flow perturbations are transmitted directly through internal volumes instead of being convected along the external surface. Once generated, such internal waves become trapped within the closed volume and multiple reflections take place which induce unrealistic oscillations in the external pressure distributions. As a consequence, supersonic calculations require control of both the external and the internal flow characteristics, which becomes possible by superposition of source and doublet distributions over the boundary surface.

2.3 Boundary conditions.

As already stated in 2.1, a set of boundary conditions must be enforced at the selected control points in order to determine the singularity parameters and thus make the function ϕ unique. For panels carrying sources and doublets, one set of boundary conditions is used to suppress the disturbances inside the configuration by using a Dirichlet type specification (i.e. $\phi = 0$ or Φ = const) while the known permeability of the surface is used to determine the value of $\partial\phi/\partial n$ on the exterior surface. For impermeable portions of the configuration, the natural requirement is to have zero normal velocity at the control points, i.e.

$$\partial\phi/\partial \mathbf{n} + \mathbf{V}_\infty \times \mathbf{n} = 0, \tag{6}$$

which, remembering the significance of $\partial\phi/\partial \mathbf{n}$, can be put into the equivalent form

$$\phi_x n_x + \phi_y n_y + \phi_z n_z + \mathbf{V}_\infty \times \mathbf{n} = 0. \tag{7}$$

For compressible flow, however, this condition is often not enforced, either intentionally or unintentionally. It is well known that it is possible to derive the compressible flow field around a given body from the incompressible solution around an affinely transformed body. To do this it is sufficient to compute the potential $\phi' = \beta^2 \phi$ in a transformed domain where $x'= x$, $y'= \beta y$ and $z'= \beta z$. Once the solution has been obtained, the incompressible perturbation velocity components are transformed to the compressible ones by means of the following relationships

$$\phi_x = \phi'_{x'}/\beta^2 \;,\; \phi_y = \phi'_{y'}/\beta \text{ and } \phi_z = \phi'_{z'}/\beta. \tag{8}$$

Due to the anisotropy of the transformations, it is immediately recognizable that different results will be obtained whether the boundary condition (6) is fulfilled at the transformed body or at the real one. In fact, if equation (6) is enforced on the transformed body - a common interpretation of Goethert's rule in references before [6] - one obtains

$$\phi'_{x'} n'_{x'} + \phi'_{y'} n'_{y'} + \phi'_{z'} n'_{z'} + \mathbf{V}_\infty \times \mathbf{n}' = 0, \tag{9}$$

which, transformed back to physical domain by means of (8) and remembering that $\mathbf{n}'= (\beta\, n_x, n_y, n_z)$, gives

$$\beta^2 \phi_x n_x + \phi_y n_y + \phi_z n_z + \mathbf{V}_\infty \times \mathbf{n} = 0. \tag{10}$$

Comparing expression (10) to (7) it is found that the residual normal velocity at the real body is

$$R = M^2_\infty \; \phi_x n_x \,, \tag{11}$$

that is, the residuum is proportional to the square of the freestream Mach number and is essentially concentrated near forward- or rear facing panels, where the term $\phi_x n_x$ is not negligible. For an impermeable panel perpendicular to the flow application of condition (4) at the transformed body gives a residual velocity equal to M^2_∞/β^2. At Mach 0.80, for example, a velocity of magnitude 1.777 coming out of the surface would be obtained at the leading edge of an unswept wing or at the nose of a bluff body.

Such inconsistencies are avoided when the condition (6) is fulfilled at the surface of the real body, as suggested by Kraus in ref. [6]. To do this it is sufficient to use the backward transformations (8) so that (6) can be recasted using the transformed variables into

$$\phi'_{x'} n'_{x'}/\beta_2 + \phi'_{y'} n'_{y'} + \phi'_{z'} n'_{z'} + \mathbf{V}_\infty \times \mathbf{n}' = 0. \tag{12}$$

This expression is known as Goethert's rule 2 and is used in most european Panel methods and in the USSAERO [7].

An apparently different approach is taken in PANAIR. The Prandtl-Glauert equation (1) is interpreted as a Laplace equation of the quantity

$$\rho \mathbf{w} = \rho_\infty (\beta_2 \phi_x , \phi_y , \phi_z), \tag{13}$$

obtained by linearisation of the mass-flux vector $\rho\mathbf{V}$. Further, the boundary condition depending on the surface permeability is expressed in term of mass flux instead of in term of velocity. For impermeable surfaces this means that

$$\rho(\partial\phi/\partial n + \mathbf{V}_\infty \times \mathbf{n}) = 0. \tag{14}$$

However, when the linearized mass flux $\rho\mathbf{w}$ is used, this expression transforms into (10), so that this boundary condition is fully equivalent to the kinematic boundary condition applied at the surface of the transformed body (see also ref. [8]). At panels inclined to the flow, therefore, enforcement of zero-mass flux does not imply that the local velocity is tangential to the surface, an apparent paradox deriving from the non-alinement of the linearized mass-flux vector to the perturbation velocity. In fact, Melnik and Mason have shown that an additional term should be taken into account when linearizing the mass-flux vector, [9]. In this case, the expression (14) transforms into (7), since mass-flux and velocity vectors are now parallel. The main reason why the "too"-linearized mass flux boundary condition is still used is to be found in the possibility to reduce the computational effort when panels carrying source and doublet distributions are used. In this case, in fact, application of the 3rd Green's theorem states that the source strengths are equal to the normal component of the freestream velocity and therefore are 'a priori' known. The unknown doublet distribution is then determined by enforcing zero perturbation potential inside the configuration. It is inferred that the computational cost associated with the solution of the linear system is largely decreased. But all panel methods using this formulation (so called Morino's boundary condition) can not and do not fulfill the condition of tangential flow at panels inclined to the flow at Mach numbers different from zero. In order to improve the solution, some artifices have been suggested. In PANAIR an empirical correction is done when the term ϕ_x becomes negative so that the velocity vector is realigned to the mass-flux vector. A modification of the internal boundary condition $\phi = 0$ has been proposed in [10], by which it is possible to approximately fulfill the expression (12) while still having the source strength predetermined by the Green's theorem.

3. THE NUMERICAL CODE.

Consideration of the unique features of linearized supersonic flow discussed in the previous section has been the leading factor in the formulation of the present method. Surface distributions of both source and doublet singularities have been used to enable enforcement of the internal Dirichlet boundary condition necessary to suppress the interior wave propagation problem. A layout of quadratically varying doublets distributed over triangular subpanels have been selected to yield continuity of the singularities over the whole configuration. Analytical integration of the induced potential function and perturbation velocities has been provided to fulfill the exact tangential velocity condition at the actual configuration surface. Furthermore, integration over surfaces more inclined to the flow than the Mach cone - so called superinclined panels - has been carried out to render the geometrical capabilities independent of the calculation Mach number. Some of the features necessary to comply with the supersonic requirements increase largely the computational effort at subsonic Mac numbers but would not have a significant impact on the result accuracy. For this reason, proper options have been provided which allow to use more efficient procedures at subsonic speeds.

The development of the numerical code emboding the aforementioned concepts took advantage of the availability of the subsonic higher-order panel method code of ref. [11] which has been extended to cover the supersonic case and upgraded in the subsonic part using [2] and [3] as main references. A schematic description of the most important features of the newly generated code is given in the following sections.

3.2 Geometry and singularity modelling.

The network concept is used to discretize the whole configuration into a convenient number of subdomains. A network is a homogeneous collection of panels which share the geometrical treatment, the type(s) of singularity distributions and the boundary condition specification. Each network is a logically independent entity since according to the type and the number of singularity distributions a set of control points is defined in such a way that the singularity distributions are uniquely determined by the associated boundary conditions. HISSS has a number of different types of networks carrying different singularity distributions and having different control point locations.

To represent solid portions of the configuration three types of networks can be used, i.e. source, doublet or composite networks. The most general type is the composite network which consists of both a source and a doublet singularity. This network can be used for all components of a configuration, allowing to avoid the sometimes artificial distinction between lifting and non-lifting surfaces typical of some earlier methods. To determine both singularity strengths, two boundary conditions must be enforced at control points located at panel centroids: a range of options is available and possible choices will be discussed in a next section. Additional control points located along the network edges are used to enforce doublet continuity across neighbouring networks. Source networks are used to represent non-lifting bodies or, in conjunction to composite networks, to simulate smooth base flows at thick trailing edges or at fuselage afterbodies. Doublet networks can be used to model lifting surfaces without thickness, a cost-effective useful approach to analyse load distributions.

Vorticity generated at the configuration surface must be carried downstream by using special networks carrying doublicity distributions which are constant streamwise and quadratically varying in the transverse direction or constant in both directions. Edge control points are used to obtain zero net vorticity along the edge abutting composite or doublet networks. Unlike other panel methods - in which vorticity trailing to downstream infinity is implicitly taken into account in the integration scheme - wake geometry must here be specified in the input data.

Additionally, control networks are available which can be used for - as their name suggests - controlling the flow characteristics over a specified surface. Of course, such networks do not carry any singularity.

The network geometry is specified by the coordinates of a rectangular array of MxN grid points. Four points belonging to two adjacent m-rows and n-columns identify a panel. On each panel, positive and negative sides are defined by the sequence of the four corner points (positive anti-clockwise) which determines the orientation of the normal vector **n** too. Allowable network geometry is completely arbitrary for subsonic calculations but networks to be used at supersonic Mach numbers must consist of elementary surfaces (panels and subpanels) which are all sub- or super-inclined. Triangular panels are not allowed inside a network, but can be used on a network edge, provided that the whole edge collapse into a single point. Internal checks are built in the code to abort the computation for configurations not matching these geometrical requirements.

With regard to the treatment of the panel geometry, three options are available. Panels can be treated as flat, curved or subdivided into subpanels. The latter option is mandatory at supersonic speeds, since quadrilateral flat panels cannot be used to fit an arbitrarily curved surface unless the corners are moved away from the surface itself. In this way, gaps between adjacent edges are generated, which imply an inadmissable jump to zero doublet in the empty space. Instead, splitting up the panel into eight triangular subpanels allows to obtain planar elements without relocating the corner points. The curved representation - which is restricted to the subsonic case - is obtained by taking into account the local curvature in the integration of expression (2) and (3). However, the singularity distributions are located in the panel average plane in this case too.

Singularity distributions used in this method are of higher-order, namely linearly varying sources and quadratically varying doublets. Let ξ, η and ζ be the coordinates of a local cartesian reference system having its origin in the center of the panel (or subpanel), the ζ axis parallel to the local normal and the ξ axis aligned to the compressibility axis. Then the singularity distributed over each element (panel or subpanel) can be expressed in the following form

$$\sigma(\xi,\eta) = \sigma_o + \sigma_\xi \xi + \sigma_\eta \eta \tag{15}$$

for the source and

$$\mu(\xi,\eta) = \mu_o + \mu_\xi \xi + \mu_\eta \eta + \mu_{\xi\xi} \xi^2 + \mu_{\xi\eta} \xi\eta + \mu_{\eta\eta} \eta^2 \tag{16}$$

for the doublet. The coefficents σ_i and μ_i of distributions (15) and (16) are not considered as independent variables but instead are taken to linearly depend on the values λ_j assumed by the singularity distributions at a prescribed set of control points. In a mathematical sense, two relationships of the type

$$[\ \sigma_i\ ] = [\ SSF\ ] \times [\lambda_j] \text{ with } i=1,3 \tag{17}$$

and

$$[\ \mu_i\ ] = [\ DSF\ ] \times [\lambda_j] \text{ with } i=1,6 \tag{18}$$

are to be established in some way. Smoothness of distributions between neighbouring elements is assumed here as the driving criterion in defining the relationships. Actually, matrices [SSF] and [DSF] are defined in two different ways, depending upon the panel geometry. For flat and curved panels, continuity of the distributions can be enforced only in a least square sense, by mimimizing the following weighted residual functions

$$ES = \Sigma\, W_k \left\{ \sigma_k - \sigma(\ \xi_k , \eta_k\ ) \right\}^2 \tag{19}$$

and

$$ED = \Sigma\, W_k \left\{ \mu_k - \mu(\ \xi_k , \eta_k\ ) \right\}^2 \tag{20}$$

where σ_k, $\sigma(\xi_k,\eta_k)$ and μ_k, $\mu(\xi_k,\eta_k)$ respectively are the values of the actual singularity strength and of that obtained by extrapolating the spline function at the neighbouring control point k and W_k is a weighting parameter. In this case, the wanted matrices are the coefficients of the linear system obtained by differentiating (19) and (20) with respect to σ_i and μ_i. When the subpanels are used, the two matrices are built in two steps. First, the coefficients of the distributions (15) and (16) are expressed in terms of the values assumed by the singularities at the vertices of the subpanels belonging to the same panel. By assuming different distributions on each subpanel it is possible to make the distributions exactly continuos along the common edges. Vertex values are then correlated to the adjacent singularity parameters. Since the subpanel vertices lying on the edges of two adjacent panel coincide, it follows that the singularity distributions match together even between panels. In that way the doublet continuity requirement is exactly fulfilled within a network. The maximum order of the matrices [SSF] is 3x5 for the panel-based spline and 3x9 for the subpanel-based spline. The respective order of the [DSF] matrices are 6x9 and 6x21. The actual size of the matrices depends on the dimensions of the network and on the position of the panel relevant to the network edge.

3.2 Calculation of the aerodynamic influence coefficients.

Once the spline functions and the geometrical quantities have been computed, the program proceeds to compute and assemble the influence coefficient matrix (AIC) whose elements, together with the doublet matching conditions, are used to determine the unknown singularity distributions. In the present method the matrix AIC is of order 4xMxN since the three components of the perturbation velocity and the potential function induced by each of the N singularity parameters λ onto the the M midpanel control points are computed. Unlike zero-th order methods, where the integration over each element gives directly the influence coefficients, a two step procedure is necessary here. The integrals resulting from substitution of (15) and (16) in expressions (2) and (3) are first evaluated. Thus, the contribution of each elementary surface (panel or subpanel) is obtained as function of the coefficients of the singularity distributions σ_i and μ_i. It is then expressed in terms of the singularity parameters λ_j by multiplication of the matrices (17) and (18). The computational effort required to recast the influence coefficients as function of λ makes up the larger amount of the additional computing time associated with the use of higher-order singularities. In fact, all the integrals deriving from the higher order terms are evaluated using recursion formulae which are started using two fundamental integrals involved in the calculation of the zero-th order terms. Using the notations of [2], all the integrals required can be put in the following form

$$H(m,n,k) = \iint_{\Sigma} \frac{(x-\xi)^{m-1} \ (y-\eta)^{n-1}}{\rho^k} \, d\xi \, d\eta, \qquad (21)$$

where

$$\rho = \{(\xi-x)^2 + s[\ (\eta-y)^2 + h^2\ ]\}^{1/2}$$

with $s = 1$ for $M_\infty = 0$ and $s = -1$ for $M_\infty = \sqrt{2}$. Note that a local compressible coordinate system has been associated to the element Σ to limit the integration to the two fundamental cases addressed in the section 2.1. These integrals are recursively computed using auxiliary line integrals of the form

$$F(m,n,k) = \iint_L \frac{(x-\xi)^{m-1} \; (y-\eta)^{n-1}}{\rho^k} \, dl, \tag{22}$$

extending over the perimeter L of the integration surface Σ. The recursive procedure starts using the integrals H(1,1,3) and F(1,1,1) as fundamental solutions. These two integrals involve the computation of inverse trigonometric functions and their evaluation is the most time-consuming part in the set up of the AIC matrices.

To save computing time, approximate integration formulae are used when the quantities $1/\rho^k$ are sufficiently small so that they can be expanded about the centroid of Σ using a Taylor series. In this case, the integrals (2) and (3) depend only on the local variables ξ and η, since the coefficients of the Taylor expansion can be brought out of the integral. Thus, the integrals can be computed once and stored together with the other geometrical quantities. Following [12], two different sets of approximate formulas are defined by retaining all expansion terms through the second order (multipole expansion) or truncating the Taylor serie to the zeroth term (monopole), whose range of applicability is assumed to be dependent on the ratio of the distance ρ between the induced control point and the centroid of the inducing element to the maximum diagonal t of the element itself. In HISSS three characteristic values for ρ/t are considered which defines four ranges where different procedures are used. In the first two regimes, the exact integration is used to obtain the needed accuracy at points close to the elements: integration over the subpanels (if any) is limited to values of $\rho/t < (\rho/t)_1$, while the entire panel is considered for values $\rho/t < (\rho/t)_2$. The approximate formulae obtained using the multipole expansion are used in the intermediate field where $(\rho/t)_2 < \rho/t < (\rho/t)_3$ and the monopole expansion is employed in the far field where ρ/t is larger than $(\rho/t)_3$. Clearly, the values chosen for $(\rho/t)_i$ should maximize computing efficiency and still yield an acceptable level of accuracy. Although an analysis can be carried out to determine the error introduced by the approximate formulae into the influence coefficients, this leaves out of consideration the fact that the actual loss of accuracy depends upon the solution itself through the values of the singularity strengths and therefore cannot be known a priori. Experience has shown, for example, that supercritical and internal flows are more sensitive to the used extent of the intermediate and far field procedures than external subcritical flows. For this reason, user's specified values can be introduced in the calculation. Default values for $(\rho/t)_i$ provided in the program have been defined to optimize subcritical external flows while at supersonic speeds an absolute minimum application range - based on a characteristic length of the configuration - has been introduced for the near field procedure.

From the analysis of the behaviour of the potential and of the velocity components induced by a flat element it is known that these functions are continuous and finite everywhere except over the element itself and, in the supersonic cases along the characteristic surfaces originating at the element contourings the induced point approaches the element edges, both tangential and anormal velocities can become unbounded. Logarithmic singularities are induced by the edge source strength and by doublet derivative normal to the edge. Edge doublet strength - which is equivalent to a vortex line - induces a singular term inversely proportional to the distance from the edge. At supersonic speeds, these singular terms are propagated by the characteristic lines into the whole downstream flowfield. Actually, the line vortex term is not computed at supersonic speeds, since the method yields strict doublet continuity when the subpanels are used. Inside the element, bounded discontinuities appear when the induced point is approaching the inducing singularity surface from the two sides. Normal velocity induced by a source distribution is discontinuous with a jump equal to the local singularity strength. Doublet distributions induce discontinuities in both the potential and in the tangential velocity components, the jumps being proportional to the local strength and to its local gradient respectively. Existence of such discontinuities implies that a positive or negative side must be specified when assigning the boundary conditions and calculating the flow solution at a given control point. To preserve generality in boundary condition specification, the discontinuous terms are computed and stored separately from the continuos terms. Due to the local character of the discontinuities, this additional storage is usually a small fraction of the total AIC space requirement. In addition, as shown in Section 3.5, the knowledge of the jump values can be used to calculate the induced velocity using the induced potential only instead of the whole AIC matrices.

3.3 Doublet matching at network edges.

Continuity of doublet across networks is obtained by enforcement of a matching algorithm at edge control points. The algorithm exploits the singular terms induced by the local doublet strength and its derivative at the edge of the panel. By requiring the velocity to remain finite at the edge implies that the doublet strength and its derivative must be continuous at the edge. It is known that the unbounded terms behave as μ/ρ and $\mu' \log(1/|\rho|)$ where ρ is the distance from the edge to the point where the velocity is induced and μ' is the derivative normal to the edge. Then a linear equation can be set up for each edge control point by moving the control point inside the panel at a distance ρ from the edge and considering all the contributions from the networks abutting the edge. The logarithmic singularity induced by the doublet derivative discontinuity is quite mild and could be tolerated but it must be included in the algorithm to avoid formation of linearly dependent equations. The algorithm proved to be reliable and robust also in awkward situations, as for example at network corner points where multiple network joints occur and at partially abutted edges. When the program computes an edge control point, it scans every composite, doublet and wake network edge to check if the control points lies within a prescribed distance from it. If no other edge than its own is found, the algorithm implies zero local doublet strength. When a wake network edge is found, its potential jump is correctly transmitted to the adjacent networks. At wake control points, an implicit Kutta condition is obtained. At supersonic edges, however, the algorithm has to be partially modified since discontinuities in the doublet derivatives are physically meaningful and must be retained in the solution. For this purpose, two different algorithms are applied at network edge control points. Mid- edge control points found to lay upstream of a supersonic edge are relocated a small distance ahead and treated as midpanel control point, i.e. the aerodynamic influence coefficients are computed and an aerodynamic boundary condition is later enforced here. Similar points lying downstream of a supersonic edge are then used to match the doublet strength only. To avoid definition of linearly dependent equations, a bookkeeping must be set up. Doublet continuity is enforced only at the first control point found to be downstream of a given edge, while any other control point abutting the same edge is relocated downstream and treated as a midpanel point. Finally, aerodynamic influence coefficients are computed also for control points downstream of supersonic edges lying in the symmetry plane. The basic algorithm is used at corner points but the logarithmic singularity is neglected for upstream supersonic edges and both singular terms are discarded for downstream supersonic edges. Thus, perturbations are not allowed to propagate upstream across the supersonic edges. The identification of supersonic edges and the selection of the proper matching algorithms are performed automatically by the program. However, the user can override the standard procedure and specify directly the type of matching algorithm to be applied to a given network abutment. Of particular interest are the possibilities to use different forms of Kutta condition (e.g. explicit flow-tangency condition) or to prescribe a specified amount of circulation.

3.4 Boundary condition specification and linear system solution.

A system of simultaneous linear equations is used to determine the strengths of the unknown singularity parameters λ. The whole set is formed by the doublet matching constraints - which are defined automatically by the program - and by the user's specified boundary condition equations. Separate parameters in the input-list are used to direct the formation of the left-hand side of these equations and to specify the value for the right-hand side term. Formally, each boundary condition equation can be conceived as a formal equation of the type

$$g(LHS) = G(LHS)*VRHS + QRHS + PRHS, \tag{23}$$

where LHS, VRHS, QRHS and PRHS are the user's specified parameters, g is the array of the perturbation functions which may be specified at the control points and G contains the analogous values assumed by the equivalent freestream variables. A wide variety of boundary condition specifications is implemented in the method. It includes specification of normal mass-flux (LHS from 1 through 4), of normal velocity (LHS = 5-8) and of potential function values (LHS=9-12). These conditions may be applied on the positive or the negative side of the panel - for values of the Fortran function MOD(LHS,4) respectively equal to 1 or 2 - or prescribed in terms of the mean or jump value between the two sides - MOD(LHS,4) equal to 3 or 0 -. The parameters LHS, VRHS and QRHS are constant for each network, while PRHS is used to specify boundary condition values variable from panel to panel, necessary - for example - when simulating boundary layer effects through the transpiration technique.

Generality of equation (23) coupled with the multiple surface singularity modelling make of the method a very flexible tool that can model almost all types of linearized potential flows. Some care, however, must be exercised when associating the boundary conditions to a given singularity type since not all of the possible combinations determine a mathematically well-posed problem. A detailed discussion of this topic can be found in [13] and [14]. Only a few example of well-posed combinations are presented here. In the next, for sake of simplicity but without loss of generality, the positive side will be considered to be the panel face wetted by the flow of interest.

A nonlifting configuration may be modeled by source panels only. For impermeable body wall, the flow tangency condition is enforced by selecting LHS = 5, VRHS=-1., QRHS = PRHS = 0. . If LHS = 1 is used, for compressible flow the Goethert rule 1 is enforced (see Section 2.3).

A lifting configuration without thickness may be represented using doublet panels. Since ϕ_n is continuous, each one of the three specifications LHS = 1, 2 or 3 together with VRHS=-1., QRHS = PRHS = 0. will give identical results, i.e. impermeability in terms of linearized mass flux. At $M_\infty \neq 0.$, however, it is impossible to model the configuration as a stream surface, unless the surface be entirely parallel to the compressibility axis.

A thick lifting configuration may be represented by doublets only or, as more usual, by a combination of sources and doublets. When doublet panels only are used, an internal Dirichlet b.c. must be enforced, i.e. LHS = 9, VRHS = -1., QRHS = PRHS = 0. This implies $\phi_n = 0$ on the outer surface and therefore only Goethert rule 1 can be enforced here again. Note that since a constant doublet distribution induces null velocity, a Neumann's type b.c. specification would lead to an ill-posed problem. Although theoretically feasible, the doublet-only formulation is very seldom used in practice since the large doublet strength necessary to produce the external velocity tends to blow up truncation errors making the results quite sensitive to the local panelling characteristics. Instead, superposition of source panels to control the value of ϕ_n on the outer side of the panels results in a substantial decrease of the doublet strengths. In the present method, this modelling is obtained using composite panels. Two boundary conditions must be specified at the midpanel control points. A range of different specifications may be applied. The most widely used type is a straightforward application of Green's theorem. The local source strength is used to cancel the normal component of the freestream velocity while the doublet distribution is determined by assigning zero perturbation potential on the interior side of the configuration. In equation (23) this combination is obtained by setting LHS = 4, VRHS = -1. for the source and LHS = 10, VRHS = 0. for the doublet specification. This is an indirect or implicit way to eliminate the linearized mass flux through the surface of the configuration. The main attraction of this type of boundary condition specification - known in the literature as Morino's type - is that the source strengths depends uniquely on the geometry of the surface and therefore are known 'a priori' so that the doublet strengths only must be solved for. At supersonic flows, propagation of interior waves is prevented in an average sense since the inner perturbation potential is zero at each control point. A slight deviation from the Morino's type specification is obtained by considering the jump of the normal velocity between the two sides instead of the jump of the linearized mass-flux. This specification - obtained by substituting LHS = 8 to LHS = 4 - is an implicit way to enforce impermeability in terms of velocity. Applying the jump operator Δ to the equation (10) of Section 2.3, this condition can be recasted as

$$\Delta(\phi_n) + (\beta^2 - 1)\, n_x\, \Delta(\phi_x) + \mathbf{V}_\infty \times \mathbf{n} = 0, \tag{24}$$

which shows the appearance of influence of the local gradient of the doublet distribution. Excluding the incompressible case ($\beta^2 = 0$) and the case of panels parallel to the compressible axis ($n_x = 0$), the source strength can not be determined a priori. The resulting equation is sparse, since it contains only the local source strength and the doublet singularity parameters involved in the computation of the matrix [DSF], see formula (18).

The external boundary condition may be also specified explicitly. In this case, mass-flux or velocity boundary conditions are enforced by selecting LHS = 1 or 5., respectively. The resulting equations contain, at least at subsonic speeds, all the singularity parameters. The choice between implicit and explicit boundary condition specification is a questionable matter. For a given configuration, both should give - provided that a sufficiently large number of panels be used - identical results. In practice, with the panel density currently affordable in the calculations different results are produced, with more pronounced discrepancies spotted in regions of strongly perturbated flows. In such regions and especially at supersonic speeds, the explicit b.c. specifications give better results. A cost effective compromise can be obtained by exploiting the HISSS capability to specify the boundary condition at a network level thereby limiting the use of the more costly direct specification to regions of strongly perturbated flow.

Once all the boundary condition equations and the doublet matching constraints have been set up, a linear system of the type

$$[\ A\ ]\ \{\ \lambda\ \}\ =\ \{\ B\ \} \tag{25}$$

must be solved for. Before starting the solution procedure, all the equations which imply a direct specification of an element of the solution vector $\{\lambda\}$ are solved independently, so that the order of the linear system can be reduced. This situation occurs always at network control points where a boundary condition of the type LHS = 4 or LHS = 12 is enforced - so that the local source or doublet strength is directly specified - and at control points lying on free edges, where the doublet strength is identically zero. For the case LHS = 8, the magnitude of the off-diagonal coefficients a_{ij} deriving from the ϕ_x term of equation (18) must be checked. In practice, when

$$\Sigma\ |a_{ij}|\ <\ \varepsilon\ |a_{ii}|\ \text{with}\ \varepsilon = 10.e{-4} \tag{26}$$

a direct specification for the element λ_i is assumed in the program.

To solve the system, consideration of the characteristics of the matrix [A] recommends the use of direct matrix solvers rather than iterative procedures. In fact, doublet matching constraints produce sparse, not diagonalized equations; at supersonic speeds, large off-diagonal values can be obtained whenever a control point lies close to a Mach surface of an upstream panel. In the program, three methods are available: a matrix solver based on the GAUSS-elimination procedure with pivoting and two matrix inversion solvers based on a Gauss inversion and an LU decomposition, both without pivoting. For one solution, the direct solver requires less computational effort, but the inversion of the matrix allows the computation of additional right-hand side vectors at virtually no-extra cost. Clearly, generality of equation (23) - which allows to investigate different combinations of angles-of-attack and of-sideslip, variation of surface permeability due to boundary layer effects (transpiration modelling) or of engine ingestion/exhaust ratios by changing the right-hand-side vector - is exploited at best when an inversion procedure is used. From a computational point of view, the LU decomposition solver has been found to be more efficient then the Gauss one. Optimized for the solution of very large systems, the procedure uses a sophisticated management of the data transfers from and to the mass-storage disks so that all the available central memory can be used while retaining the I/O operations at a minimum.

3.5 Calculation of results.

Multiplication of AIC matrices by the solution vector gives directly the 'mean' velocity and potential induced by the singularities at the control points. Half of the local 'jump' values and the freestream terms must be added to obtain the global velocity and potential on the selected side of the configuration. When the composite networks have been used to determine the value of the potential function inside the configuration, the external velocity may be computed by adding the 'jump' values to the derivative of the potential function on the interior side. If the kinematic boundary conditions have been specified in terms of the 'jump' values - i.e. LHS = 4 or 8 - it would then be possible to avoid the computation and the storage of the velocity AIC submatrices. It is to note, however, that although theoretically equivalent, the results obtained by the two procedures do not coincide exactly. To retain logical compatibility, the full AIC procedure must be applied when the external flow tangency condition has been specified directly - LHS = 1 or 5 and the second method used together with indirect specifications. Still, the two sets of results will not be identical. The differences are however negligible, provided that the panel density is adequate to the model the flow perturbations. Thus, these residuals can be used to monitor the accuracy of the solution and detect regions where the panelling should be refined.

Pressure coefficients are computed from the velocities by means of either the isentropic or a 2nd-order formula. The latter is derived using the small perturbation assumptions and is therefore considered to be more appropriate for a method solving the Prandtl-Glauert equation. Its use tends to exagerate the effects of small-perturbation theory on panels inclined to the flow. Better comparison with Euler code results have been obtained by employing the isentropic formula, especially at supersonic speeds.

Finally local load distributions and global aerodynamic force and moment coefficients are obtained by integration of surface pressures. Additional quantities, such as mass-flux and momentum, are computed for completeness.

4. CHARACTERISTIC OF THE CODE.

The method is coded using the language FORTRAN 77. The code consists of about 24,000 lines, of which 15,000 are executable statements and the rest are comment used for internal documentation. The entire code follows a modular structure with one main module managing the flow of the calculation, six modules made up of several subroutines contributing to perform a well-defined task and one module grouping utility procedures used throughout the program. Several options are available, which allow to stop the execution of the calculation at different levels - e.g. geometry generation or assembling of the AIC matrices - or to run the program in different back-up modes. These allow to skip parts of the calculation already performed - and whose relevant data have been stored - for the same configuration at the same Mach number. Interrupt modes are useful for checking the input data during the definition phase of a new configuration. Back-up modes can be used for respecifying the boundary condition types without recomputing the AIC matrices or for calculating further combinations of angle-of-attack/sideslip by retrieving the inverse matrix.

Initially developed on the VAX-780/11 computer system, the code is now implemented even on the MBB IBM mainframe computers, on different CDC machines operating under NOS/VE and, with only minor modifications, on the Cyber 205 and CRAY 1S vector processors. CPU requirements depend on the particular installation and on the array sizes declared in the code. Upper bound values of array dimensions are set by Fortran PARAMETER statements which specify the maximum allowable number of networks, panels and unknown singularity parameters. Present PARAMETER values used in the code installed on the IBM computer are:

- max. number of network .... NNETTX = 120,
- panels .... NPANTX = 5000,
- control points .... NSNGTX = 6000,

which correspond to a CPU storage of about 500 Kbytes. On this machine, a substantial reduction in I/O operations is gained by storing part of the panel ands subpanel information data and of the solving matrix into large COMMON blocks which are allocated at execution times within the additional storage space available through the MVS/XA operating system. PARAMETER statements define the quantity of data which can be dynamically allocated, while the rest - if any - is automatically stored onto mass-storage disks. Current values are:

- number of panel information ...... NPANSX = 2000,
- number of subpanel information ... NSPNSX = 2000,
- order of solving matrix .... NEQLU = 900,

which correspond to an additional storage of about 30 Mbytes. On machines without virtual CPU storage the parameters must be reset to appropriate values which compromise at best geometrical requirements with installation capabilities.

5. APPLICATIONS.

5.1 Cone-cylinder-cone configuration.

A classical testcase for studying the internal wave propagation problem is the simple body of figure 1. The configuration consists of two 15 deg.s cones joint by a circular cylinder. The longitudinal pressure distributions obtained using the present method have been compared in [15] with calculations from an Euler code and from NLRAERO - a low-order panel method [16]. Figure 2 shows the comparison for the symmetrical flow condition at $M_\infty = 2.0$. Use of composite panels associated with specification of internal perturbation potential inside the body is instrumental in damping out the severe pressure fluctuations exhibited by the source-only modelling of NLRAERO in the aft-cone. Further, the HISSS solution agrees closerly with the Euler calculation even on the forward cone and on the cylindrical part. For this example, smooth results are obtained even using the low-order formulation, provided that the source be replaced by triplet, a combined source-vortex singularity developed by Woodward [17]. However, since, the triplet does not guarantee continuity of geometry, its effectiveness in eliminating the internal wave propagation problem is restricted to simple geometries [18],[1].

Fig. 1 - Cone-cylinder-cone testcase

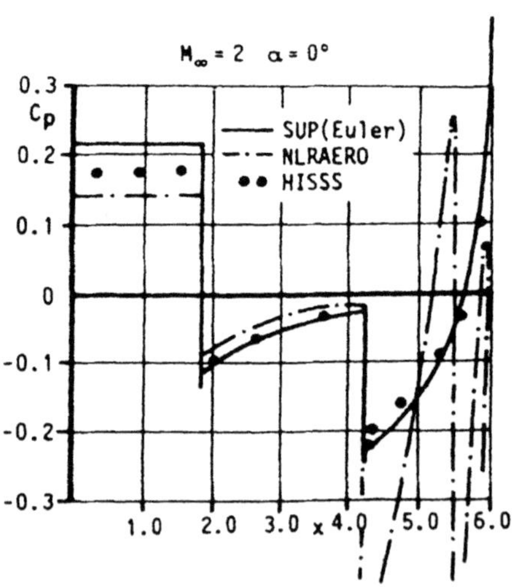

Figure 2 - Comparison of surface pressures

5.2 Supersonic cruise wing-body configuration of NASA TM 78792.

The geometrical panelling of this configuration typical of a supersonic cruise airplane is presented in figure 3. A total number of 795 composite panels grouped in 9 networks have been used to model the external surface of the configuration. Explicit external normal velocity and interior perturbation potential have been specified as boundary conditions, leading to a linear equation system of 1896 unknown singularity parameters. Perturbation velocities have been computed from the full AIC matrices and pressure coefficients derived by the isentropic formula. In figure 4 the comparison between the calculations and the experimental data of [19] is produced for the longitudinal characteristics at $M_\infty = 2.3$ and 3.3. The agreement between computation and experiment is remarkably good, considering the relatively high Mach numbers. For comparing the lift-to-drag L/D characteristic, the theoretical drag curve have been shifted so that the values at $\alpha = 0$ concide with the relevant experimental data. The increment is mainly due to skin friction drag which is - of course - neglected in the potential flow calculation. The surface pressure distributions computed at an angle of attack close to the maximum L/D condition are shown in form of isobar curves - with a pressure step equal to 0.01 - in figure 5.

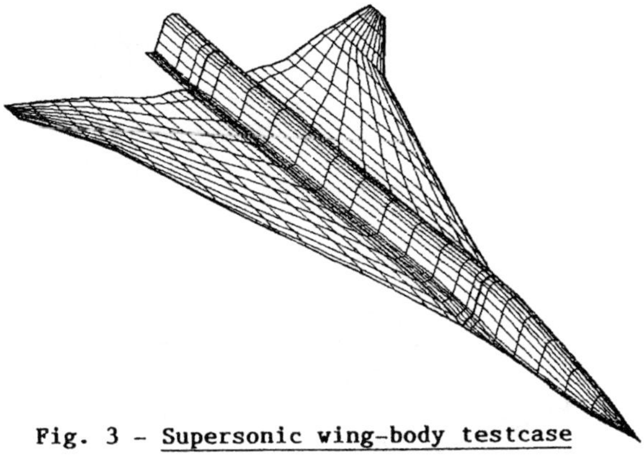

Fig. 3 - Supersonic wing-body testcase

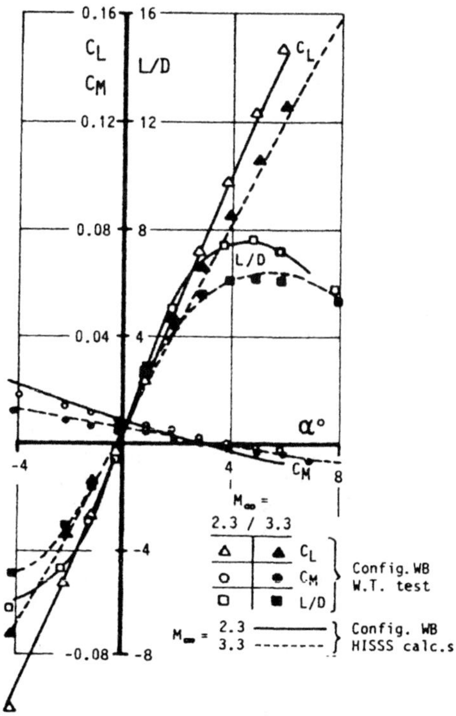

Fig. 4 - Theory vs. experiment comparisons

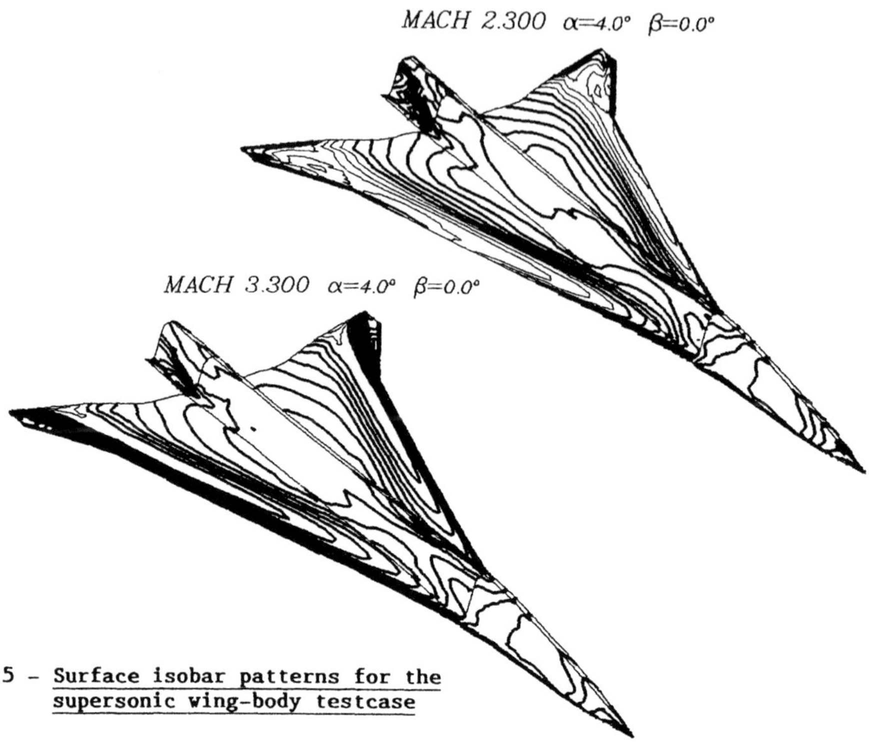

Fig. 5 - Surface isobar patterns for the supersonic wing-body testcase

CONCLUDING REMARKS.

In the technical literature the formulation followed here is usually referred to as h i g h e r order - in opposition to l o w or l o w e r order ones used by other panel methods - although there are many more differences between HISSS - and perhaps PANAIR - and other panel methods than the mere order of singularity distributions. In fact, the singularity distributions used by the two classes of methods differ even in location and in type and different boundary conditions are enforced for determining the singularity strengths. Among others, there are two most important features which characterise the present formulation.

The first feature is the capability to make use of the mathematical properties of the singularity discontinuities for modelling the physics of the flow. This feature is mandatory for calculations in supersonic flow, where due to the hyperbolic nature of the governing equation singular terms induced by singularity discontinuities are transmitted into the influence domain along characteristic surfaces. Discontinuities in source distributions and in doublet derivatives induce jumps in the perturbation velocity which are the linearized part of recompression shocks or of Prandtl-Mayer expansions. Discontinuities in the doublet strength induce very large velocity jumps which have no physical meaning and therefore must be avoided. In this formulation, continuity of the doublet strength is ensured over the entire configuration while source and doublet derivative jumps can develop along the boundaries of networks - i.e. computational subdomains - into which the whole configuration is subdivided.

The second feature is the wide range of options available for defining the modelling of each particular flow problem. In contrast to other methods, where type of singularity distributions and boundary conditions specifications are coupled 'ab initio' together and sometimes even associated once for ever with specific configuration components - as for example source on body and vortex on lifting surfaces - the present method allows the user to define for each network number and type of singularities and to specify type, location and values of the boundary conditions necessary to determine the singularity strengths. Coupled with very few restrictions for the allowed network geometries, the method is a very flexible tool which allows the modelling of rather different flow problems, ranging from analysis of external flows around impermeable bodies to simulation of internal channel flows. Moreover, flexibility in modeling is of paramount importance in the calculation of supersonic flows. Here, use of double distributions of singularities makes available one additional boundary condition which can be used to suppress the transmission of non-physical perturbation waves inside the configuration, a negative feature of methods using a single surface singularity.

As should be expected, both features have a large impact on the practical use. To take full advantage of the capabilities of the method, singularity panels should be located on the actual surface of the configuration. Enforcement of continuous doublicity distributions implies continuity of surface geometry which must be guaranteed by the input data. As a result, the preparation of the geometrical input is often cumbersome because the panelling must be defined without generating gaps or overlaps and intersection lines between different configuration components must be computed. For applications to complex geometries involving the definition of some thousands of panels, the use of interactive preprocessors capable to access a CAD-supplied geometry data base represents the only efficient way to solve the task [4].

On the other hand, even the freedom in defining the singularity distributions and in selecting the boundary condition specifications can become a burden for a potential user not fully aware of the background theory. In an effort to increase user's friendliness of their method, PANAIR developers are evaluating the use of an expert system [20] which should be able to prompt the user with the most appropriate choices relevant to his flow problem. A compromise has been obtained in HISSS by assigning to the variables controlling the various options some default values and declaring these variables into NAMELIST Fortran statements. In this way, the user can omit to specify all or part of the options in which case the program will assume the default values - which are eventually dependent on the Mach number. But he is also able to override the default values and make his own selection when running non-standard problems.

R E F E R E N C E S

[1] Fornasier, L., "Treatment of Supersonic Configurations by an Updated Low-Order Method", J. of Aircraft, Vol. 21, Apr. 1984.

[2] Johnson, F.T., "A General Panel Method for the Analysis and Design of Arbitrary Configurations in Incompressible Flows", NASA CR-3079, 1980.

[3] Ehlers, F.E., et al., "A Higher Order Panel Method for Linearized Supersonic Flow", NASA CR-3062,1979.

[4] Fornasier, L., and Heiss, S., "Application of HISSS Panel Code to a Fighter-Aircraft Configuration at Subsonic and Supersonic Speeds", AIAA Paper 87-2619.

[5] Eberle, A., "3-D Euler Calculations Using Characteristic Flux Extrapolation", AIAA Paper 85-119.

[6] Kraus, W., "The MBB Subsonic Panel Method - Part 1", MBB-UFE 632-70, 1970.
[7] Woodward, F. A., "An Improved Method for the Aerodynamic Analysis of Wing-Body-Tail Configurations in Subsonic and Supersonic Flow", Nasa CR-2228, 1973
[8] Melnik, R., and Mason, W., "Mass Flux Boundary Conditions in Linear Theory", AIAA Journal, Vol. 22, Nov. 1984.
[9] Butter, D.J., "Compressibility Corrections used in Panel Methods", HSA-MAE-R-FDM-0039 Report,1978.
[10] Nathman, J.K., "The Solution of the Prandtl-Glauert Equation around Non-slender Bodies", Analytical Method Report, 1987.
[11] Lucchi, C.W., "MBB Higher Order Panel Method - First Results", MBB-UFE122-AERO-MT-509,1981.
[12] Hess, J.L., "Calculation of Potential Flow about Arbitrary Three Dimensional Lifting Bodies", MDC-J0546 Report, 1969.
[13] Carmichael, R.L., and Erickson, L.L., "PAN AIR - A Higher Order Method for Predicting Subsonic or Supersonic Linear Potential Flows About Arbitrary Configurations, AIAA Paper 81-1255, 1981.
[14] Towne, M.C., et al., "PAN AIR Modeling Studies", AIAA Paper 83-1830, 1983.
[15] Mortel, P., et al., "Calcul d'ecoulements supersoniques autour de missiles tactiques", Proceed.s of the 23ème Colloque d'aerodynamique appliquee, AAAF,1986.
[16] Hoeijmakers, H.W.M., "A Panel Method for the Determination of the Aerodynamic Characteristics of Complex Configurations in linearized Subsonic or Supersonic Flow", NLR Report 80124 U, 1981.
[17] Woodward, F.A., and Landrum, E.J., "The Supersonic Triplet - A New Aerodynamic Panel Singularity with Directional Properties", AIAA Journal, Vol. 18, No. 2, 1980.
[18] Woodward, F.A., and Fornasier, L., "Investigation of the Triplet Concept Using a Higher-Order Panel Method", ICAS Paper 84-001, 1984.
[19] Shrout, B.L., and Fournier, R.H., "Aerodynamic Characteristics of a Supersonic Cruise Airplane Configuration at Mach Numbers of 2.30, 2.96 and 3.30", NASA TM 78792, 1979.
[20] Conner, R.S., et al., "PAN AIR Consultation Using Expert System Techniques", AIAA Paper 85-4094, 1985.

CHARACTERISTICS AND BOUNDARY ELEMENTS FOR THREEDIMENSIONAL NONSTATIONARY NAVIER STOKES FLOWS

F.K. Hebeker
Universität-GHS Paderborn, Fb. 17
D-4790 Paderborn, West Germany

SUMMARY

We report on a new numerical approach based on the boundary element method to compute nonstationary incompressible Navier Stokes flows in a threedimensional container. Asymptotic error estimates are presented as well as numerical examples, in particular the secondary vortex flow in a spherical tank has been computed.

INTRODUCTION

The numerical prediction of threedimensional Navier Stokes flows is a challenge as well for mathematicians as for applied scientists. The now "classical" approaches lead to huge practical difficulties: finite difference methods are principally easy to handle but are hampered by curved geometries, whereas finite element methods are fairly flexible but are even more complicated to get implemented on the computer. In the present paper we report on a new approach based on a boundary element method combined with a spectral solver and characteristics (to handle the convective term) as an alternative.

Let us consider the initial boundary value problem of the Navier Stokes equations governing viscous homogeneous flows in a bounded cavity $\Omega \subset \mathbb{R}^3$ (with smooth boundary $\partial\Omega$):

$$v_t + v \cdot \nabla v - \nu \nabla v + \nabla p = F \quad , \quad \text{div } v = 0 \quad \text{in} \quad \Omega,$$
$$v_{|\partial\Omega} = 0, \qquad v_{|t=0} = v_o \quad . \tag{1}$$

Here the vector field $v(t,x)$ or the scalar field $p(t,x)$ denotes the velocity (at time t at the point $x \in \Omega$) or the pressure function of the flow, respectively, and $\nu \sim \frac{1}{Re}$ is the dynamic viscosity of the medium. The exterior forces are summarized up to the vector field $F(t,x)$.

Our numerical procedure is as follows: On short time intervals we compute the characteristics (of the acceleration terms of the differential equations). This corresponds to a stabilized time-discretization of the convective term, if the Navier Stokes equations transformed on characteristic ("Lagrangean") coordinates are discretized implicitly. Here the characteristics are computed backwards in time [11], consequently in each time step we are left with a stationary Stokes-like problem which can be treated immediately by using boundary elements and fast spectral solvers [2]. Each part of this method is mathematically justified by asymptotic error analysis. The algorithm is tested by threedimensional numerical computations.

This paper is organized as follows: First, the nonstationary Navier Stokes equations are discretized (semi-) implicitly in time by use of the characteristics approach. Second, the linear stationary problem appearing at each time step is solved by the combined boundary element spectral method. Third, we present the results of a numerical test series for a large period. Finally, this algorithm is applied to study secondary vortex flows in a spherical cavity.

A more detailed version of this paper will be published elsewhere [6].

TIME-DISCRETIZATION AND CHARACTERISTICS

In general unique solvability of (1) is asserted mathematically on possibly small time intervals only. Assume that a unique solution exists on the time interval $[o,T]$ being sufficiently smooth there. A time grid $t_k = k\tau$ ($\tau = T/N$ step length; $k = o,1,\dots,N$) is introduced. Assume that an approximate v_k has been computed for $v(t_k,.)$. Then we approximate the characteristics corresponding to the hyperbolic part

$$v_t + v_k \cdot \nabla v \qquad \text{for the time interval } [t_k,t_{k+1}],$$

of the differential equations by solving the characteristic differential system

$$\dot{x}(t) = v_k\,(x(t)) \quad , \quad x(t_{k+1}) = x, \tag{2}$$

(on the time interval $[t_k,t_{k+1}]$) *backwards* in time. Let

$$X_k(t,t_{k+1},x) \quad \equiv \quad x(t) \quad , \tag{3}$$

denote the general solution of (2), depending on the given data. Then for the flow fields $\hat{v}(t,x) = v(t,X_k(t,t_{k+1},x))$, $\hat{p}$ analogously, in characteristic ("Lagrangean") coordinates we obtain a new system of PDEs. where the convective term has dropped out, due to the fact that the partial derivative $\hat{v}_t$ just corresponds to the total derivative of $v(t,X_k(t,t_{k+1},x))$. If we discretize now the new PDE's by means of a fully implicit scheme, then we are left with this linear stationary boundary value problem.

$$\frac{v_{k+1}(x) - v_k(X_k(t_k,t_{k+1},x))}{\tau} - \nu\Delta v_{k+1}(x) + \nabla p_{k+1}(x) = F(t_{k+1},x), \tag{4}$$

$$\operatorname{div} v_{k+1} = 0 \text{ in } \Omega \quad , \quad v_{k+1}|_{\partial\Omega} = 0 \quad ,$$

to be solved for each time step.

This way of taking the Navier Stokes problem has the major advantage that we get a *stabilized* algorithm (by semi-implicit time-discretization) but are still left with a boundary value problem with PDE of constant coefficients: hence efficient boundary element methods are applicable!

The characteristic differential system (2) does not cause serious troubles: for any internal node x it is solved by means of a simple Euler-Cauchy procedure, hence the first-order accuracy of our implicit time-stepping method is not afflicted.

Use of Lagrangean coordinates has a long tradition in hydrodynamics, as well for analytical as for numerical investigations (see [8], [12], [4], [14], for instance). Our algorithm is closely related to that by Pironneau [11] (called "transport-diffusion algorithm"). It has been investigated there theoretically in a finite element setting, but it might prove even more advantageous in connection with boundary element methods: this is indicated by the result of our numerical examples below.

The convergence analysis for this algorithm is summarized in

Theorem 1: Assume that $v_o \in H_o^1(\Omega) \cap H^2(\Omega)$ be divergence-free and F sufficiently smooth. Let

$$\nabla v \in L^\infty(\bar{\Omega} \times [o,T]) \tag{5}$$

hold. Assume further that the characteristic system (2) is solved exactly. Then the characteristics method (2), (3), (4) is convergent of the first order ($k=1,\dots,N$):

$$\| v_k - v(t_k,\cdot) \|_{L^2(\Omega)} \le \text{const. } \tau \ , \quad \text{with} \quad \tau \to 0 \ . \tag{6}$$

The proof (to be published in [6]) by energy analysis is based partly on sharp a priori estimates of the solution of (1) (which follow from (5), cf. [7]: note that we do not require the nasty "compatibility condition" at $t = 0$!) , partly on sharp estimates of the gradient of the general solution of the characteristic differential system (following [12]). Pironneau's proof requires essentially stronger regularity assumptions, on the other hand the additional error due to finite element approximations has been taken into account there.

THE BOUNDARY ELEMENT SPECTRAL METHOD

We are concerned now to the stationary BVP. (4), written shortly as

$$\begin{aligned} &\lambda u - \nu \Delta u + \nabla q = f \ , \quad \operatorname{div} u = 0 \quad \text{in } \Omega \ , \\ &u = 0 \qquad \text{on the boundary } \partial\Omega. \end{aligned} \tag{7}$$

The relatively simple shape of these differential equations strongly suggests to use boundary elements to solve (7), but we are confused by the nonhomogeneous term f: usually it is handled by "volume potentials" (e.g. [14],[3],[13]), but these are extremely time-consuming tools to be circumvented for numerical purposes. Hence in [2] the idea of using a fast spectral solver to compute a particular solution of the nonhomogeneous differential equations has been introduced. Consequently we result in a "boundary element spectral method" which is summarized as follows.

We split the problem (7):

$$u = u_1 + u_2 \quad , \quad q = q_1 + q_2 \quad , \tag{8}$$

where (u_1,q_1) is any solution of the differential equations of (7), but (u_2,q_2) solves (7) with $f = 0$ and the boundary condition

$$u_2 = -u_1 \quad \text{on} \quad \partial\Omega.$$

Let us consider both subproblems separately:

A particular solution (u_1, q_1) is obtained as follows. The flow region Ω is covered by a finite number of blocks B_i. Now the nonhomogeneous field f is extended smoothly to $\cup B_i$, and then localized to each B_i using a partition of unity, so that each localized field is periodic with respect to its block. This process of "extension and localization" is particularly simple in those cases where it would suffice numerically to include Ω in just one block and to extend f by zero (see the numerical examples below).

In each block the periodic solution of the nonhomogeneous differential equations is obtained as a (surprisingly simple) *Fourier series* [1]. Here the vector Fourier coefficients are computed by means of the product trapezoidal rule, and the series is evaluated by using the Fast Fourier Transform. Let (u_1^n, q_1^n) be an approximation by truncating the series ($3 \times 8n^3$ terms).

The core of the problem consists of solving then (7) with $f = 0$ and boundary condition $u_2 = -u_1^n$ on $\partial\Omega$. It is solved efficiently, namely in complicated geometries, by means of a *boundary element method*. For this the boundary is divided into a finite number of surfaces S_i, each of it being mapped on a domain $D_i \subset \mathbb{R}^2$ by a smooth application [15]. Assume that each D_i is rectangular, for example. It is decomposed in small squares, immediately leading to a decomposition of the boundary into quadrangular panels, the "boundary elements". This construction is particularly simple if Ω may be mapped smoothly onto a ball, e.g., hence polar coordinates may be adopted to describe the boundary $\partial\Omega$ by means of just one global coordinate frame (see the example below).

A fundamental solution of (7) is simply constructed, hence the solution of the BVP. may be obtained by a corresponding double layer potential, the vector surface source of it satisfies a boundary integral equations' system of the second kind [14]. We are looking for an appriximate solution ψ_h which is piecewise bilinear on each D_i and globally continuous. The unknown coefficients are obtained from a collocation-type boundary element procedure, resulting in a linear algebraic system with a dense but relatively small matrix. The boundary integrals are evaluated by a product Gaussian formula.

A convergence analysis of this boundary element spectral method has been carried out in [5] for the boundary element part and in [1] (see [9], too) for the spectral part. Let us summarize these results.

Theorem 2: Assume that $f \in H^s(\Omega)$, $s > \frac{3}{2}$, be smoothly extended and localized to the blocks. Let (u_{nh}, q_{nh}) denote the approximate solution of (7) by the boundary element spectral method. Then this interior uniform convergence result

$$\| u_{nh} - u \|_{C(\bar{G})} \leq \text{const.}\ (n^{-p} + h \cdot \log \frac{1}{h}) \tag{10}$$

(when $n \to \infty$, $h \to 0$) holds, where $p = s + \frac{1}{2} - \varepsilon$ ($\varepsilon > 0$ arbitrary). Here $G \subset \Omega$ is any domain with positive distance of $\partial\Omega$. The constant generally depends on $\lambda = 1/\tau$.

A complete proof is given in [6]. It should be pointed out that this convergence result even takes into account the error due to numerical quadrature. Actually, this is responsible for the fact that we get an *interior* asymptotic error estimate only: currently we are missing more accurate asymptotic expansions of the boundary integrals near the boundary.

NUMERICAL EXAMPLES

This new approach has been tested in case of Navier Stokes flow in a spherical cavity (radius = 1). It is covered by a cube (edge length = 16/7), and the data f are extended by zero for the spectral method. The sphere is represented by polar coordinates $(\Theta,\varphi) \in D \equiv [0,\pi] \times [0,2\pi]$, D is decomposed in quadratic elements for the boundary element procedure.

In our first example we study the quantitative behaviour of the algorithm, namely its stability properties, in case of an artifical flow

$$v(t,x) = (t+1)(e^{-|x|^2} - e^{-1})(z-y,\ x-z,\ y-x)^T , \quad p(t,x) = P(t)$$

(due to the lack of a nontrivial analytical solution of the Navier Stokes problem). The viscosity is $\nu = 1$. We carried out 50 time steps of the step length $\tau = 0.1$. We used 12,288 Fourier coefficients, 200 boundary elements (= 546 degrees of freedom), and the midpoint rule to evaluate the boundary integrals. The CPU time was about 3,800 secs. per time step on a PRIME 750. From the Fig. 1 namely the high degree of stability of the computations should be noticed (cf. [14] for a scheme explicit in the convective term).

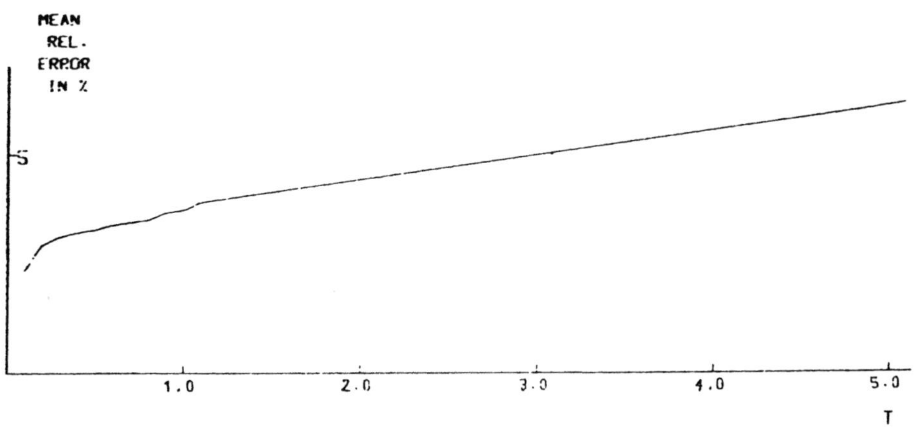

Fig. 1 Mean relative error in case of a test computation

In our second example we compute a Navier Stokes flow resulting from a rigid body rotation ($\omega = 100$) when the container is suddenly brought to rest. Toroidal vortices are developing, as it is seen in the following figures showing a vertical cross section of the spherical container. The Reynolds number is about 600, the time step $\tau = 0.01$. See Fig. 2 below.

ACKNOWLEDGMENT

The author is indepted to Prof. Dr. E. Krause, Aachen, for providing our team for challenging physical examples. Thanks are going to Mrs. E. Rohde who wrote the plot programs, and also to the Deutsche Forschungsgemeinschaft for some financial support.

REFERENCES

[1] BORCHERS, W.: "Eine Fourier-Spektralmethode für das Stokes-Resolventenproblem", U. of Paderborn 1985, submitted, 13 pp.

[2] BORCHERS, W., HEBEKER, F.K.: "The boundary element spectral method and applications in 3-D viscous hydrodynamics", C.A. BREBBIA et al. (eds.), Boundary Elements VIII, Berlin 1986, 823-828.

[3] FISCHER, T.M., ROSENBERGER, R.: "A boundary integral method for the numerical computation of the forces exerted on a sphere in viscous incompressible flows near a plane wall", U. of Darmstadt 1985, submitted, 30 pp.

[4] HEBEKER, F.K.: "An approximation method for the Cauchy problem to the 3-D equation of vorticity transport", Math. Meth. Appl. Sci. 5 (1983), 439-475.

[5] HEBEKER, F.K.: "Efficient boundary element methods for 3-D viscous flows", Numer. Meth. PDE 2 (1986), 273-297.

[6] HEBEKER, F.K.: "Characteristics and boundary elements for three-dimensional Navier Stokes flows", in preparation.

[7] HEYWOOD, J.G., RANNACHER, R.: "Finite element approximation of the nonstationary Navier Stokes problem, Part I", SIAM J. Numer. Anal. 19 (1982), 275-311.

[8] KATO, T.: "On classical solutions of the two-dimensional nonstationary Euler equation", Arch. Rat. Mech. Anal. 25 (1967), 188-200.

[9] MADAY, Y., QUARTERONI, A.: "Spectral and pseudospectral approximations of the Navier Stokes equations", SIAM J. Numer. Anal. 19 (1982), 761-780.

[10] ORSZAG, S.A.: "Numerical simulation of incompressible flows within simple boundaries, Part I", Stud. Appl. Math. 2 (1971), 293-326.

[11] PIRONNEAU, O.: "On the transport-diffusion algorithm and its applications to the Navier Stokes equations", Numer. Math. 38 (1982), 309-332.

[12] RAUTMANN, R.: "Ein Näherungsverfahren für spezielle parabolische Anfangswertaufgaben mit Operatoren", Lecture Notes in Math., 267 (1972), 187-231.

[13] TOSAKA, N., KAKUDA, K.: "Numerical simulations for incompressible viscous flow problems using the integral equation methods", C.A. BREBBIA et al. (eds.), Boundary Elements VIII, Berlin 1986, 813-822.

[14] VARNHORN, W.: "Zur Numerik der Gleichungen von Navier-Stokes", PhD. thesis, U. of Paderborn 1985.

[15] WENDLAND, W.L.: "On some mathematical aspects of boundary element methods for elliptic problems", J.R. Whiteman (ed.), The Mathematics of Finite Elements and Applications V, London 1985, 193-227.

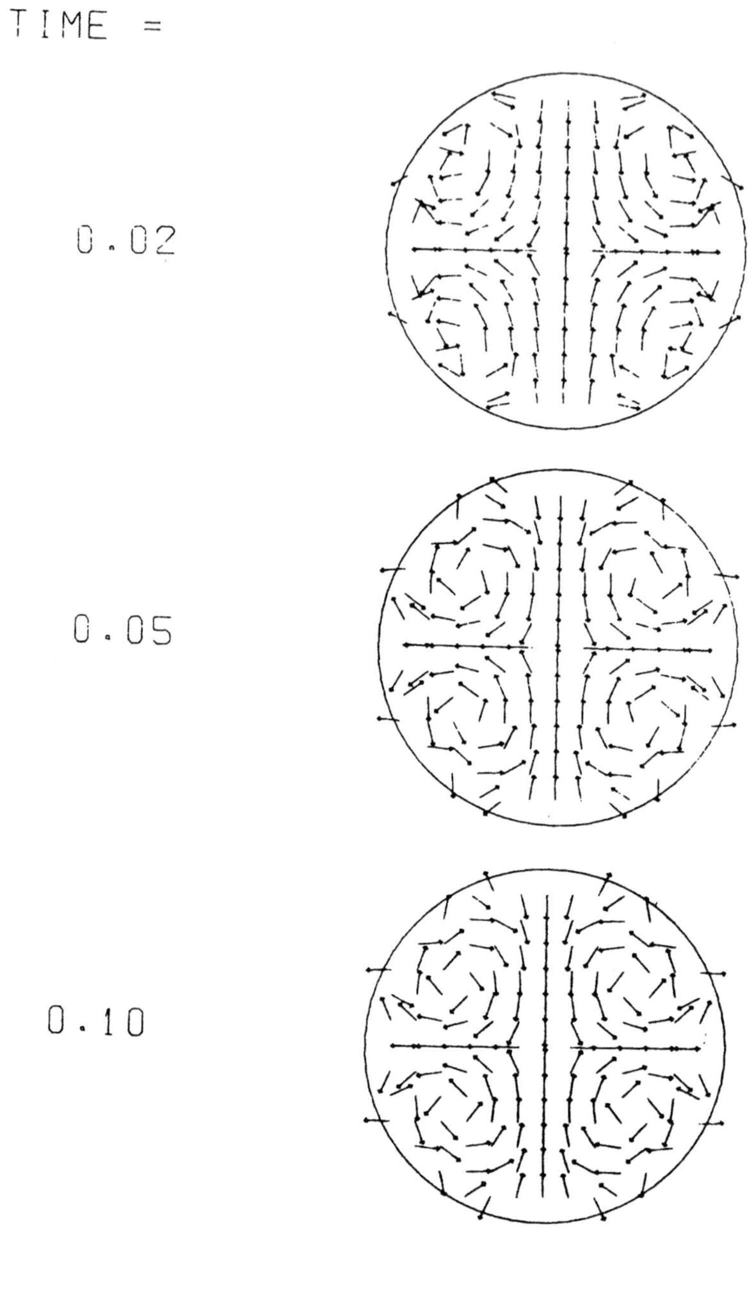

Fig. 2 Vortex development of spherical flow (Re = 600, time increment τ = 0,01)

APPLICATION OF AN ADVANCED PANEL METHOD TO AERODYNAMIC PROBLEMS OF AIRCRAFT DESIGN

John L. Hess and Walter O. Valarezo
Aerodynamics Research and Technology
McDonnell Douglas Corporation, Douglas Aircraft Company
Long Beach, California 90846, USA

SUMMARY

Panel methods are the most commonly used flow computation techniques for analyzing complicated three-dimensional geometries. The well-known surface source method developed at DAC over the years has recently been extended to treat more general geometries and more complicated flow situations. This paper describes the basic features of the method and discusses some of the newer capabilities: a higher-order formulation, improved iterative solver, interactive boundary-layer method, and a propeller capability that predicts not only blade performance but also the mutual interference between a propeller and the airframe on which it is mounted. Many sample results are compared with experimental data and specific cases of application to aircraft design problems are included.

BASIC SURFACE SOURCE METHOD

The potential field due to the body can be written as an integral over the body of the form

$$\phi = \int \left[\frac{1}{r}\,\sigma + \frac{\partial}{\partial n}\left(\frac{1}{r}\right)\mu\right]ds \,. \tag{1}$$

The first term represents a source distribution of variable strength σ, and the second term is a dipole distribution of strength μ. The function given by Eq. (1) automatically satisfies Laplace's equation and approaches zero at infinity regardless of the nature of the functions σ and μ. These functions are adjusted to satisfy the condition of zero normal velocity over the body surface. The present method utilizes a surface source technique where the source densities are adjusted to satisfy a condition of zero normal velocity on the exterior of the body surface, and the dipole is used an as auxiliary variable to produce lift. The basic method assumes that each surface panel is planar and that the source density on each panel is constant. This combination has come to be known as a first-order surface panel method.

On each panel a control point is selected where the normal-velocity boundary condition is applied and where velocity and pressure are eventually calculated. By differentiating Equation 1, the velocity induced by the source density on a panel at a point in space is

$$\vec{v}(\text{panel}) = [\iint_{\text{panel}} \text{grad}(1/r)dS]\sigma(\text{panel}) , \qquad (2)$$

where the constant value of σ has been taken outside the integral whose value therefore depends only on geometry. Suppose N panels are used to define the body, and let σ_j denote the value of source density on the jth panel. The velocity induced by the jth panel at the control point of the ith panel is defined as $\vec{V}_{ij}\sigma_j$, where $\vec{V}_{ij}$ is an integral of the form of Equation 2. The corresponding normal velocity at the ith control point is

$$A_{ij}\sigma_j = \vec{n}_i \cdot \vec{V}_{ij}\sigma_j , \qquad (3)$$

where $\vec{n}_i$ is the unit normal vector to the ith panel. Applying the normal-velocity boundary condition at all control points then yields the following set of linear algebraic equations:

$$\sum_{j=1}^{N} A_{ij}\sigma_j = -\vec{n}_i \cdot \vec{V}_\infty . \qquad (4)$$

This is the numerical approximation of the integral equation that expresses the zero normal-velocity boundary condition. Once the σ_j have been determined, the disturbance velocities at the control points due to the body are given by

$$\vec{v}_i = \sum_{j=1}^{N} \vec{V}_{ij}\sigma_j \qquad i = 1, 2, ..., N . \qquad (5)$$

While the addition of lift in two dimensions causes no significant increase in the complexity of the problem, the problem of three-dimensional lifting flow is not only considerably more complicated than nonlifting flow but requires assumptions that are somewhat arbitrary [1]. The main features are illustrated in Fig. 1. Wings or other lifting portions of the configuration are characterized by having trailing edges from which issue trailing vortex wakes. So-called bound vorticity is hypothesized to lie on or within the wing surface, with strength varying in the direction parallel to the trailing edge. The bound vorticity strength is adjusted to satisfy a condition of smooth flow off the entire trailing edge, and the trailing vorticity has constant strength in the stream direction and an initial strength at the trailing edge equal to the local spanwise derivative of bound vorticity strength. The location of the wake once it leaves the trailing edge is initially unknown, which introduces a nonlinearity into the problem. In the present method the wake shape is assumed or input by the user. This restriction seems to be of little consequence since calculated results are not sensitive to the details of the wake shape. However, the location of the wake may be determined by iterating the calculation if necessary. The variation of bound vorticity in the chordwise direction is assumed and the strengths at a number of spanwise locations are the unknown values that are adjusted to satisfy the Kutta condition at an equal number of locations along the trailing edge.

A key issue in any lifting procedure is the method of applying the Kutta condition. Classically, the Kutta condition is stated as the avoidance of an infinite velocity at the trailing edge. Such a condition cannot be enforced numerically, and some other criterion, which is a consequence of the classical Kutta condition, must be used. Physically, it is necessary that upper and lower-surface pressures on a lifting section must approach a common value at the trailing edge. This is the form of the Kutta condition employed in the present method. Since pressure can be expressed in terms of the square of the velocity, the Kutta condition equations are nonlinear.

Another important feature of the present panel method is its use of an iterative matrix solution. At Douglas Aircraft, a large panel number case such as that of Fig. 2 is run approximately every forty-five minutes by one of the various design teams. Numerical efficiency is required to minimize computational expense and maximize computer throughput (turnaround time). The cases to which the present method is applied regularly have panel numbers up to 4000 panels. For such cases an iterative solution of the linear equations that express the boundary condition is an order of magnitude faster computationally than a direct elimination solution. As is well known, the computational effort for an iterative solution is proportional to the square of the number of equations, i.e., to the square of the panel number, while that of the direct solution is proportional to the cube. Thus the advantage of the former becomes greater as the panel number increases and as the computation time becomes a more important factor. For the iterative solution to realize this advantage, it must converge reliably in a relatively small number of iterations,

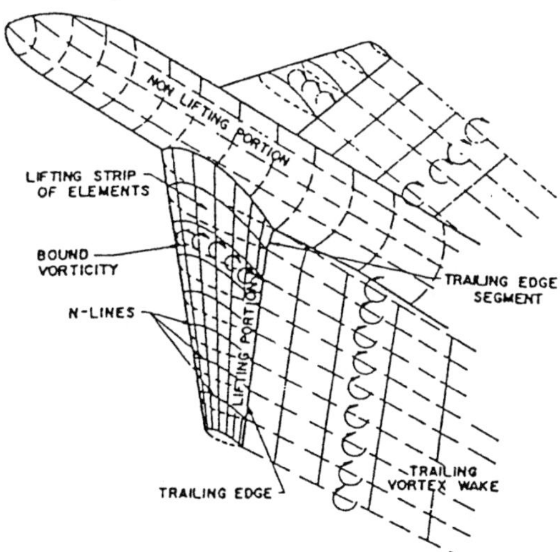

Fig. 1: Panel-method nomenclature.

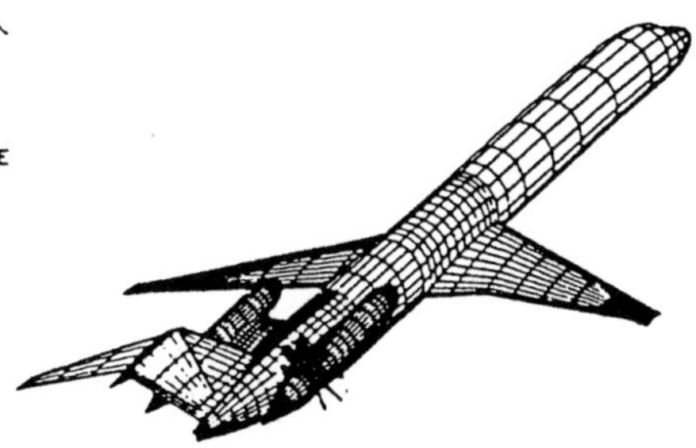

Fig. 2: Complex geometry.

say 10-20. As reported in [2], an accelerated, block Gauss-Siedel achieved this efficiency in the nonlifting panel method. The addition of lift produces a major complication in that the values of bound vorticity must be included among the unknowns and the set of equations must contain those expressing the Kutta conditions at various span locations along the trailing edges. As mentioned above, these last are nonlinear. The standard "Newton-type" method for solving sets of nonlinear equations consists of successive linearizations followed by iteration. That is, in each iteration a set of linear equations is solved, but the coefficient matrix changes with each iteration. This alternative type of iteration was incorporated into the basic block Gauss-Seidel scheme, and after a restructuring of the acceleration procedure, a reliable method was obtained. The number of iterations required for convergence is increased about 30% compared to the non-lifting case. Further details of this scheme are given in [3]. Figure 3 shows the convergence history for a lifting solution about a circular ring wing at 10° angle of attack. For this case, both unaccelerated block Gauss-Siedel and the simplified acceleration schemes had failed to converge after 30 iterations, while the new acceleration scheme had converged after 15 iterations.

HIGHER-ORDER IMPLEMENTATION

While the basic method uses a constant source density on flat panels, the method has been extended to account for source-derivative and surface-curvature effects [4-6]. The higher-order implementation has been shown to yield a large gain in accuracy for certain types of surfaces, particularly those having concave regions and interior flows.

The key integrals for the velocity components induced by a panel, Eq. (2), can be obtained in closed analytic form if the panel is flat even if the singularity strength is variable. This has led many investigators to use variable singularities on flat panels, but, as will be seen below, this approach is mathematically inconsistent. While the exact effect of a curved panel cannot be obtained in closed form, an expansion can be used to obtain this effect to an order consistent with other approximations. Suppose it is desired to find the velocity potential at the point P due to a variable source density σ on a small curved portion S of the body surface (Fig. 4). This is obtained as an expansion about the effects due to a flat panel S_F that represents the projection of S into the tangent plane. A coordinate system, ξ, η, is constructed in the

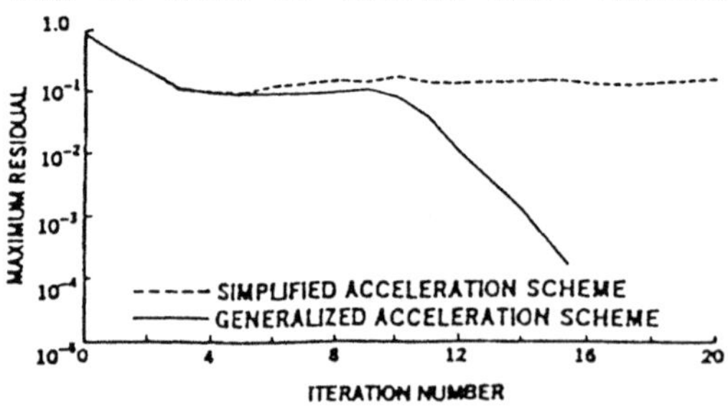

Fig. 3: Convergence for ring wing.

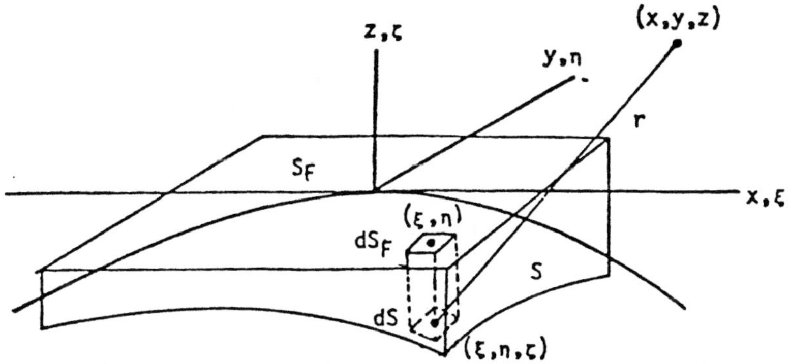

Fig. 4: Curved surface panel and its projection in the tangent plane

the tangent plane with the origin at thepoint of tangency. The key assumption of the higher-order formulation is that the surface shape and source density have rapidly convergent Maclaurin series in tangent-plane coordinates so that higher-order terms are small compared to lower-order terms. Specifically,

$$\sigma = \sigma_o + (\sigma_x \xi + \sigma_y \eta) + (\text{2nd order}) \ , \qquad (6)$$

$$\zeta = (P\xi^2 + 2Q\xi\eta + R\eta^2) + (\text{3rd order}) \ , \qquad (7)$$

where ζ is the vertical distance from the curved surface to the tangent plane and where the coefficients σ_x, σ_y, P, Q, and R are constants. Then the potential at P can be expanded in the form

$$\phi(P) = \iint_S \frac{\sigma}{r} dS$$

$$= \phi^{(o)}\sigma_o + [\phi^{(P)}P + 2\phi^{(Q)}Q + \phi^{(R)}R + \phi^{(1x)}\sigma_x$$

$$+ \phi^{(1y)}\sigma_y] + (\text{higher order}) \ , \qquad (8)$$

where each individual potential on the right side of Eq. (8) is an integral over the flat tangent panel S_f and can be written in closed analytic form. Velocity components are obtained by differentiation. The term $\phi^{(o)}\sigma_o$ represents a flat panel with a constant source density, i.e., the first-order implementation. The five second-order terms are those in the square bracket of Eq. (8). Three are proportional to second derivatives of the surface and two to first derivatives of the source density. Mathematically, it is inconsistent to account for the σ_x and σ_y terms without accounting for the P, Q and R terms. Similarly, it is inconsistent to include second derivatives of σ without including third derivatives of the surface. In the program, certain surface fits are used to assign definite numerical values to P, Q and

R for each panel. Numerical differentiation formulas are used to express σ_x and σ_y in terms of σ_0 and the values of σ on the surrounding panels, all of which are initially unknown. Collecting all the terms proportional to each σ_j then gives a calculation procedure similar in form to that defined by Eqs. (3) - (5).

STEADY PROPELLER FLOWFIELD CALCULATION

A general method for making detailed calculations of flow about propellers with centerbodies [7] was constructed by adapting the lower-order basic method [1]. Features of the method include allowance for the input of an axisymmetric nonuniform onset flow, provision for blade symmetry, generation of a constant radius helical wake, and a special far-wake approximation to improve computational efficiency. While the flow about propellers is inherently unsteady, the flow can be reduced to a steady flow if the propeller operates in an axisymmetric flowfield whose symmetry axis is identical with the propeller's axis of rotation. For this case, the flow is independent of time in a coordinate system fixed in the blades. The propeller onset can thus be uniform or nonuniform but it must be axisymmetric. Paneled segments are used to model the wake but numerical experimentation has revealed that only few "paneled" wake segments are required to converge on a blade loading level due to the use of a far-wake approximation representing the cylindrical wake of an infinitely-many-bladed propeller.

This steady propeller method for isolated propeller/nacelle geometries has been extended to treat the case of an installed propeller configuration [8]. While the basic propeller method is a unique capability to computationally explore the details of the flow about propeller/nacelle geometries, the problem of interest to the aircraft aerodynamicist is that of estimating the propeller-airframe interference effects for a geometry such as that shown in Fig. 2. In its exact form this problem is unsteady and will require an unsteady flow analysis capability. To provide a near-term capability to address the interference problem, fluctuating velocity fields resulting from the rotating propeller are time-averaged and mutual interference is obtained as equivalent steady flows.

VISCOUS EFFECTS AND SEPARATED FLOWS

The source panel method has been combined with an inverse boundary-layer method by Cebeci et al. [9]. Unlike direct boundary-layer methods, the inverse method is capable of calculating flows with separation. Figure 5 compares calculated results with experiment. The interactive procedure yields marked improvement over the purely inviscid results at the higher angles of attack and in fact agrees with

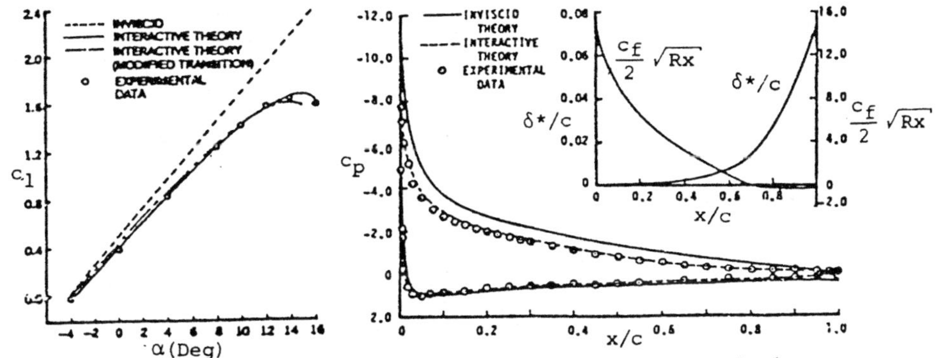

Fig. 5: Use of interactive boundary-layer method

experimental data up to the stall condition for this case. The computed skin-friction coefficient changes sign at the 70% x/c location (indicating separation) and the computed pressure distribution agrees well with test data at this high angle of attack condition.

APPLICATIONS

The various capabilities of the method and the generality of the surface panel technique have been key factors in the extensive application of the present method to a variety of aerodynamic design problems. Figure 6 displays a sample of full-aircraft geometry from those routinely analyzed using the present method. The more densely paneled areas of the geometry often indicate the surfaces where more detailed information about the flow is desired by the designer. Figure 7 shows computed pressure distributions compared to experimental results for an aft-fuselage geometry. The computed results show good agreement with test data at the pressure rows on the nacelle, pylon and fuselage surfaces.

Recently, much effort has been directed at the computation of propeller flowfields. Figure 8 shows a four-bladed propeller

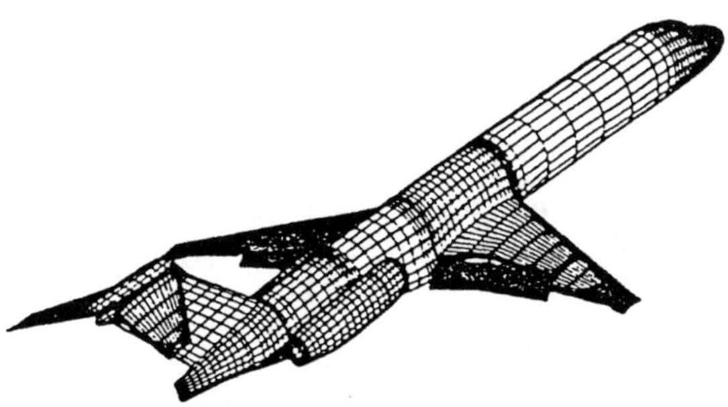

Fig. 6: Full Aircraft Configuration

for which experimental surface pressure data were available. Comparisons of computed chordwise pressures with experiment are shown in Figs. 9 and 10 for the design and an off-design case, respectively. Agreement with test data is good and for the off-design case the predicted suction peak on what would otherwise be the pressure side of the blade was confirmed by two-dimensional analysis run at an equivalent angle of attack. Figure 11 shows the two-dimensional results and verifies all aspects of the panel method pressure distribution qualitatively. These comparisons demonstrate the capability of the propeller panel method to yield valuable information about the flow in regions of

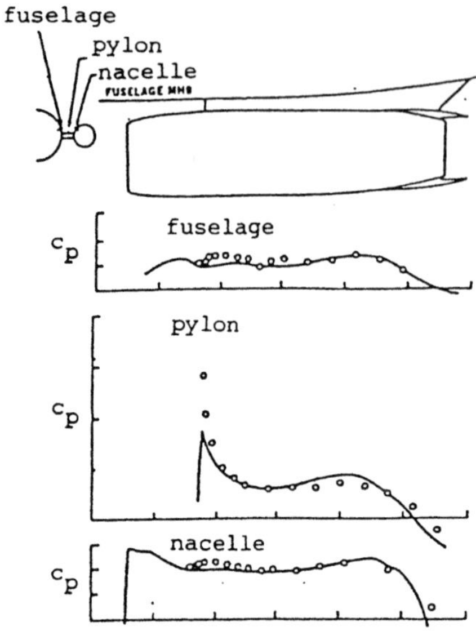

Fig. 7: Pressure comparparison

Fig. 8: Four-bladed propeller

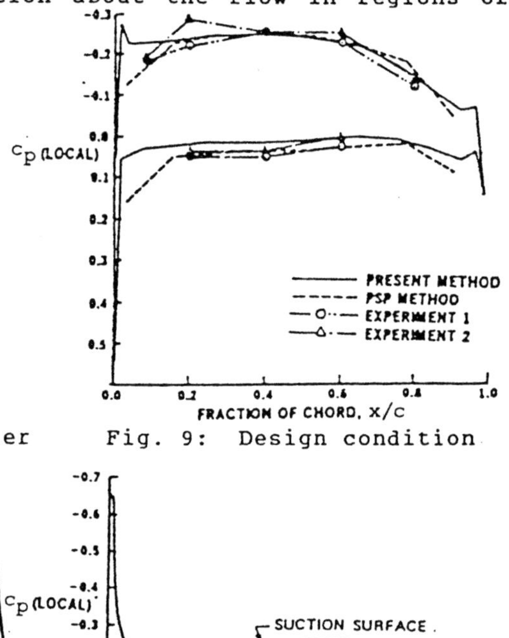

Fig. 9: Design condition

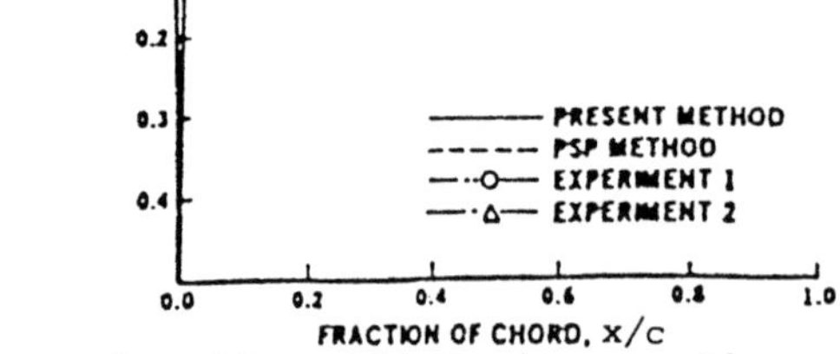

Fig. 10: Off-design condition

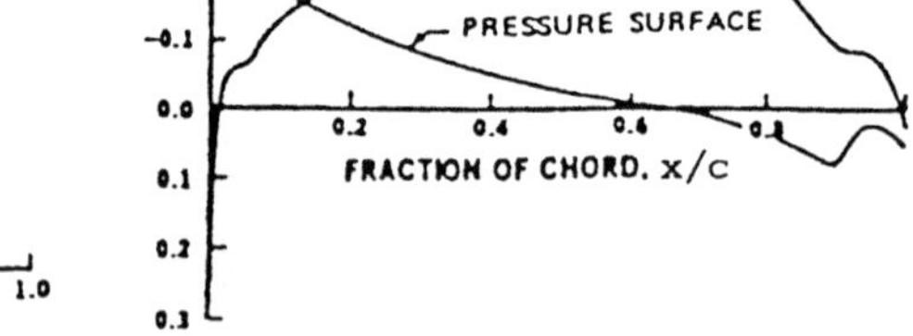

Fig. 11: 2-D pressure distribution

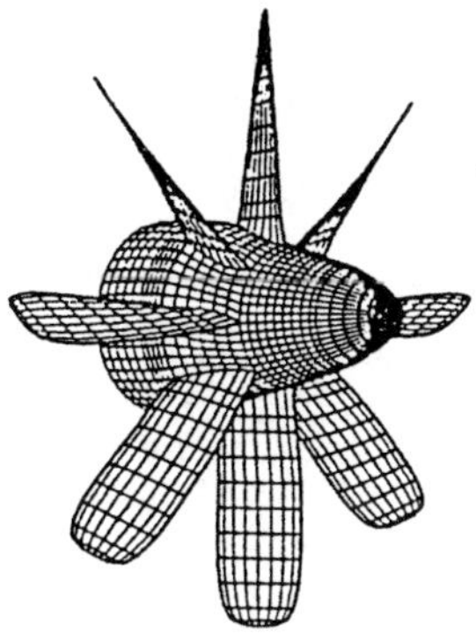

Fig. 12: Eight-bladed propeller

the blade where less sophisticated methods (PSP-vortex lattice method [10]) are limited and where pressure instrumentation of a model might be impractical. An advanced eight-bladed propeller/nacelle geometry is shown in Fig. 12. Figure 13 shows computed versus experimental results for thrust and power coefficients and efficiency for a range of advance ratios. The calculations are close to the experimental data and differ by an amount consistent with the neglect of viscosity. The present method has been employed in a time-

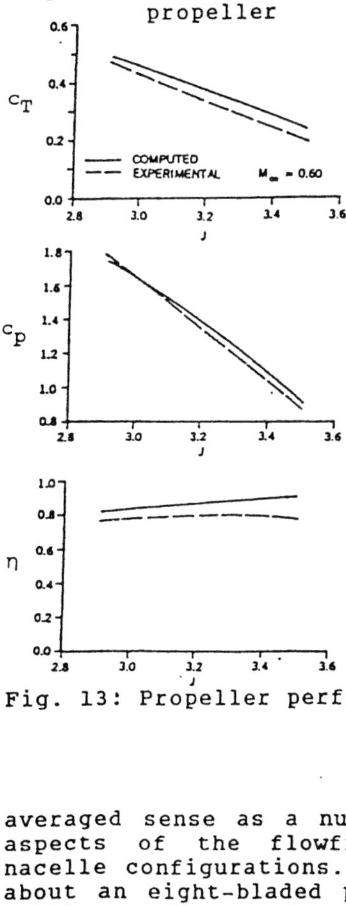

Fig. 13: Propeller performance

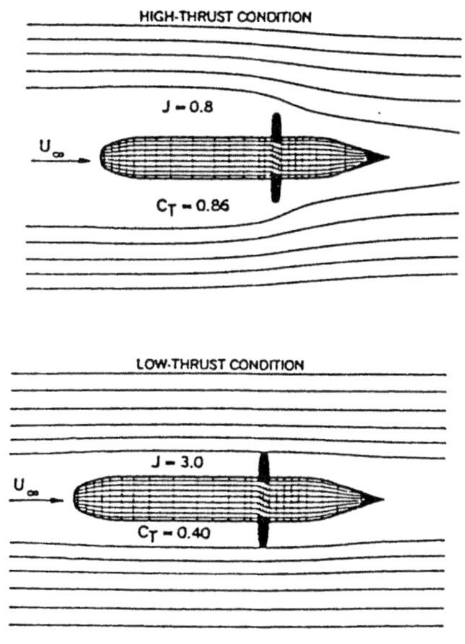

Fig. 14: Computed streamlines in the field

averaged sense as a numerical wind tunnel to survey various aspects of the flowfield surrounding advanced propeller/ nacelle configurations. Figure 14 shows computed streamlines about an eight-bladed propeller at both high and low thrust settings. It can be seen that at the lower advance ratio (higher thrust) there is significant convective acceleration of the fluid into the propeller, while at the higher advance ratio (lower thrust) the freestream is virtually undisturbed by the slowly rotating propeller. This is the kind of flow behavior described in textbooks and now readily demonstrated by the present method. Computed axial velocities were compared with laser velocimeter measurements for the eight-bladed propeller of Fig. 15. Computed versus experimental axial velocity distributions as functions of nondimensional radial distance from the axis of rotation are shown in Fig. 16. The agreement with experimental data is excellent and shows the expected result of increasing axial velocity with increasing distance along the propeller axis of rotation. The present method has also been successfully applied to the problem of computing the influence of an installed propeller on the remainder of the aircraft. Wind tunnel test data were available for the wing/body/nacelle/pylon/propeller configuration shown in Fig. 2 for both power off and power on cases. Figure 17 shows computed versus experimental surface pressure

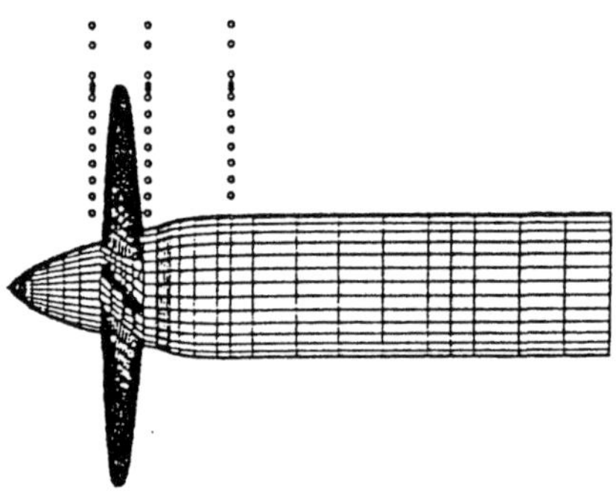

Fig. 15: Propeller and probe locations

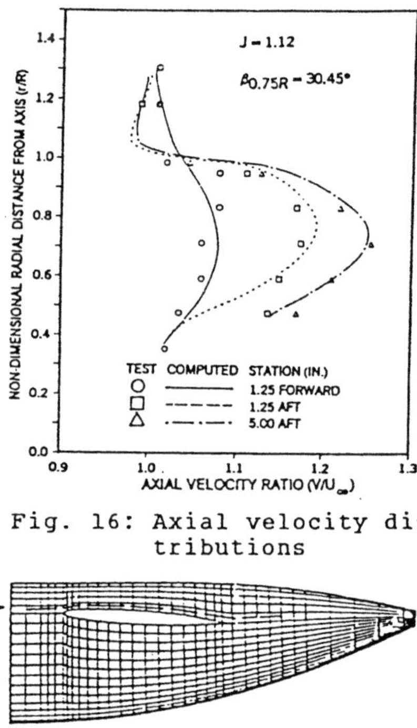

Fig. 16: Axial velocity distributions

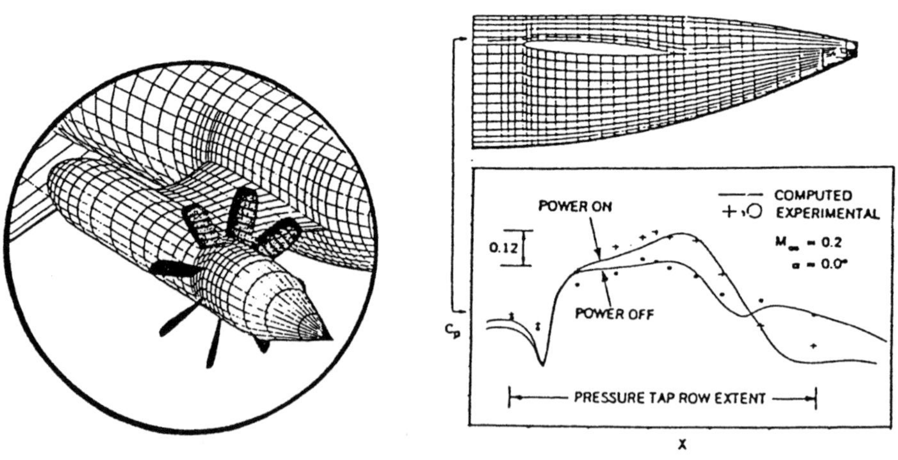

Fig. 17: Surface pressures on aft-fuselage section

distributions on the aft-fuselage section just above the pylon. The agreement with test data is excellent and the effects of power are captured in trend as well as in magnitude.

CONCLUSION

A review of current capabilities of the DAC source panel method has been presented. The various capabilities have proven so useful that they are virtually indispensible tools used by aircraft aerodynamicists to address design tasks daily. To increase the range of problems to which the code applies, many aspects of the method are currently being refined or extended and new areas of application are under current research.

REFERENCES

[1] HESS, J.L.: The Problem of Three-Dimensional Lifting Flow and Its Solution by Means of Surface Singularity Distribution. Computer Meth. in Appl. Mech. and Eng., Vol. 4, No. 3, Nov. 1974.

[2] HESS, J.L. and FRIEDMAN, D.M.: Analysis of Complex Inlet Configurations Using a Higher-Order Panel Method. AIAA Paper No. 83-1823, Jul. 1983.

[3] CLARK, R.W.: A New Iterative Matrix Solution Procedure for Three-Dimensional Pane Methods. AIAA Paper No. 85-0176, Jan. 1985.

[4] HESS, J.L. and STOCKMAN, N.O.: An Efficient User-Oriented Method for Calculating Compressible Flow About Three-Dimensional Inlets. AIAA Paper No. 79-0081, Jan. 1979.

[5] HESS, J.L.: A Higher-Order Method for Three-Dimensional Potential Flow. Rept. No. NADC-77166-30, Jun. 1979.

[6] HESS, J.L. and FRIEDMAN, D.M.: An Improved Higher-Order Panel Method for Three-Dimensional Lifting Flow. Rept. No. NADC-79277-60, Dec. 1981.

[7] HESS, J.L. and VALAREZO, W.O.: Calculation of Steady Flow About Propellers Using a Surface Panel Method. AIAA J. of Propulsion and Power, Vol. 1, No. 6, Nov.-Dec. 1985.

[8] VALAREZO, W.O. and HESS, J.L.: Time-Averaged Subsonic Propeller Flowfield Calculations. AIAA Paper No. 86-1807-CP, Jun. 1986.

[9] CEBECI, T., CLARK, R.W., CHANG, K.C., HALSEY, N.D. and LEE, K.: Airfoils with Separation and the Resulting Wakes. J. Fluid Mech., Vol. 163, 1986, pp. 323-347.

[10] GREELEY, D.S. and KERWIN, J.E.: Numerical Methods for Propeller Design and Analysis in Steady Flow. Trans. of SNAME, Vol. 90, 1982.

THE PREDICTION OF TRANSONIC INTERFERENCE FLOW BY MEANS OF A HYBRID METHOD

H. Jakob

Messerschmitt-Bölkow-Blohm GmbH, Bremen
Hünefeldstr. 1-5, 2800 Bremen, Germany

SUMMARY

A combination of panel method and finite difference method is proposed to calculate inviscid transonic interference flow in space. The underlying idea of this so-called "hybrid method" is to solve approximately the full potential equation governing transonic spatial flow in a specified wing section. For this purpose a computational grid of H-type is established around a wing section profile. In all nodes the cross flow is evaluated in advance by means of a panel method, including interference effects from fuselage or nacelle. Using these quantities as boundary conditions, a finite difference method for planar flow finally yields transonic flow results, taking into account oblique shocks on the contours. Special features of the panel method involved, like appropriate compressibility corrections and computation of second derivatives of the velocity potential, are discussed. Some computational results of the hybrid method are given in the case of a wing-fuselage configuration and compared to wind tunnel experiments.

INTRODUCTION

In aerodynamics, surface singularity or panel methods are widely used to predict interference flow around complex configurations. Unfortunately, they are restricted to pure subsonic or supersonic flow. In the design process of a modern transport aircraft, however, we are highly involved in transonic flow conditions, where supersonic flow regions are enclosed in a subsonic onset flow. The shape of the supercritical flow region, which may be terminated by a shock front and which is crucial to estimation of drag, cannot be predicted by surface singularity methods, even not, if results are corrected for compressibility. Instead, in this case, one has to apply a so-called field method, using field panels, finite differences or finite volumes, which will raise up computing time by an order of magnitude. Furthermore, an appropriate computational grid in space has to be established, whose generation around complex configurations may be a tedious task. Therefore, in the following, a procedure called "hybrid method" is proposed, which combines the obvious advantages of panel methods in spatial flow and finite difference methods in planar flow to get transonic flow predictions around wing sections including strong interference effects from fuselage or nacelle.

OUTLINE OF THE HYBRID METHOD

The full potential equation governing transonic flow in space, reads in cartesian coordinates (x, y, z):

$$(a^2-u^2)\,\Phi_{xx}+(a^2-v^2)\,\Phi_{yy}+(a^2-w^2)\,\Phi_{zz}-2uv\,\Phi_{xy}-2uw\,\Phi_{xz}-2vw\,\Phi_{yz}=0\,, \qquad (1)$$

where

$\Phi(x,y,z)$ = velocity potential in (x, y, z),

$\begin{pmatrix}u\\v\\w\end{pmatrix}$ = velocity vector,

a = local speed of sound.

The partial differential equation (1) is to solve for the velocity potential Φ together with boundary conditions of tangent flow on the wetted surface and undisturbed onset flow at infinity:

$$\nabla\Phi\cdot\vec{n} = 0 \quad \text{on the surface,} \qquad (2)$$

$$\nabla\Phi|_{\infty} = \begin{pmatrix}u_{\infty}\\v_{\infty}\\w_{\infty}\end{pmatrix} \quad \text{at infinity,} \qquad (3)$$

where

$\nabla\Phi = \begin{pmatrix}\Phi_x\\\Phi_y\\\Phi_z\end{pmatrix}$ = gradient vector of Φ ,

$\vec{n}$ = outer normal vector on the surface.

The subscript ∞ denotes onset conditions.

A wing is assumed to be placed in the cartesian coordinate system (x, y, z) according to Fig. 1 with x as main flow or chordwise direction and +y as spanwise direction.

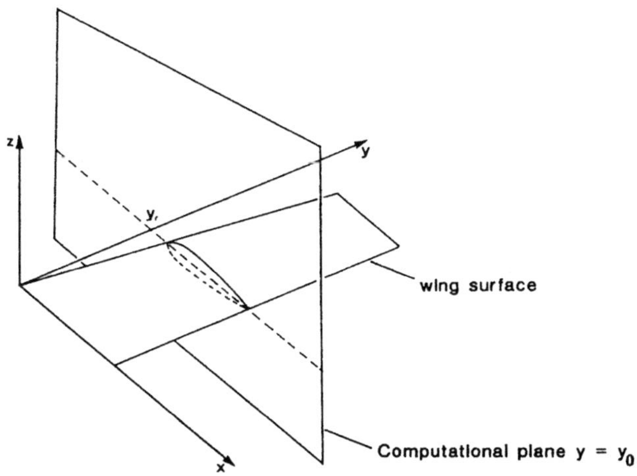

Fig. 1 Computational plane $y = y_0$ of the hybrid method

Let us consider a fixed wing section $y=y_0$ (Fig. 1). For the moment, all derivatives of Φ with respect to y, i.e. v, Φ_{xy}, Φ_{yy} and Φ_{yz} are assumed to be known in advance. Then (1) can be reordered by combining all terms of these derivatives to a quantity R on the right hand side:

$$(a^2-u^2)\,\Phi_{xx} + (a^2-w^2)\,\Phi_{zz} - 2uw\,\Phi_{xz} = R\,. \tag{4}$$

Equation (4) formally coincides with the full potential equation for planar flow, except that, generally, R is not equal to zero and a, the local speed of sound, contains an additional velocity component in y-direction. Equation (4) is to be solved by a finite difference method [1] for planar flow in the plane $y=y_0$. For this purpose, we choose a computational grid of H-type around the section profile, whose coordinate lines are formed by the potential- and streamlines (φ - and ψ - lines) of incompressible flow around the wing section profile at $y=y_0$ (Fig. 2).

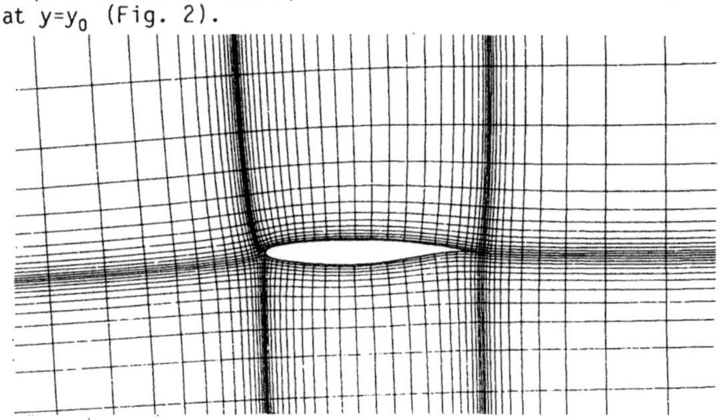

Fig. 2 Orthogonal H-type coordinate system in a wing section $y = y_0$ (formed by potential- and streamlines of incompressible flow)

Furthermore, to be more versatile, especially in handling oblique shocks on the wing surface, the direction of cross flow derivatives is allowed to deviate from the y-direction. The new direction η may vary along chord and should be aligned along local sweep at the leading and trailing edges and approximately along the shock front on the wing surface.

Transforming (1) and (2) into the profile oriented (φ, η, ψ) coordinate system, we get:

$$\begin{aligned}
&\left(a^2 \beta_\varphi - f^2 \bar{\Phi}_\varphi^2\right) \Phi_{\varphi\varphi} + \left(a^2 \beta_\psi - f^2 \bar{\Phi}_\psi^2\right) \cdot \Phi_{\psi\psi} \\
&= 2\left(f^2 \bar{\Phi}_\varphi \bar{\Phi}_\psi + a^2 \frac{\varphi_x \varphi_z}{f^2} \sin^2\gamma\right) \Phi_{\varphi\psi} \\
&+ \frac{1}{2}\left[\bar{\Phi}_\varphi^2 + \bar{\Phi}_\psi^2\right] \cdot \left[\Phi_\varphi \left(f^2\right)_\varphi + \Phi_\psi \left(f^2\right)_\psi\right] - a^2 \frac{\sin^2\gamma}{f^2} \left(\Phi_\varphi \varphi_{xx} - \Phi_\psi \varphi_{xz}\right) \\
&- \frac{1}{f^2}\left(a^2 - \Phi_y^2\right) \Phi_{\eta\eta} + 2\Phi_y \left(\bar{\Phi}_\varphi \Phi_{\eta\varphi} + \bar{\Phi}_\psi \Phi_{\eta\psi}\right) \\
&+ 2 \frac{\sin\gamma}{f^2}\left[a^2 \Phi_{\eta x} - \frac{1}{2} \Phi_y \Phi_z \left(\bar{\Phi}_\psi \left(f^2\right)_\varphi - \bar{\Phi}_\varphi \left(f^2\right)_\psi\right)\right] ,
\end{aligned} \tag{5}$$

where

$$a^2 = a_\infty^2 - \frac{\varkappa - 1}{2}\left[(\nabla\Phi)^2 - \left(u_\infty^2 + v_\infty^2 + w_\infty^2\right)\right] ,$$

$$f^2 = \varphi_x^2 + \varphi_z^2 ,$$

$$\beta_\varphi = 1 - \frac{1}{f^2} \varphi_z^2 \sin^2\gamma , \quad \beta_\psi = 1 - \frac{1}{f^2} \varphi_x^2 \sin^2\gamma ,$$

$$\bar{\Phi}_\varphi = \frac{1}{\cos\gamma} \Phi_\varphi - \frac{\tan\gamma}{f^2}\left(\Phi_\eta \varphi_x + \varphi_z \sin\gamma \, \Phi_z\right) ,$$

$$\bar{\Phi}_\psi = \frac{1}{\cos\gamma} \Phi_\psi + \frac{\tan\gamma}{f^2}\left(\Phi_\eta \varphi_z - \varphi_x \sin\gamma \, \Phi_z\right) ,$$

$$\Phi_y = \frac{1}{\cos\gamma}\left[\Phi_\eta - \Phi_x \sin\gamma\right] ,$$

γ = angle between the η-direction and the y-direction.

The boundary condition reads:

$$\Phi_\psi = \frac{\mp n_y}{\sqrt{f^2 \left(1 - n_y^2\right)}} \cdot \Phi_y . \tag{6}$$

Here n_y denotes the y-component of the normal vector, the negative sign is valid for the upper side of the profile. To solve (5) together with (6) by a finite difference method in two dimensions φ and ψ the following quantities must be known in advance:

1) the derivatives Φ_η and $\Phi_{\eta\eta}$ of the velocity potential Φ with respect to the cross flow direction η .

2) the potential Φ itself as a starting solution and as a boundary condition on the outer border of the H-type grid, where, however, it has to be updated due to changes in circulation during the solution process.

In all nodes of the computational grid (Fig. 2), these quantities will be evaluated approximately by a panel method, containing all interference effects in the specified wing section. For this purpose, the panel method must have some special features, which will be treated in the next chapter.

OUTLINE OF THE PANEL METHOD

1. The compressibility correction

The cross flow quantities generated by the panel method should be corrected for compressibility effects.

However, applying various compressibility rules to profiles of different pressure type, it is next to impossible to find a single rule which yields best results in all cases. Here we adapt the so-called "Göthert-rule 2" [2] for two reasons. Firstly, we have to evaluate all three components of a corrected velocity vector and not only a corrected pressure coefficient. A simple scaling procedure for the components of the velocity vector based on the ratio of the compressible and incompressible total velocity, obviously yields wrong results even on the surface of an infinite swept wing. The "Göthert rule 2", on the other hand, evaluates a corrected perturbation velocity potential from which a velocity vector and higher derivatives are easily deduced.

Secondly, the cross flow quantities around the leading edge of the wing are crucial to the method, because in this region 3D-effects are dominant. Although the Göthert-rule 2 generally underestimates the flow velocity on the suction side, it was found to give reliable results near the leading edge (Fig. 3).

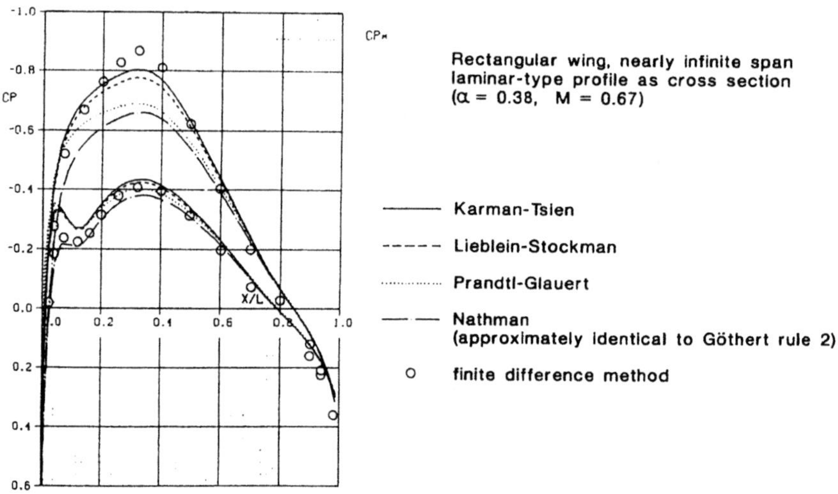

Fig. 3 Comparison of compressibility correction rules implemented in program VSAERO (AMI)

2. Evaluation of second derivatives of the velocity potential

In all nodes of the H-type grid (Fig. 2) around a section profile, the panel method has to generate the second derivative of the velocity potential Φ with respect to η .

Although numerical differentiation is possible, it would be more convenient to get all derivatives directly from the singularity distribution. Therefore the computational grid is adapted to the panelling in the cross section, in the sense, that the endpoint P of every coordinate line φ = constant on the contour coincides with a control point of the panel method, i.e. centroid of a panel (Fig. 4).

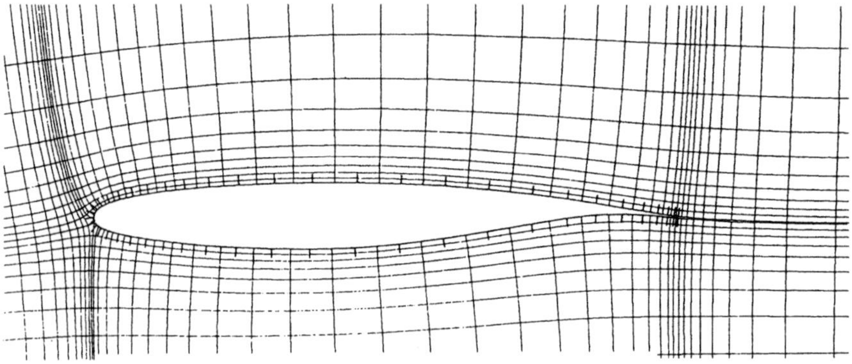

Fig. 4 H-Type grid around section profile, adapted to panelling

If we assume a source singularity distribution on the contour, it can be shown, that, to get correct second derivatives of the velocity potential, we have to apply a piecewise linearly varying source density and that we have to take into account surface curvature.
This is demonstrated by the example of an infinite swept cylinder with circular cross section normal to the sweep direction (Fig. 5).

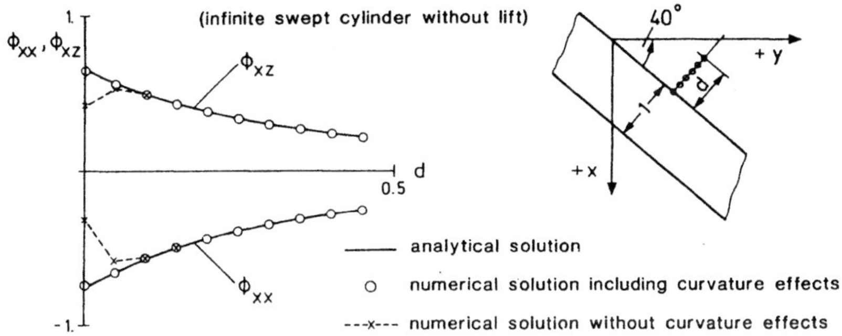

Fig. 5 Second derivatives of velocity potential near a contour

A series of off-body points is choosen, which approach a fixed control point along its outer normal vector. Using the distance from the off-body point to the surface as independent variable, the results of the second derivatives Φ_{xx} and Φ_{xz} are shown in Fig. 5. Compared to the analytical solution, we only get satisfactory results, if surface curvature effects are included, which is performed here by the method of Hess [3] .

EXAMPLES

As an example, results calculated by the hybrid method are shown at various wing stations of a wing-fuselage configuration (Fig. 6). The influence of the tailplane to the wing pressure is assumed to be small, so it is omitted from the geometry for reasons of simplicity. Because the wing is designed according to an isobars concept, the η-direction, that is the direction for which cross flow is taken from a panel calculation, coincides with the line of constant chord. Comparing the numerical results to experiments at four wing stations (Fig. 7) we must keep in mind, that viscous effects are not included which may explain the deviations in lift coefficient and trailing edge pressure.
Furthermore, all calculations are performed using a finite difference scheme in nonconservative form, which usually results in pressure coefficients too low immediately behind shocks and in shock locations shifted upstream. The effect of a nacelle installation is given in the last picture of Fig. 7, which shows the computed pressure distribution close to the engine.

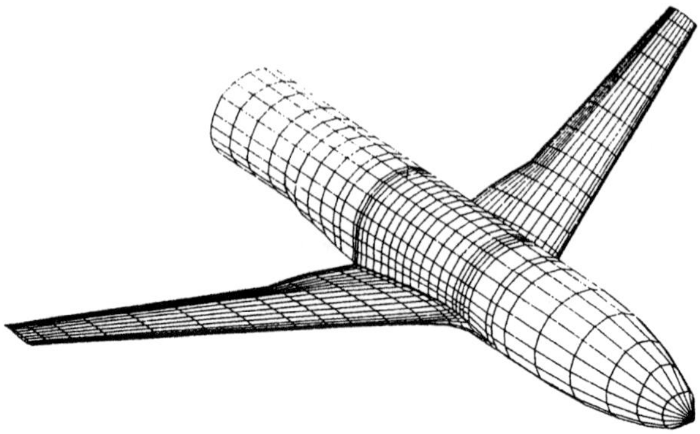

Fig. 6 Panel model of a wing-fuselage configuration (tailplane omitted for simplicity)

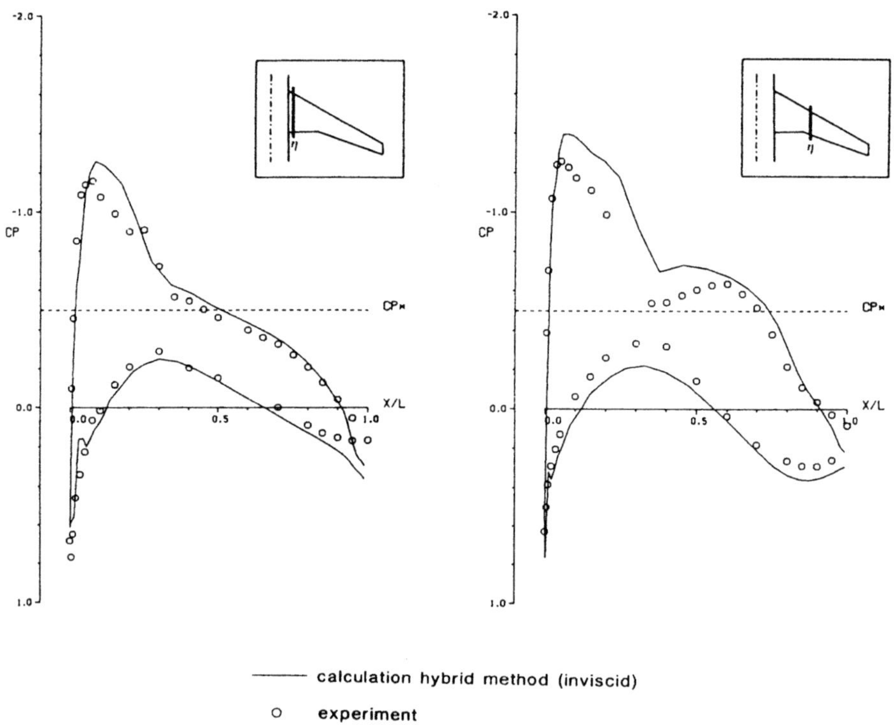

Fig. 7 Calculated pressure distributions in various wing stations compared to experiment (α = 2.24°, M = 0.78)

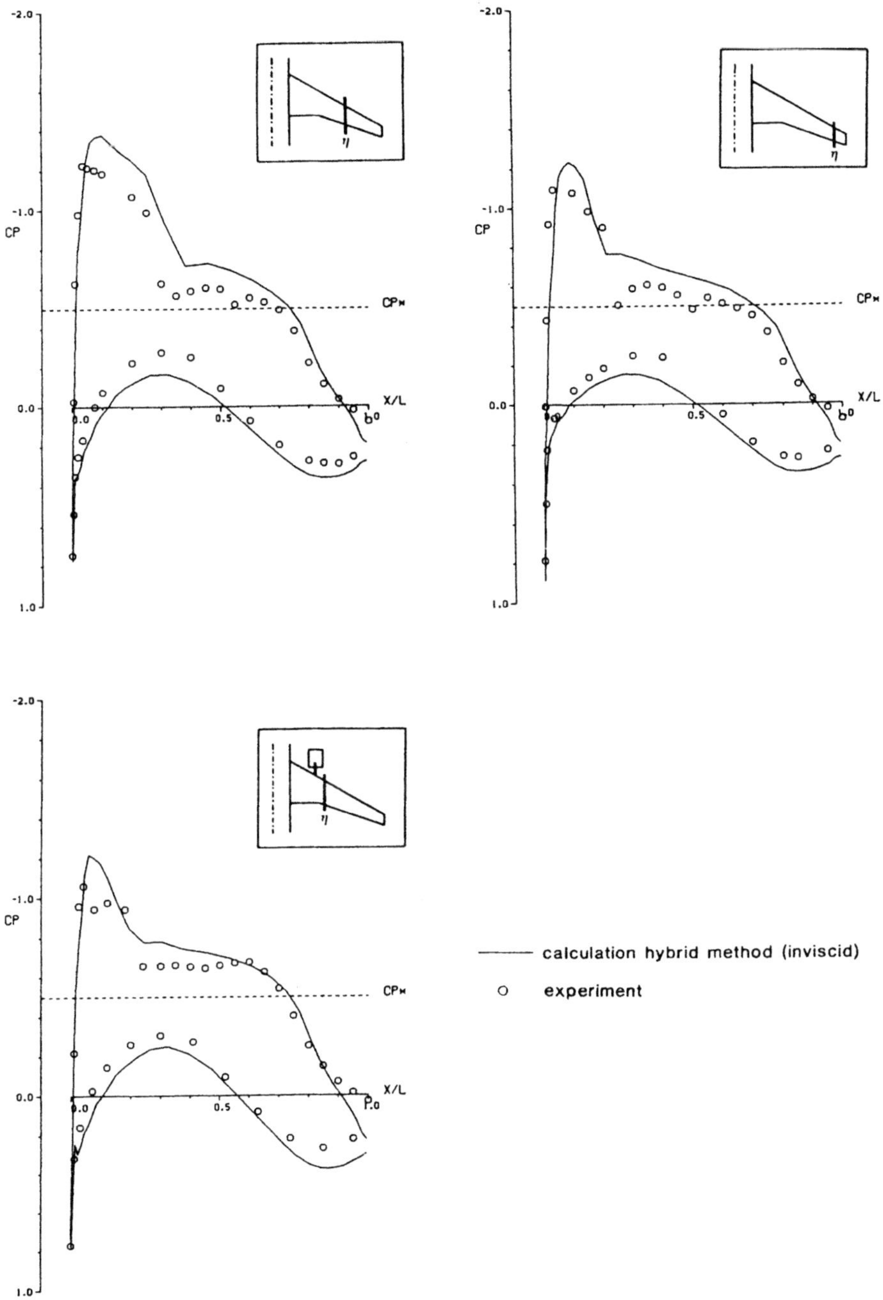

Fig. 7 Calculated pressure distributions in various wing stations compared to experiment (α = 2.24°, M = 0.78)

CONCLUDING REMARKS

A hybrid method is presented, which allows the prediction of pressure in specified wing sections including interference effects. The combination of 3D-panel method and 2D-finite difference method is intended to yield results of engineering accuracy by reasonable computing times.

Further extensions are possible:

a) Implementation of a correct 3D-Rankine-Hugoniot shockpoint operator[4].

b) Iterative coupling to a boundary layer code handling infinite swept wings.

c) Specification of sectional lift without changing the angle of attack by slightly modifying the cross flow derivatives.

REFERENCES

[1] Jakob, H.
Ein Verfahren zur Berechnung der ebenen transsonischen Strömung in Stromlinienkoordinaten
ZKP Ergebnisbericht Nr. 48, LFK 8111 7, Dezember 1984.

[2] Brettauer, N., Kraus, W., Sacher, P.
Das MBB-Unterschall-Panel-Verfahren
Teil 1: Das Verdrängungsproblem ohne Auftrieb in kompressibler Strömung
MBB-Bericht Nr. UFE 632-70 (Ö), 31. Dez. 1970.

[3] Hess, J. L.
Consistent Velocity and Potential Expansions for Higher-Order Surface Singularity Methods
Report No. MDC J6911,
Douglas Aircraft Company, June 1975 .

[4] Mertens, J., Klevenhusen, K. D., Jakob, H.
Accurate Transonic Wave Drag Prediction Using Simple Physical Models
AIAA 24th Aerospace Sciences Meeting
January 6-9, 1986 / Reno, Nevada .

INTEGRAL EQUATION SOLUTION FOR TRANSONIC AND SUBSONIC AERODYNAMICS

Osama A. Kandil and Hong Hu

Department of Mechanical Engineering and Mechanics

Old Dominion University, Norfolk, Virginia, USA

SUMMARY

Two methods are presented to solve for the subsonic and transonic flows around airfoils. The first method is based on the integral solution of the full-potential equation with a shock-capturing technique only or with shock capturing-shock fitting technique. In the second method, the integral solution of the full potential equation is coupled with an embedded region of Euler equations around the shock location. The second method is a computationally efficient technique for flows with strong shocks where the entropy increase and vorticity production across the shock are not small. Several numerical examples are presented and compared with the experimental data and other computational results.

INTRODUCTION

Currently, the available computational inviscid methods for aerodynamics prediction can be divided into two types: the integral equation (IE) methods and the finite-difference (FD) methods including the finite-volume techniques.

Starting in 1970, a great deal of progress has been made in solving transonic flow problems by using the FD methods. Notable contributions are due to Murman and Cole [1], Ballhaus and Bailey [2], Steger and Lomax [3], Garabedian and Korn [4] and Jameson [5, 6]. Although the FD methods are successful in dealing with transonic flows over a wide range of Mach numbers, there are several drawbacks associated with these methods. In the FD methods, fine grid points are needed over a large computational domain around the source of disturbance and a special treatment is required to satisfy the far-field boundary conditions. Moreover, there are major technical difficulties in generating suitable grids adaptable to complex aerodynamic configurations. Other problems include the effects of the numerical dissipation and the large number of iteration cycles to obtain converged solutions.

Panel method for subsonic and supersonic aerodynamics computations have been in use since the 1960's. These methods use the linearized potential equation and special forms of the IE method. For transonic flows and in the pre-computer era, the transonic small-perturbation equation was studied by using approximate IE methods which are based on special partial integration forms. Because of the approximating assumptions on the decay of the perturbation velocity in the far-field and the shock-fitting nature of these approximate IE methods, they were not competitive with the FD methods.

With the exception of the work given in reference [7] and [8], several investigators have recently used the IE method for the full-potential equation to solve for the transonic flow around three-dimensional wings [9] and airfoils [10]-[12]. The work by Piers and Slooff [7] and Tseng and Morino [8] uses the transonic small disturbance equation. In all these recent papers, the potential equation is rewritten as a Poisson's equation which is solved using the Green's theorems. In addition to the surface integral terms appearing in the classical IE solution, a field integral term appears in the IE solution for transonic flows. This term corresponds to the nonlinear compressibility terms. Type-dependent finite differencing [8]-[12] or artifical viscosity [7] are used to capture shock waves in the flow.

In the present paper, we expand on the work done by Kandil and Yates [9] by introducing two additional methods: the first is a shock capturing-shock fitting method and the second is an IE solution with an embedded computational Euler domain around the shock location. The two methods are applied to NACA 0012 and the results show good agreements with experimental data and other FD results.

FORMULATION AND INTEGRAL-EQUATION SOLUTION

The dimensionless governing equations of the two-dimensional, steady, compressible, potential flow are given by

$$\Phi_{xx} + \Phi_{yy} = G \ , \qquad (1)$$

$$G = -\frac{1}{\rho}\,(\rho_x\,\Phi_x + \rho_y\,\Phi_y) \ , \qquad (2)$$

$$\rho = [1 + \frac{\gamma-1}{2}\,M_\infty^2\,(1 - \Phi_x^2 - \Phi_y^2)]^{1/\gamma-1} \ . \qquad (3)$$

where Φ is the total velocity potential, ρ the density, γ the gas index and M_∞ the freestream Mach number. The characteristic parameters are the freestream velocity and density U_∞ and ρ_∞, and the airfoil chord length ℓ. The boundary conditions consist of the flow tangency condition and the infinity condition, and for lifting airfoils, a Kutta condition is enforced at the trailing edge. These are given by

$$\nabla\,\Phi \cdot \hat{n} = 0 \qquad \text{on } g(x,y) = 0 \ , \qquad (4)$$

$$\nabla\,\Phi \rightarrow \bar{e}_\infty \qquad \text{away from } g \ , \qquad (5)$$

$$\Delta C_p\big|_{TE} = 0 \ . \qquad (6)$$

where $\hat{n}$ is the unit normal to the airfoil surface g, $\bar{e}_\infty$ a unit vector parallel to the freestream velocity, ΔC_p the pressure jump and TE refers to the trailing edge. The formal integral solution of equation (1) in terms of the velocity field with explicit line integral for the shock surface is given by

$$\nabla\Phi(x,y) = \bar{e}_\infty + \frac{1}{2\pi}\oint_g q_g(s)\,\frac{(x-\xi)\,\bar{i} + (y-\eta)\,\bar{j}}{(x-\xi)^2 + (y-\eta)^2}\,ds$$

$$+ \frac{1}{2\pi}\oint_g \gamma_g(s)\,\frac{(y-\eta)\,\bar{i} - (x-\xi)\,\bar{j}}{(x-\xi)^2 + (y-\eta)^2}\,ds$$

$$+ \frac{1}{2\pi}\iint G(\xi,\eta)\,\frac{(x-\xi)\,\bar{i} + (y-\eta)\,\bar{j}}{(x-\xi)^2 + (y-\eta)^2}\,d\xi d\eta$$

$$+ \frac{1}{2\pi}\oint_S q_S(s)\,\frac{(x-\xi)\,\bar{i} + (y-\eta)\,\bar{j}}{(x-\xi)^2 + (y-\eta)^2}\,ds \qquad (7)$$

where q_g and γ_g are the surface source and vortex distribution, respectively, S refers to the shock surface and q_S is the source strength of the shock. It should be noted that one does not need to use both of the first and second integral terms for the velocity field calculation. For symmetric flows, either the first or second integral can be used, while for asymmetric flows, either the second integral or both integrals can be used. The third term is the field integral term corresponding to the total (in the present formulation) flow compressibility. The fourth term explicitly represents the contribution of the shock surface. For the shock-capturing method, the fourth integral is dropped from equation (7) since the third term implicitly includes the shock surface contribution [9]. For the shock capturing-shock fitting method, this term is explicitly included to sharpen the shock. Since the details of the shock-capturing method have been covered in reference [9], here we concentrate on the shock-capturing-shock fitting technique.

SHOCK CAPTURING-SHOCK FITTING METHOD

In this method, equation (7), without the last integral term, is used to calculate the total velocity $\nabla\Phi$ during the shock-capturing part of the method. Since G in the third integral is a nonlinear term, an iterative procedure is required. For implementation, the airfoil surface is divided into a number of straight panels with linear sources $q_g(s)$ and/or vortex $\gamma_g(s)$ disribution. The field, outside of the airfoil, is divided into a number of rectangular elements except at the airfoil surface where trapezoidal elements are used, Figure 1. The strength of these field elements $G(\xi,\eta)$ is assumed constant and equal to the centroidal value. Starting with a linear G expression, equations (4) and (6) are enforced to obtain q_g and/or γ_g. The local Mach number M is determined from

$$M = M_\infty \nabla\Phi/(\rho)^{\frac{\gamma-1}{2}} \qquad (8)$$

where ρ is found from equation (3). A type-dependent differencing is used to find ρ_x and ρ_y of the nonlinear G expression of equation (2). The conditions behind the shock are determined by using the Rankine-Hugoniot relations

$$V_{2n} = \frac{(\gamma-1)\,M_{1n}^2 + 2}{(\gamma+1)\,M_{1n}^2}\,V_{1n} \;, \qquad (9)$$

$$V_{2t} = V_{1t} \;, \qquad (10)$$

$$\rho_2 = \frac{(\gamma+1)\,M_{1n}^2}{(\gamma-1)\,M_{1n}^2 + 2}\,\rho_1 \;, \qquad (11)$$

where the subscripts 1 and 2 refer to the conditions ahead and behind the shock, respectively, while the subscripts n and t refer to the normal and tangential directions to the shock, respectively. The iteration procedure is repeated between the surface panels and the field elements until q_g, γ_g and G converge. Once the shock is captured, shock panels are introduced to sharpen the shock. The fourth integral term of equation (7) is now considered. The strength $q_S(s)$ is found from

$$q_S = -\,(V_{1n} - V_{2n}) = -\,\frac{2\,V_{1n}}{\gamma+1}\,(1 - \frac{1}{M_{1n}^2}) \;, \quad M_{1n} > 1 \,. \qquad (12)$$

The orientation of the shock panels, β, is iteratively determined from

$$\beta = \sin^{-1}\left[\frac{1.2\,\sin\beta\,\sin\theta}{\cos\,(\beta-\theta)} + \frac{1}{M_1^2}\right]^{1/2} \qquad (13)$$

where θ is the relative direction of the flow behind the shock to that ahead of the shock. With the fitted shock panels taken into account, the iteration procedure is repeated until convergence is achieved.

INTEGRAL EQUATION-EMBEDDED EULER DOMAIN METHOD

In order to account for strong shocks and still use a small computational domain, we combine the IE solution with a small embedded computational Euler domain, Figure 1. The dimensionless conservation form of the unsteady Euler equations for two-dimensional flows are given by

$$\frac{\partial q}{\partial t} + \frac{\partial E}{\partial x} + \frac{\partial F}{\partial y} = 0 \qquad (14)$$

where the flow vector field q and the flux components E and F are given by

$$q = [\rho, \rho u, \rho v, \rho e]^t , \tag{15}$$

$$E = [\rho u, \rho u^2 + p, \rho uv, \rho uh]^t , \tag{16}$$

$$F = [\rho v, \rho uv, \rho v^2 + p, \rho vh]^t . \tag{17}$$

The total energy and enthalpy per unit mass are given by

$$e = \frac{p}{(\gamma+1)\rho} + (u^2 + v^2)/2, \quad h = e + p/\rho . \tag{18}$$

Since we are interested in the steady flow solution only, the energy equation [last elements in equations (15)-(17)] which is a differential equation is replaced by the algebraic steady form which states that the total enthalpy is constant. Hence, the energy equation is replaced by

$$p = (\rho/\gamma)\ [1/M_\infty^2 + (\gamma-1)\ (1 - u^2 - v^2)/2] . \tag{19}$$

The basic finite-volume equation is obtained by integrating equation (14) over x and y to obtain

$$\iint \frac{\partial q}{\partial t}\, dA + \oint (E\, dy + F\, dx) = 0 . \tag{20}$$

Equation (20) is then applied to each cell of the embedded grid of the Euler domain. The resulting difference equation is

$$\left(\frac{\partial q}{\partial t}\right)_{ij} \Delta A_{ij} + \sum_{r=1}^{4} (E\, \Delta y_r + F\, \Delta x_r) = 0 \tag{21}$$

where ΔA is the cell area, r refers to the cell-side number and the integer subscripts i, j refer to the centroidal values. The Euler solver is a central-difference finite-volume method which uses Runge-Kutta time stepping with explicit second- and fourth-order dissipation terms. The details of this solver are given in reference [6], and hence it will not be covered here. The solution procedure for the combined method starts with the IE computation for several iterations to establish the shock location and the boundary and initial conditions for the Euler domain. The Euler solver is then used to capture the shock and to calculate the flow vector field. Fixing the field values of the Euler solution, the IE computation is performed to update the boundary conditions for the Euler domain. The procedure is repeated until convergence is achieved.

NUMERICAL EXAMPLES

Figures 2-8 show samples of the computational results with application to the NACA 0012. Figure 2 shows the results for shock-free flows using the IE method with vortex panels only and with source panels only along with comparison with the Euler solution of reference [13]. Figures 3 and 4 give the results of the effect of the computational domain size for nonlifting and lifting shock-free cases. In all these test cases, a total number of 140 panels has been used on the airfoil surface. According to these results, we used in the subsequent examples a computational domain size of 2x1.5 (where unity represents the chord length) and 140 vortex panels on the airfoil surface.

Figure 5 shows a comparison between the shock-capturing results and the shock capturing-shock fitting results. Obviously, the latter method is effective in sharpening the shock. In Figure 6, we show a comparison of the shock capturing-shock fitting results with the results of Garabedian [14] and the experimental data [15]. For a lifting case, Figure 7 shows the comparison of the shock capturing-shock fitting results with those of Steger and Lomax [3] and the experimental data of reference [16]. A typical number of iteration cycles to achieve convergence is 30 cycles. Finally, we show in Figure 8 a comparison of the results of IE-embedded Euler domain method with those of Jameson's Euler method [6]. This case took 15 IE iteration cycles and 300 Euler cycles to achieve a residual error of 10^{-3}.

CONCLUDING REMARKS

The IE formulation and solution for transonic and subsonic external aerodynamics have been presented. In addition to the previously developed shock-capturing method, two methods have been developed: the first is a shock capturing-shock fitting method and the second is the IE-embedded Euler domain method. The former method effectively sharpens the shock while the latter method is capable of dealing with strong shocks efficiently. The results are in good agreement with the experimental data and other existing FD solutions which use the potential and Euler equations. These methods are currently extended to treat unsteady flows.

ACKNOWLEDGEMENT

This research work is supported by NASA Langley Research Center unde Grant No. NAG-1-648.

REFERENCES

[1] MURMAN, E. M. and COLE, J. D.: "Calculation of Plane Steady Transonic Flows," AIAA Journal, 9 (1971) pp. 114-121.

[2] BALLHAUS, W. F. and BAILEY, F. R.: "Numerical Calculation of Transonic Flow about Swept Wings," AIAA Paper 72-677 (1972).

[3] STEGER, J. L. and LOMAX, H.: "Transonic Flow about Two-Dimensional Airfoils by Relaxation Procedures," AIAA Journal, 10 (1972), pp. 49-54.

[4] GARABEDIAN, P.R. and KORN, D.: "Analysis of Transonic Airfoils," Comm. Pure Appl. Math., 24 (1972), pp. 841-851.

[5] JAMESON, A.: "Iterative Solution of Transonic Flows over Airfoils and Wings," Comm. Pure Appl. Math., 27 (1974), pp. 283-309.

[6] JAMESON, A.: "Transonic Airfoil Calculations Using the Euler Equations," in Numerical Methods in Aeronautical Fluid Dynamics, Edited by ROE, P. L., Academic Press, (1982), pp. 289-308.

[7] PIERS, W. J. and SLOOF, J. W.: "Calculation of Transonic Flow by Means of a Shock-Capturing Field Panel Method," AIAA Paper 79-1459 (1979).

[8] TSENG, K. and MORINO, L.: "Nonlinar Green's Function Methods for Unsteady Transonic Flows," in Transonic Aerodynamics, edited by D. NIXON, AIAA, New York (1982), pp. 565-603.

[9] KANDIL, O. A. and YATES, E. C., JR: "Computation of Transonic Vortex Flow Past Delta Wings - Integral Equation Approach," AIAA Paper 85-1582 (1985) also AIAA Journal, 24 (1986), pp. 1729-1736.

[10] OSKAM, B.: "Transonic Panel Method for the Full Potential Equation Applied to Multicomponent Airfoils," AIAA Journal, 23, (1985), pp. 1327-1334.

[11] ERICKSON, L. L. and STRANDE, S. M.: "A Theoretical Basis for Extending Surface--Paneling Methods to Transonic Flow," AIAA Journal, 23, (1985), pp. 1860-1867.

[12] SINCLAIR, P. M.: "An Exact Integral (Field Panel) Method for the Calculation of Two-Dimensional Transonic Potential Flow Around Complex Configurations," Aeronautical Journal, June/July (1986), pp. 227-236.

[13] SELLS, C. C. L.: "Plane Subcritical flow Past a Lifting Airfoil," Proceedings of the Royal Society, London, 308 (Series A) (1968), pp. 377-401.

[14] GARABEDIAN, P., KORN, D. G. and JAMESON, A.: "Supercritical Wing Sections," Lecture notes in economic and mathematical systems, 66 (1972).

[15] HALL, M. G.: "Transonic Flows," IMA; Controller HMSO London (1975).

[16] United Aircraft Corp., Sikorsky Aircraft Division: "Two-Dimensional Wind Tunnel Tests of an H-34 Main Rotor Airfoil Section," TRECTR 60-53 (1960).

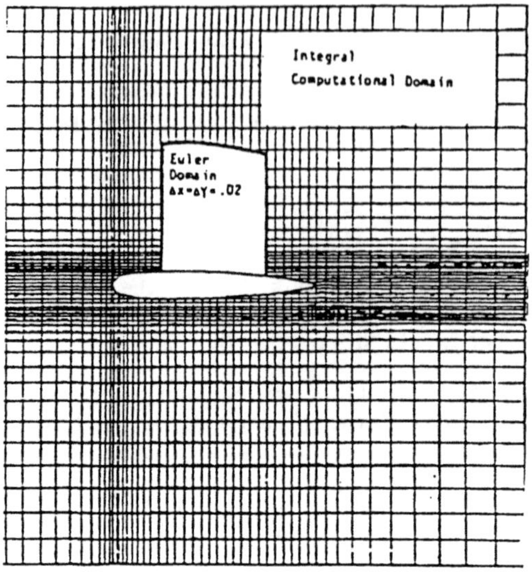

Fig. 1 Computational Domain of Integral Equation Solutions with Embedded Euler Domain

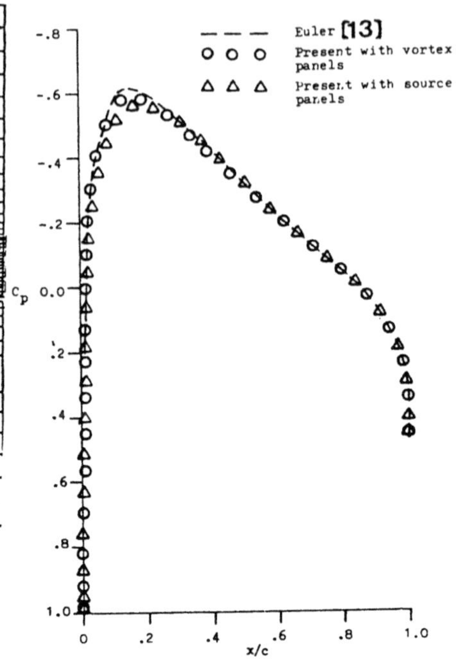

Fig. 2 NACA 0012 aerofoil, M_∞ = .72, α = 0.0, computational domain: 2 x 1.5

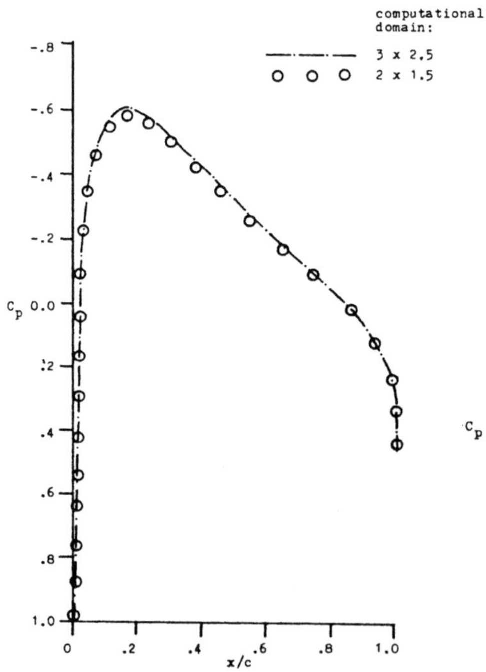

Fig. 3 NACA 0012 aerofoil, M_∞ = .72, α = 0.0, with vortex panels, effect of the size of the computational domain

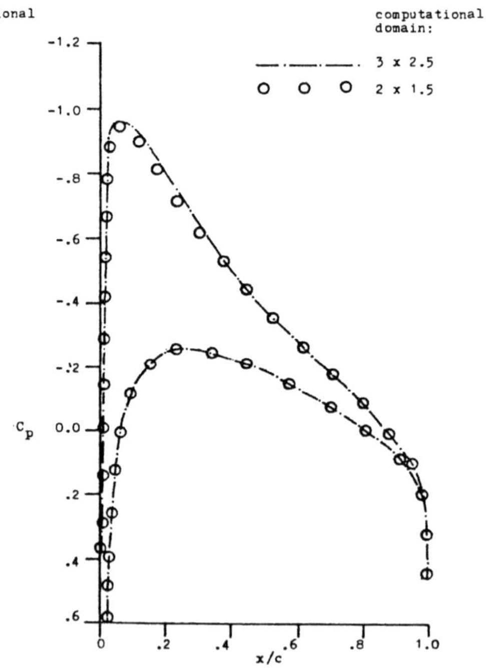

Fig. 4 NACA 0012 aerofoil, M_∞ = .63, α = 2.0, with vortex panels, effect of the size of the computational domain

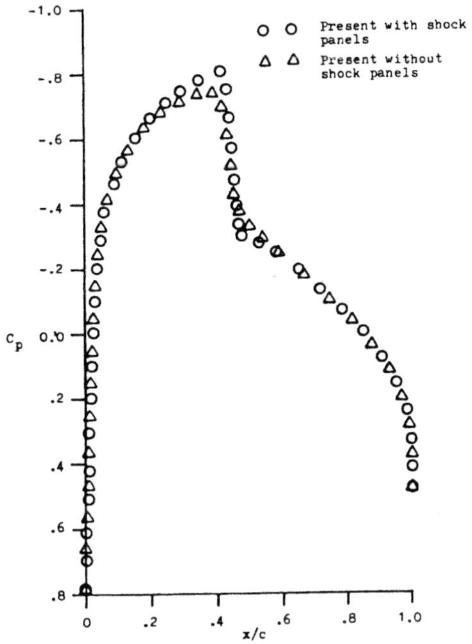

Fig. 5. NACA 0012 aerofoil, M_∞ = .80, α = 0.0, with vortex panels, computational domain: 2 x 1.5, effect of shock panels.

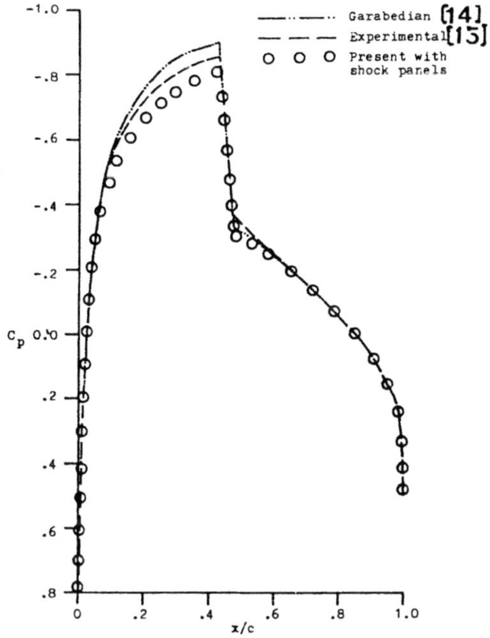

Fig. 6. NACA 0012 aerofoil, M_∞ = .80, α = 0.0, with vortex panels, computational domain: 2 x 1.5.

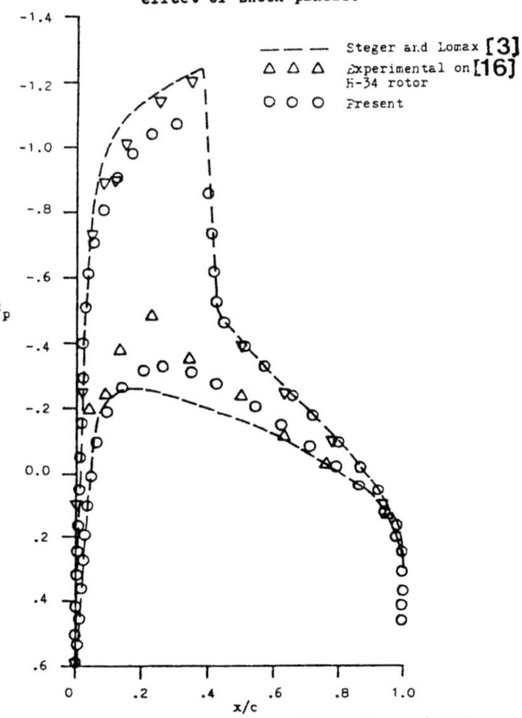

Fig. 7. NACA 0012 aerofoil, M_∞ = .75, α = 2.0, with vortex panels, the computational domain: 2 x 1.5.

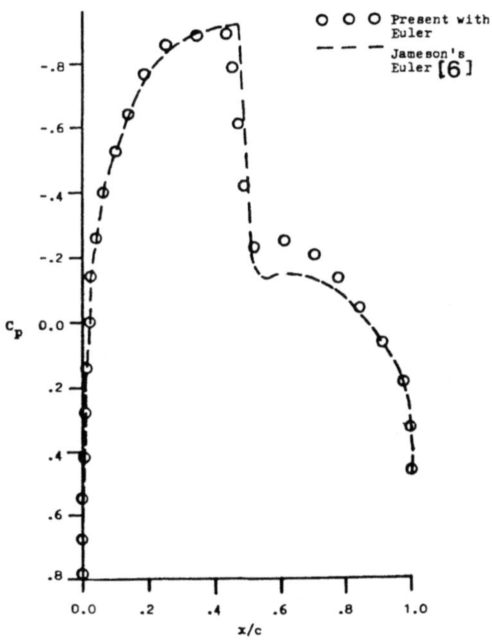

Fig. 8. IE Solution with Embedded Euler Domain for the case: NACA 0012, M_∞ = 0.8, α = 0.0.

STEADY AND UNSTEADY POTENTIAL FLOWS AROUND AXISYMMETRIC BODIES AND RING AIRFOILS

E. Katzer

DFVLR Institute of Aeroelasticity

Bunsenstraße 10, D-3400 Göttingen, Germany

SUMMARY

A higher order boundary element method is applied for the calculation of steady and unsteady incompressible potential flows around axisymmetric bodies and thin ring airfoils. As a major feature, a Fourier decomposition of the flow field leads to a reduction of the problem by one dimension and results in a very efficient algorithm. A new fast and stable algorithm is derived for the calculation of special elliptic integrals introduced by Riegels [1] which appear in the kernel functions. For the integration of the strong and weak singularities a quadrature formula is used which is exact for the leading terms of the expansion. The exclusive representation of the flow by a vorticity distribution results in a simple implementation of the Kutta condition. The unsteady flow is calculated by a time marching procedure. The present analysis is extendable to flows around bodies with small deviations from axisymmetry, to engine flows including jet simulation and to wing/engine interactions.

INTRODUCTION

The prediction of unsteady flows is of great practical importance for the design of airplanes and high-speed trains. The knowledge of the unsteady aerodynamic forces is mandatory for flutter analyses to prevent catastrophic oscillations of the wing. The tendency toward larger aircraft engines increases the importance of flow prediction around engines. For the steady case, Geißler et al. [2] show that the flow around a ring airfoil is a good representation of the engine flow. The present paper extends the investigation to unsteady flows which are calculated by a time marching procedure. This allows the incorporation of arbitrary body motions. In a first step, the engine is represented by a thin axisymmetric ring airfoil. The steady axisymmetric flow around an ellipsoid is used to prove the accuracy of the analysis.

In contrast to many panel methods, where the flow is represented by source and vorticity singularities, we will solely apply a vorticity distribution of higher order. This will allow a simple implementation of the Kutta condition. Usually when panel methods are applied the entire body surface is subdivided into small elements even if the body is axisymmetric (e.g. Haberland and Sauer [3]). A much more economic method has been obtained by Riegels [4], Weissinger [5] and Geißler [6] by applying a Fourier decomposition of the flow in circumferential direction. We will take advantage of this technique which can also be extended to bodies with small deviations from axisymmetry.

FOURIER EXPANSION OF THE GOVERNING EQUATIONS

Let us consider arbitrary motions of a rigid body B immersed in an incompressible fluid. We introduce cylindrical coordinates, figure 1, and the body surface ∂B is parametrised by the arc length s and the circumferential angle φ. The velocity v_{rel} on the body surface relative to the body motion at time t can be decomposed into the prescribed kinematic velocity of the body v_{kin} and the velocity disturbances v induced by the body motion:

$$v_{rel(p,t)} = v_{(p,t)} - v_{kin(p,t)} , \qquad p \in \partial B . \tag{1}$$

Especially for steady flows, the kinematic velocity is constant, $v_{kin} = -v_\infty$.

The induced velocity is harmonic:

$$\begin{gathered} \operatorname{div} v = \operatorname{curl} v = 0 , \\ p \to \infty \;\Rightarrow\; v \to 0 , \quad \text{for all } t \end{gathered} \tag{2}$$

and can be represented by a distribution of sources and vortices on the body surface and on a presumably existent wake W. In order to incorporate flows with circulation, we do not use sources but solely apply a vorticity distribution γ and thus obtain a representation of the induced velocity field by the Biot-Savart law

$$v_{(p,t)} = \frac{-1}{4\pi} \int\limits_{p' \in \partial B \cup W} \frac{[p-p', \gamma_{(p',t)}]}{|p-p'|^3} \mp \frac{1}{2} [n, \gamma_{(p,t)}] , \tag{3}$$

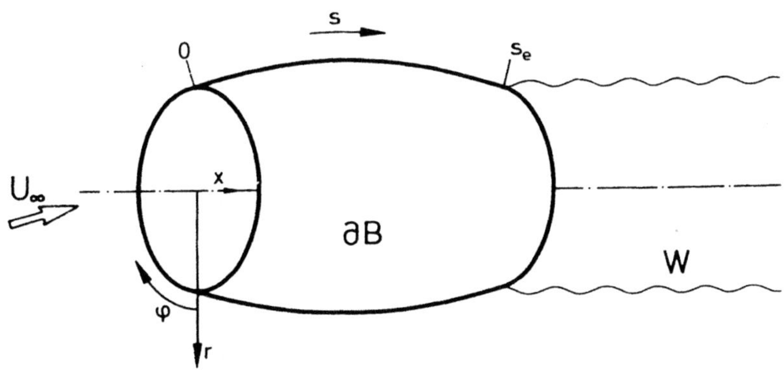

Fig.1 Flow around a thin ring airfoil: coordinate system

where

$$\mathrm{div}\ \gamma = 0 , \qquad p \in \partial B \tag{4}$$

and [*,*] denotes the cross product. The vortex distribution γ is determined by the kinematic flow condition

$$\langle n, v_{rel(p,t)} \rangle = 0 ; \qquad p \in \partial B \tag{5}$$

where n is the outer normal unit vector and <*,*> denotes the inner product.

For determination of the flow field around a ring airfoil a model for the wake is needed. The wake is formed by the vorticity generated on the surface and leaving the airfoil at the trailing edge. Restricting the analysis to small disturbances of a uniform flow U_∞ parallel to the x direction, the transport of the vorticity is approximated by a linarisation of the vorticity transport equation:

$$\frac{\partial}{\partial t} \gamma + U_\infty \frac{\partial}{\partial x} \gamma = 0 . \tag{6}$$

This leads to a wake model consisting of a cylinder extending from the trailing edge to infinity where the vorticity is convected with constant velocity U_∞. To ensure the uniqueness of the flow field, a Kutta condition is necessary. It can be formulated in terms of the circumferential component γ_φ of the vorticity at the trailing edge s_e. For steady flow we apply

$$\gamma_\varphi(s_e) = 0 \tag{7}$$

and in the unsteady case we require that

$$\frac{\partial}{\partial t} \Gamma_{(t)} + U_\infty \frac{\partial}{\partial x} \gamma_\varphi(s_e,t) = 0 \tag{8}$$

where

$$\Gamma_{(t)} = \int_0^{s_e} \gamma_\varphi(s',t)\, ds' . \tag{9}$$

For the model of a thin ring airfoil we require that no vorticity emmanates from the leading edge. This is formulated in terms of the vorticity component in streamwise direction γ_s

$$\gamma_{s(s=0)} = 0 ; \qquad \varphi \in [0,2\pi] . \tag{10}$$

In the steady case, equations (1)-(5) and (7) lead to an integral equation of the first kind. In the unsteady case a time marching method is applied which starts with the steady solution. The derivatives in equations (6) and (8) are approximated by the trapezoid rule. Together with equations (1)-(5) an integral equation for each time step has to be solved. The computational efforts are reduced by the fact that the influence matrix can be calculated and inverted once for all time steps.

Many of the commonly used panel methods panelise the total body surface into small elements and solve the discrete analogue of the integral equations (1)-(5). If the surface on which the singularities are distributed is axisymmetric, a much more efficient approach may be applied. (Note that the flow field itself must not be axisymmetric which is for example the case for wing/engine interaction). Following the ideas of Riegels [4], Weissinger [5] and Geißler [6] we apply a Fourier decomposition in circumferential direction φ to the boundary condition (5)

$$\langle n, v_{kin} \rangle = \sum_k e^{ik\varphi} c^k_{kin(s)} \qquad (11)$$

and the flow field (3)

$$\langle n, v \rangle = \sum_k e^{ik\varphi} c^k_{BS(s)} \qquad (12)$$

where

$$c^k_{BS(s)} = \int_0^{s_e} ds' \, \gamma^k_{(s')} \, K^k_{(s,s')} \, . \qquad (13)$$

c^k_{kin}, c^k_{BS} and γ^k are the Fourier coefficients of the boundary condition, the Biot-Savart law and the vorticity distribution. This approach decomposes the integral equation on a two-dimensional surface into several one-dimensional integral equations along the arc length s, one for each Fourier coefficient (figure 2). Details will be reported by Katzer [7].

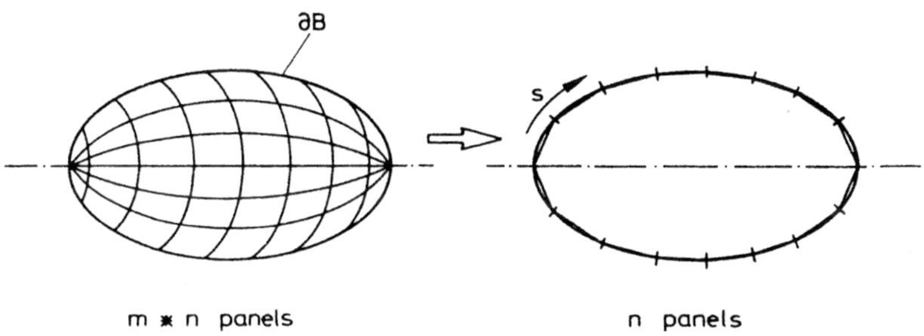

Fig.2 A Fourier decomposition reduces the two-dimensional problem into a one-dimensional problem

A comparison of the numerical efforts shows the advantages of this approach. A panelisation into $(m \cdot n)$ panels at the surface with $(m \cdot n)$ degrees of freedom results in numerical operations of the order $(m \cdot n)^2$ for calculation of the influence matrix and of the order $(m \cdot n)^3$ for solution of the linear equations. The decomposition with the same degrees of freedom leads to m real and imaginary Fourier coefficients and n panels in a meridial section. This results in an order of $m \cdot n^2$ operations for the influence coefficients and $m \cdot n^3$ operations to solve the equations which is an order of m and m^2 less than the customary approach. In practice m is approximately of an order of ten.

The efficiency is further increased by a second-order approximation of the vorticity distribution. Due to the conditions (4) and (10) the vorticity on the body surface is uniquely defined by one component, γ_φ, in circumferential direction alone. We assume that the Fourier coefficient of a transformed vorticity distribution

$$j^k_{(s)} = r_{(s)}\,\gamma^k_{\varphi(s)} \tag{14}$$

is approximated by a piecewise linear function given by the n+1 values at the panel edges. In the same way, the body surface ∂B is approximated piecewise by n conical panel elements. The resulting integral equations are solved by a collocation method where the boundary conditions are fulfilled at the n panel centers. As there are n+1 degrees of freedom for the vorticity distribution, one degree of freedom is left to fulfill the Kutta condition.

INTEGRATION OF THE KERNEL FUNCTIONS AND TREATMENT OF THE SINGULARITY

Lengthy computations (Katzer [7]) show that the kernel functions, K^k, appearing in equation (13) depend on

$$K^k_{(s,s')} \quad \text{depend on} \quad (\text{Geometry } \partial B,\ \kappa^2,\ G_{k-1},\ G_k,\ G_{k+1}) \tag{15}$$

where

$$\kappa^2 = \frac{4rr'}{(x-x')^2 + (r+r')^2}\ , \qquad \kappa^2 \in [0,1]\ . \tag{16}$$

Specific elliptic integrals are essential parts of the kernels

$$G_k(\kappa^2) = (-1)^n \int_0^{\pi/2} \frac{\cos 2k\Theta}{(1-\kappa^2\sin^2\Theta)^{3/2}}\, d\Theta\ , \tag{17}$$

which were first analysed by Riegels [1]. The efficiency of the Fourier decomposition approach would be lost if these "Riegels' integrals" could not be calculated efficiently.

Riegels [1] and Weissinger [8] reduced these integrals to the complete elliptic integrals of the first and second kind, K and E:

$$G_0(\kappa^2) = E(\kappa^2)\ /\ (1-\kappa^2)\ , \tag{18}$$

$$G_1(\kappa^2) = G_0(\kappa^2) - 2\Big(K(\kappa^2) - E(\kappa^2)\Big)\ /\ \kappa^2\ , \tag{19}$$

$$G_{k+1} = \frac{2-\kappa^2}{\kappa^2}\,\frac{4k}{2k-1}\,G_k - \frac{2k+1}{2k-1}\,G_{k-1}\ . \tag{20}$$

Unfortunately the recursion formula (20) is numerically stable only for large values of κ^2. Rewriting equation (20), the author developed a formula to calculate G_k for given values of G_{k+1} and G_{k-1}. Iterating this procedure, the author (Katzer [9]) then derived a new recursion scheme

$$G_k(\kappa^2) = a_k^{+j}(\kappa^2)\,G_{k+j}(\kappa^2) - a_k^{-j}(\kappa^2)\,G_{k-j}(\kappa^2)\ . \tag{21}$$

The upper indices, j, are powers of two and the coefficients a_k^{+j} and a_k^{-j} are calculated recursively.

Using equation (21), G_k can be evaluated from G_{2k}, which vanishes for large k, and G_0 which is given by equation (18) and a standard procedure for E. This results in a numerically stable and very efficient algorithm for calculation of "Riegels' integrals"; see Katzer [9] for details.

Let us now discuss the numerical integration of equation (13). It is well known that the kernel (15) is singular if the integration point s' approaches the collocation point s. Near that singularity "Riegels' integrals G_k" allow an expansion in terms of a power series and logarithmic terms, just as the complete elliptic integrals E and K do. This leads to an expansion of the kernels (15)

$$K^k_{(s,s')} = \frac{A}{\varepsilon} + a_k \ln \varepsilon^2 + \alpha_k + 0(\varepsilon \ln \varepsilon^2 + \varepsilon) \tag{22}$$

where

$$\varepsilon = s'-s \,, \qquad \varepsilon \to 0 \,.$$

Subtracting the leading strong and weak singularity and integrating them analytically would produce a different formula for each k. Instead we apply a numerical quadrature formula which is exact in the leading terms of an expansion of the type

$$f_{(\varepsilon)} = \frac{A}{\varepsilon} + a^0 \ln \varepsilon^2 + \alpha^0 + \varepsilon(a^1 \ln \varepsilon^2 + \alpha^1) + 0(\varepsilon^2 \ln \varepsilon^2 + \varepsilon^2) \,. \tag{23}$$

Thus for all k the same quadrature formula can be used:

$$\int_{-h}^{+h} f = h\left\{f(h/e) + f(-h/e)\right\} + 0(h^3 \ln h) \,. \tag{24}$$

This formula is exact for the first four terms of the expansion (23) and is applied in the vicinity of the singularity. Elsewhere the integral along the arc length which appears in equation (13) is evaluated by a standard numerical integration procedure.

RESULTS

As a first test case, the steady axisymmetric flow around an ellipsoid is analysed. As neither trailing edge nor wake exist, the kinematic flow condition (5) at the upstream stagnation point is used instead of the Kutta condition to ensure a unique solution of the n+1 unknowns. The numerical solution obtained for 10 and 20 panels is compared with the analytical solution in <u>figure 3</u>. The good agreement of the zero-th Fourier mode c_p^0 of the pressure coefficients even with 10 panels is surprising. There are only small deviations near the stagnation points. Thus the good representation of the flow by the present higher order method is demonstrated.

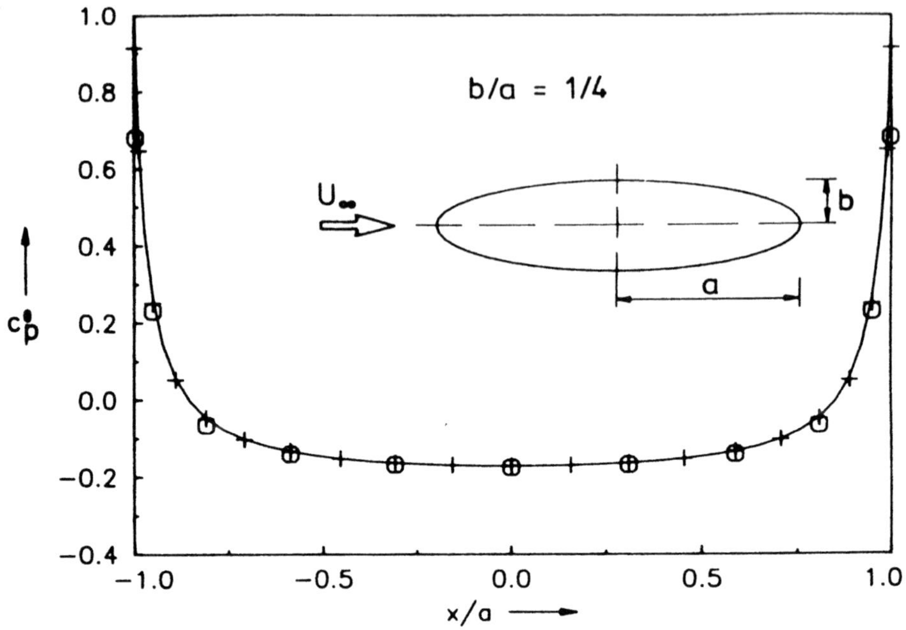

Fig.3 Axisymmetric flow around an ellipsoid: pressure;
—— analytical solution
ooo numerical solution, 10 panels
+++ numerical solution, 20 panels

A second test case is the unsteady flow around a thin cylindrical ring airfoil in harmonic heave oscillations with a reduced frequency of $\omega^* = 1$. Starting with the steady solution, figure 4 shows the time history of the vorticity on the body and in the wake for the first cycle. The real part of the first Fourier mode of the circumferential component of the vorticity, γ^1, is presented; actually the imaginary part vanishes. The convection of vorticity in the wake and its development on the body is clear. After three cycles the solution becomes periodic in time. A harmonic analysis of the pressure differences between the lower and upper side of the ring airfoil gives the in-phase (real) and out-of-phase (imaginary) parts of the pressure coefficients Δc_p^1. For large diameters, the results can be compared with the two-dimensional theory of a flat plate. The present results based on 20 panels on the body and 32 time steps per cycle compare well with this analytical theory (Försching [10]).

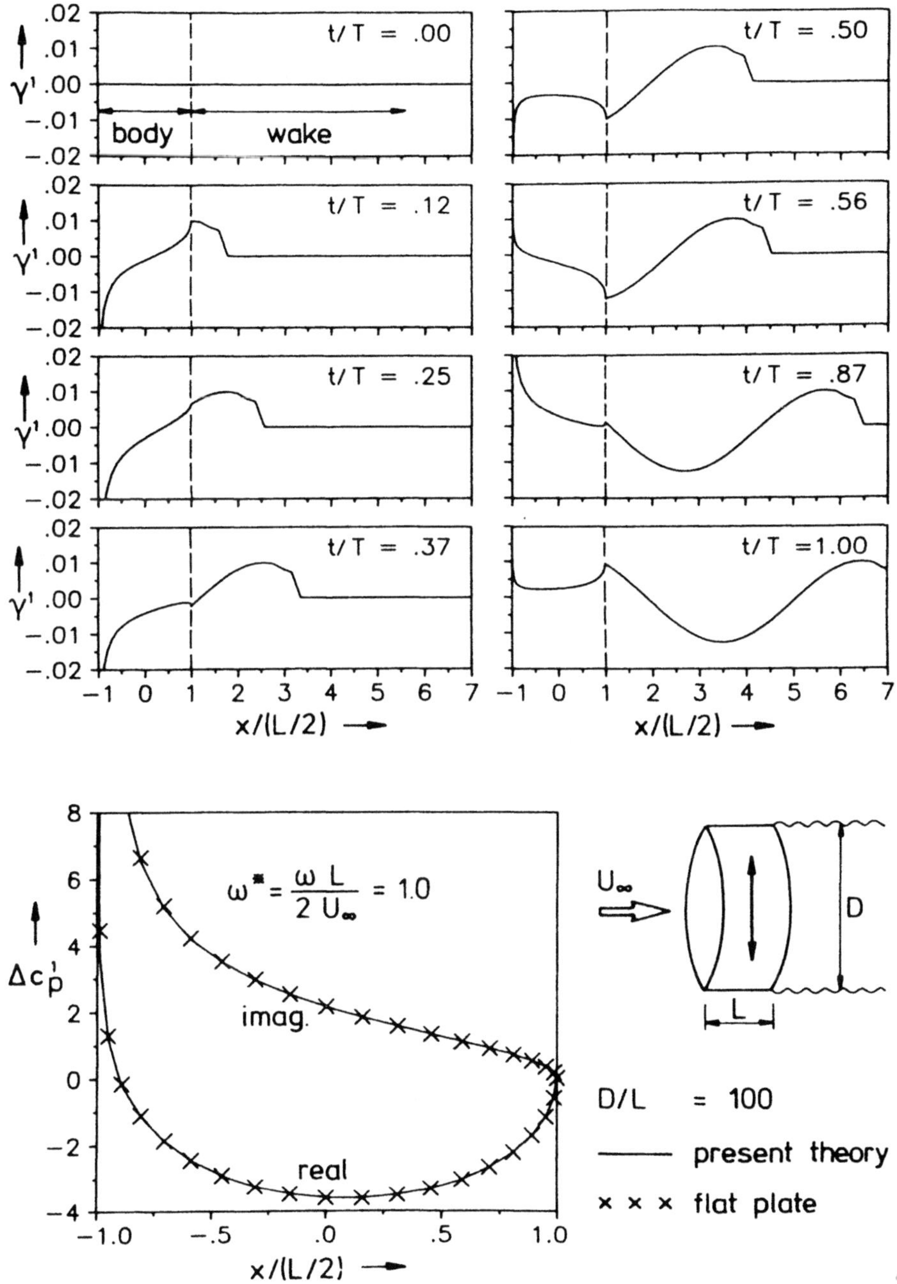

Fig. 4 Unsteady flow around ring airfoil in heave oscillations: time history and pressure compared with two-dimensional limit

CONCLUSION

A higher order panel method for incompressible potential flows around axisymmetric bodies and thin ring airfoils is presented. The flow field is represented solely by a vorticity distribution. This allows a simple implementation of the Kutta condition in the case of the ring airfoil. A greatly improved numerical efficiency compared to ordinary panel methods is achieved by two principles:

- A Fourier decomposition of the flow field and the boundary condition reduces the dimensionality of the problem from two to one.
- A higher order approximation of the vorticity by piecewise linear functions allows a reduction of the number of panels.

As a consequence of the Fourier decomposition specific kinds of elliptic integrals have to be evaluated. An efficient and numerically stable algorithm for the calculation of these "Riegels' integrals" has been developed. For the integration of the strong and weak singularity of the kernel functions a special numerical quadrature formula is presented.

A time marching procedure is applied for the calculation of unsteady flows and allows the treatment of arbitrary body motions. The reliability of the method is tested on two representative problems. The steady axisymmetric flow around an ellipsoid compares well with the analytical solution and shows that only 10 to 20 panels are needed for a good representation of the flow field. Also the test of the unsteady flow around a cylindrical ring airfoil undergoing harmonic heave oscillation show good agreement with the analytical solution. The extension of the analysis to more realistic engine geometries including center bodies and nonaxisymmetric geometries as well as the incorporation of jet effects is under development.

REFERENCES

[1] RIEGELS, F.: "Formeln und Tabellen für ein in der räumlichen Potentialtheorie auftretendes elliptisches Integral". Archiv der Mathematik 2 (1949/1950) pp.117-125.

[2] GEIßLER, W., KATZER, E., TRIEBSTEIN, H.: "Stationäre Strömung am DFVLR-Ejektor-Triebwerk. Vergleich: Theorie-Experiment". DFVLR-AVA internal report IB 232-87 J 04 (1987).

[3] HABERLAND, C., SAUER, G.: "On the Computation of Wing Lift Interference Caused by High Bypass Engines". 15th Congr. of Int. Council of Aeronautical Sci., London, 1986, Paper ICAS-86-3.9.1 (1986).

[4] RIEGELS, F.: "Die Strömung um schlanke, fast drehsymmetrische Körper". Max-Planck-Inst. für Strömungsforschung, Report Mitteilung No.5, 1952.

[5] WEISSINGER, J.: "Zur Aerodynamik des Ringflügels in inkompressibler Strömung". Z. für Flugwissenschaften 4 (1956) pp.141-150.

[6] GEIßLER, W.: "Berechnung der Potentialströmung um rotationssymmetrische Rümpfe, Ringprofile und Triebwerkseinläufe". Z. für Flugwissenschaften 20 (1972) pp.457-462.

[7] KATZER, E.: "Stationäre und instationäre Umströmung dünner Ringflügel und Triebwerke mit Simulation des Triebwerkstrahls". Will appear as DFVLR-FB report (1987).

[8] WEISSINGER, J.: "Zur Aerodynamik des Ringflügels. III. Der Einfluß der Profildicke". DVL Report No. 42 (1957).

[9] KATZER, E.: "Numerische Berechnung eines elliptischen Integrals mit Anwendung auf Potentialströmungen in Zylinderkoordinaten". DFVLR-AVA Internal Report IB 232-85 J 21 (1986).

[10] FÖRSCHING, W.: "Grundlagen der Aeroelastik". Berlin, Heidelberg, New York: Springer 1974, pp.252-257.

ON THE APPLICABILITY OF THE FREDHOLM-RADON METHOD IN POTENTIAL THEORY AND THE PANEL METHOD

J.Kral and W. Wendland*

Matematický ústav ČSAV, Žitna 25, 11567 Prague 1, CSSR
* Mathematisches Institut A, Universität Stuttgart, Pfaffenwaldring 57, D-7000 Stuttgart 80, Germany.

SUMMARY

For three-dimensional domains with piecewise Lyapounov boundaries having corners and edges one can find simple corner points where the classical assumptions for the validity of the Fredholm-Radon boundary integral equation method with continuous charges are violated. As a consequence, in this case the convergence of the panel method was not justified either. However, the use of a suitable weighted maximum norm allows the justification of the method and the convergence for rectangular domains which violate the previous assumption. Here we give a survey on these relations between geometry, solvability of the integral equations and the convergence of the panel method by introducing a more general concept for the Fredholm radius. For general domains, however, it is still open whether corresponding weighted norms can be found.

INTRODUCTION

The Fredholm-Radon method in classical potential theory goes far back in history |8|. For convex regions C. Neumann |21|, |22| already showed that the boundary Γ does not require regularity. Fredholm |7| and Radon |25| extended the applicability of the double layer boundary integral equation to non convex domains. In the two-dimensional case Γ could have corners and further singularities |14|, |25|. For three dimensions Burago and Maz'ya |3| and also |12|, |13| relaxed the regularity of Γ significantly. E.g. Γ can be a piecewise Lyapounov surface having corners and edges. The basic property for the method, however, is the requirement that the double layer potential operator on the space of continuous charges can be decomposed into a contraction plus a completely continuous operator; in other words, that its Fredholm radius is larger than 1. This condition implies a geometric restriction for admissible corners and edges of Γ. In connection with the maximum norm on $C^{0}(\Gamma)$ the restriction is sufficient and necessary for the above decomposition |3|, |13|.

The same properties also provide the solution of the exterior Neumann problem which models flows around bodies as e.g. air planes as well as the stability and convergence of the panel method with piecewise constant or piecewise linear boundary elements |2|, |11|, |29|, |30|, |31|. Unfortunately, the above geometric condition is violated at very simple corners of rectangular domains or, e.g., at the point where the trailing edge of an airplane wing meets the fuselage |15|. Hence, whenever such corner points on Γ are present, the applicability and convergence of the classical panel method is still not justified. In |15| we introduce for $C^{0}(\Gamma)$ on rectangular Lipschitz domains appropriate weighted maximum norms. With respect to these weighted norms the geometric restrictions relax so that on the rectangular domains (which violate the original assumptions) the corresponding Fredholm radius becomes larger than 1. Consequently, in this case the panel method is still stable and uniformly convergent. Although on $C^{0}(\Gamma)$ the norms have been changed, the topology of uniform convergence is still the same. Therefore we introduce corresponding new concepts of the Fredholm radius.

In |1| one finds an equivalent norm to the maximum norm associated with a non-Lipschitzian rectangular domain providing also a Fredholm radius larger than 1. This indicates the validity of the Fredholm-Radon method for rather general piecewise Lyapounov boundaries. For general non-weighted maximum norms p, however, the interpolation operators of the panel methods P_h might not maintain

$$\lim_{h\to 0} \|P_h\| = 1.$$

For the panel methods we therefore restrict ourselves to the case of weighted maximum norms. The rigorous proof for general domains still is yet to be done.

In combination with results by Dahlberg and Kenig |6| and by Maz'ya |16| we find convergence of some positive order for the panel method.

We also consider the adjoint boundary integral equation in the space of signed bounded Borel measures. For the corresponding adjoint panel method we formulate a weak convergence result.

Verchota showed in |28| that the classical boundary integral equation also is uniquely solvable in $L_2(\Gamma)$ if Γ is just a Lipschitz boundary. His analysis, however, is based on the closed graph theorem which so far does not provide a stability analysis for the panel method with collocation - in the contrary to the concept of the Fredholm radius.

If instead of the classical Fredholm-Radon integral equations one uses a variational formulation with the hypersingular normal derivative of the double layer potential and a corresponding boundary element Galerkin method then this different approach is still valid on general Lipschitz boundaries and the method converges due to Costabel |5|.

In this survey we intend to show the close relation between classical potential theory and numerical analysis of the panel method. We also hope that our conjecture for the Fredholm radius for domains with piecewise Lyapounov boundaries will be justified in the near future.

Acknowledgement: This work was partly supported by the "Stiftung Volkswagenwerk".

1. THE CLASSICAL INTEGRAL EQUATION IN POTENTIAL THEORY

Let us consider the exterior Neumann problem for the Laplacian,

$$\Delta U = 0 \quad \text{in} \quad \Omega \subset \mathbb{R}^3, \tag{1.1}$$

$$\frac{\partial U}{\partial n}\Big|_\Gamma = \psi \quad \text{on} \quad \Gamma = \partial\Omega, \tag{1.2}$$

$$U(x) = o(1) \quad \text{for} \quad |x| \to \infty. \tag{1.3}$$

Here Ω is the exterior domain to a given piecewise smooth simple closed surface Γ. The Neumann boundary condition (1.2) is to be understood in the sense of a given boundary flow |13|, i.e.

$$-\int_\Omega \nabla\varphi \cdot \nabla U \, dx =: \int_\Gamma \varphi \frac{\partial U}{\partial n} d\sigma = \int_\Gamma \varphi \, d\Psi \tag{1.4}$$

for all test functions $\varphi \in C_o^\infty(\mathbb{R}^3)$ where $\Psi \in C^*(\Gamma)$ denotes a given signed Borel measure of bounded variation on Γ, $d\Psi = \psi \, d\sigma$, defining a bounded linear functional on $\varphi \in C^o(\Gamma)$ by (1.4). For the flow without lifting, e.g., Ψ is given by

$$d\Psi = -\vec{n}\cdot\vec{v}_\infty \, d\sigma = \psi \, d\sigma$$

with $\psi \in L_\infty(\Gamma)$. Here $\vec{n}$ denotes the normal vector at the smooth points of Γ pointing into Ω. At corners and edges we may take $\vec{n} = \vec{0}$ or one of the limits from the adjacent smooth surface parts. σ denotes the 2-dimensional surface measure of Γ. The well known Green representation formula reads as

$$U(x) = \frac{1}{4\pi} \int_{y\in\Gamma} U(y) \frac{\partial}{\partial n_y} \left(\frac{1}{|x-y|}\right) d\sigma - \frac{1}{4\pi} \int_\Gamma \frac{1}{|x-y|} d\Psi \tag{1.5}$$

for $x \in \Omega$. The right-hand side in (1.5) is well defined for any fixed $x \notin \Gamma$, continuous $U_{|\Gamma}$ and $\Psi \in C^*(\Gamma)$. The crucial assumption for the Green representation theorem in the form (1.5) to remain valid for nonsmooth Γ and continuous $U_{|\Gamma}$ turns out to be

$$\sup_{x\notin\Gamma} \int_\Gamma \left| \frac{\partial}{\partial n_y} \left(\frac{1}{|x-y|}\right) \right| d\sigma < \infty. \tag{1.6}$$

Theorem 1 |12|, |13|: Let u be any given continuous function on Γ and let (1.6) be satisfied. Then the double layer potential

$$\frac{1}{2\pi} \int_{y\in\Gamma} u(y) \frac{\partial}{\partial n_y} \left(\frac{1}{|x-y|}\right) d\sigma \qquad \text{for } x\notin\Gamma$$

can be extended continuously for $x \to \Gamma$ up to the boundary Γ. Moreover, there holds the jump relation

$$\lim_{x \to x_o \in \Gamma} \frac{1}{2\pi} \int_\Gamma u(y) \frac{\partial}{\partial n_y} \left(\frac{1}{|x_o - y|}\right) d\sigma = \pm u(x_o) + Ku(x_o) \qquad (1.7)$$

with + for $x \in \Omega$ and - for $x \in \mathbb{R}^3 \setminus \overline{\Omega}$,

where the boundary integral operator K is defined by

$$Ku(x_o) = \frac{1}{2\pi} \int_{y \in \Gamma \setminus \{x_o\}} u(y) \left(\frac{\partial}{\partial n_y} \frac{1}{|x-y|}\right) d\sigma$$

$$+ \frac{1}{2\pi} u(x_o) \left(2\pi - \int_{y \in \Gamma \setminus \{x_o\}} \frac{\partial}{\partial n_y} \left(\frac{1}{|x_o - y|}\right) d\sigma\right). \qquad (1.8)$$

K is a continuous linear operator in $C^o(\Gamma)$.
Hence, from (1.5) one obtains an integral equation on Γ in $C^o(\Gamma)$ provided the single layer potential takes continuous boundary values.

Theorem 2 |3|: Let Φ be any given signed measure of bounded variation, i.e. $\Phi \in C^*(\Gamma)$, and let (1.6) be satisfied. Then the single layer potential

$$V\Phi(x) := \frac{1}{2\pi} \int_\Gamma \frac{1}{|x-y|} d\Phi$$

admits a boundary flow in $C^*(\Gamma)$ given by

$$- \int_\Omega \nabla\varphi \cdot \nabla(V\Phi)\, dx = \int_\Gamma \varphi\, d\{\mp\Phi + K^*\Phi\}$$

for all test functions $\Phi \in C_o^\infty(\mathbb{R}^3)$, where K^* is the operator in $C^*(\Gamma)$ adjoint to K,(1,8), in $C^o(\Gamma)$. For $\psi \in L_p(\Gamma)$ and $p>2$, the single layer potential $V\Phi(x)$ becomes continuous across Γ, i.e., continuous in $\mathbb{R}^3$.

In view of these results it is an essential question which surfaces would provide the estimate (1.6), which has been varified for piecewise Lyapounov boundaries |30|. In |3| one finds a rather complete discussion of (1.6) for general non-smooth boundaries. For plane domains |14| gives a precise characterization of the boundaries providing (1.6) and, in addition, assumption $\omega_1(K)<1$ below which, as shown by Netuka |20|, holds for all convex domains in $\mathbb{R}^n$. For nonconvex domains with C^1-boundaries the condition (1.6) can be violated |19|, in this case the boundary integral equations cannot any more be considered for continuous $U_{|\Gamma}$. However, (15) remains valid for Lipschitz domains and U in $H^1_{loc}(\Omega)$ (see |5|).

For a piecewise Lyapounov boundary Γ and $\psi \in L_p(\Gamma)$ with $p>2$ we find for the yet unknown boundary values $u = U_{|\Gamma}$ the classical boundary integral equation

$$u(x) - Ku(x) = -2V\psi(x) \qquad \text{for } x \in \Gamma. \tag{1.9}$$

2. CORNERS AND EDGES AND THE FREDHOLM-RADON METHOD

The mathematical justification of the panel method in potential theory rests on the applicability of the Fredholm theorems for the Fredholm integral equation of the second kind (1.9). In the space of continuous functions $C^0(\Gamma)$, this question was already tackled by J. Radon |24|, |25| if the Fredholm radius is larger than 1 (see also the survey in |17|). For the unweighted maximum norm, however, it turns out that even nonconvex simple rectangular corners must be excluded |15|. We therefore equip $C^0(\Gamma)$ with the weighted norm

$$\|u\|_{C^0_w} = \max_{x \in \Gamma} |u(x)w(x)| \tag{2.1}$$

associated with some given piecewise continuous weight function w(x) satisfying

$$0 < w_0 = \inf_{x \in \Gamma} w(x) \quad \text{and} \quad \sup_{x \in \Gamma} |w(x)| = W_0 < \infty .$$

Note that all these weighted norms are equivalent on $C^0(\Gamma)$ generating the same topology of uniform convergence. Correspondingly, the space of signed bounded Borel measures $C^*(\Gamma)$ has the weighted norms

$$\|G\|_{C^*_w} = \sup_{\|\varphi\|_{C^0_w} \leq 1} \left|\int_\Gamma \varphi \, dG\right| = \int_\Gamma \frac{1}{w} |dG| . \tag{2.2}$$

The Fredholm radius is defined by the inverse number of

$$\omega_w(K) := \inf_C \|K - C\|_{C^0_w, C^0_w} , \tag{2.3}$$

where C traces all compact linear operators acting on $C^o(\Gamma)$ and where the operator norm in (2.3) is defined as usual by

$$\|A\|_{C^o_w, C^o_w} := \sup_{0 \neq \varphi \in C^o(\Gamma)} \|A\varphi\|_{C^o_w} \; / \; \|\varphi\|_{C^o_w} .$$

As is well known, there holds

Theorem 3 |24| |27|: If $\omega_w(K)<1$ then the classical Fredholm theorems are valid for (1.9) in $C^o(\Gamma)$.

For domains satisfying assumption (1.6) one finds an explicit expression for $\omega_w(K)$, i.e.

$$\omega_w(K) = \lim_{0<\delta \to 0} \{\sup_{x \in \Gamma \setminus \gamma} \frac{1}{2\pi} \int_{0<r\leq\delta} \frac{1}{w(y)} \left|\frac{\partial}{\partial n_y}\left(\frac{1}{|x-y|}\right)\right| d\sigma \} . \quad (2.3)$$

This equation is a simple extension of the case $w(y)\equiv 1$ for which (2.3) is due to |3| and |13| (see also |30|). For "convex corners", i.e. if the cone of all half lines originating at x being tangential to Γ is convex, one finds

$$\omega_1(K) = \sup_{x \in \Gamma} \frac{1}{2\pi} \left| 2\pi - \int_{\Gamma\setminus\{x\}} \frac{\partial}{\partial n_y}\left(\frac{1}{|x-y|}\right) d\sigma \right| ,$$

hence, in this case $\omega_1(K)<1$ will be satisfied if Γ does not contain spines nor edges of zero angle |30|.

For non convex three-dimensional corners, however, $\omega_1(K)$ can easily become larger than 1. E.g. for

$\mathbb{R}^3\setminus\Omega = ([-1,1]\times[0,1]\times[0,1]) \cup ([-1,0]\times[-1,1]\times[0,1])$

we have $\omega_1(K) = \frac{5}{4}>1$, generated by the corners at the origin and at the point $(0,0,1)$ |15|. However, if we choose

$$w(y) = \begin{cases} 15/56 & \text{in } E_1 , \\ 45/56 & \text{in } E_2 , \\ 1 & \text{in } E_3 , \end{cases} \quad (2.4)$$

where

$E_1 = [-\frac{1}{2},0] \times [0,\frac{1}{2}] \times (\{0\} \cup \{1\}) \subset \Gamma,$

$E_2 = (([-\frac{1}{2},0] \times [-\frac{1}{2},0])) \cup (([0,\frac{1}{2}] \times [0,\frac{1}{2}])) \times (\{0\} \cup \{1\}) \subset \Gamma,$

$E_3 = \Gamma \setminus (E_1 \cup E_2),$

we find $\omega_w(K) \leq \frac{14}{15} < 1$.

This example shows that the definition of the Fredholm radius depends on the choice of the norm even if the topology remains the same. Therefore an invariant definition of the Fredholm radius would be far more appropriate, as e.g.

$$\omega(K) := \inf_{0<w} \inf_{C} \|K-C\|_{C^o_w, C^o_w} \tag{2.5}$$

where w traces all positive lower semicontinuous weight functions and C all compact linear operators acting on $C^o(\Gamma)$, or

$$\widetilde{\omega}(K) = \inf_{p} \inf_{C} p(K-C)$$

where p ranges over all norms in $C^o(\Gamma)$ inducing the topology of uniform convergence.

Theorem 4 |1|, |15|: Let Γ be the boundary to a so-called "rectangular set" i.e. Γ can be covered by a finite number of planes parallel to the coordinate planes. Then $\widetilde{\omega}(K)<1$, i.e. there exist a norm p inducing to topology of uniform convergence in $C^o(\Gamma)$ and a compact linear operator C acting on $C^o(\Gamma)$ such that $p(K-C)<1$. For a rectangular Lipschitz boundary Γ we have $\omega(K)<1$ |15|.

In view of Theorem 4 and the corresponding two-dimensional situation we formulate for the three-dimensional case the following

Conjecture: For a simple closed piecewise Lyapounov surface Γ we have

$$\widetilde{\omega}(K) \leq \sup_{x\in\Gamma} \frac{1}{2\pi} \left| 2\pi - \int_{\Gamma\setminus\{x\}} \frac{\partial}{\partial n_y} \left(\frac{1}{|x-y|}\right) d\sigma \right|. \tag{2.6}$$

If the conjecture is true then $\widetilde{\omega}(K)<1$ provided the boundary does not contain spines nor edges of zero angle. In two dimensions (2.6) is due to J. Radon |25| where 2π is to be replaced by π.

Now let $\omega(K)<1$. Then to Γ there exists a lower semicontinuous positive weight function w such that $\omega_w(K)<1$. If we fix an even C_0^∞-cut-off function $\varphi:\mathbb{R} \to [0,1]$ vanishing on $[1,\infty)$ such that $\varphi=1$ on $[-\frac{1}{2},\frac{1}{2}]$ then, because of (2.3), there is some $\delta>0$ such that

$$K = C_\delta + K_\delta \tag{2.7}$$

where

$$C_\delta u(x) = \frac{1}{2\pi} \int_\Gamma \left[1-\varphi\left(\frac{|x-y|}{\delta}\right)\right] u(y) \left(\frac{\partial}{\partial n_y} \frac{1}{|x-y|}\right) d\sigma(y) \tag{2.8}$$

is smoothing of every order $k\in\mathbb{N}$, i.e. $C_\delta : C^0(\Gamma) \to C^k(\mathbb{R}^3)$ is continuous for every k. The remainder

$$K_\delta u(x) = \frac{1}{2\pi} \int_{y\in\Gamma\setminus\{x\}} \varphi\left(\frac{|x-y|}{\delta}\right) u(y)\left(\frac{\partial}{\partial n_y} \frac{1}{|x-y|}\right) d\sigma(y)$$

$$+ \frac{1}{2\pi} u(x) \left(2\pi - \int_{y\in\Gamma\setminus\{x\}} \frac{\partial}{\partial n_y}\left(\frac{1}{|x-y|}\right) d\sigma\right) \tag{2.9}$$

is a contraction in C_w^0, i.e.

$$\|K_\delta\|_{C_w^0,\, C_w^0} = q < 1. \tag{2.10}$$

Both properties are essential for

i) the validity of Fredholm's theory for the integral equation (1.9) (see Nikolski's Theorem |10,(5.XIII)|) and

ii) for the damping of higher frequencies of Fourier expansions of u underlying any of the multigrid panel methods for (1.9) |9|, |23|.

It is well known that the integral equation (1.9) has at most one solution |13|. Hence, (1.9) is uniquely solvable in $C^0(\Gamma)$ for any given continuous right-hand side.

The justification of the above conjecture remains one of the fundamental open problems in classical potential theory to our current days.

3. REMARKS ON THE ADJOINT OPERATOR

In the indirect method the flow potential U is often sought in the form of a single layer potential

$$U(x) = \frac{1}{4\pi} \int_{y\in\Gamma} \frac{1}{|x-y|}\, dG \qquad (3.1)$$

For a domain having corners and edges, the single layer density will be a signed measure $G\in C^*(\Gamma)$. Here the boundary condition (1.2), (1.4) yields the boundary integral equation

$$G - K^* G = \Psi \qquad \text{in } C^*(\Gamma) \qquad (3.2)$$

for $G\in C^*(\Gamma)$ with given $\Psi\in C^*(\Gamma)$. Again, if $\omega(K) < 1$ then the adjoint operator K_δ^* will be a contraction in $C_w^*(\Gamma)$ due to duality. By definition (2.8), we find for the image measure

$$dC_\delta^* G = (\underset{\sim}{C}_\delta^* G(x))\, d\sigma\ ,$$

where the density $\underset{\sim}{C}_\delta^* G(x)$ is given by

$$\underset{\sim}{C}_\delta^* G(x) = \frac{1}{2\pi} \int_\Gamma \left[1-\varphi\left(\frac{|x-y|}{\delta}\right)\right] \left(\frac{\partial}{\partial n_x} \frac{1}{|x-y|}\right) dG\ ,$$

and where the mapping $\underset{\sim}{C}_\delta^*: C^*(\Gamma) \rightarrow C^k(\mathbb{R}^3)$ is continuous for every $k\in\mathbb{N}$. According to Theorem 2 we now have

Theorem 5 |13|: If $\tilde{\omega}(K)<1$ then the equation (3.2) admits a unique solution $G\in C^*(\Gamma)$ to any given boundary flow $\Psi\in C^*(\Gamma)$. If $V\Psi$ is contiuous in $\mathbb{R}^3$ then for the solution G also U(x), given by (4.1), will be contiuous in $\mathbb{R}^3$.

Hence, for the flow problem with $\psi(y) = -\vec{n}(y)\cdot\vec{v}_\infty$ we can construct the potential also via (3.1), (3.2) provided $\tilde{\omega}(K)<1$.

Remark: This approach is valid as long as at least (1.6) (and $\omega(K)<1$) will be satisfied. This is the case e.g. for piecewise Lyapounov surfaces. One can find an example of a surface Γ satisfying (1.6) but with

$$\sup_{x\in\Gamma} \int_{\Gamma\setminus\{x\}} \left|\frac{\partial}{\partial n_x}\left(\frac{1}{|x-y|}\right)\right| d\sigma = \infty\ . \qquad (3.3)$$

Hence, the adjoint operator K^* <u>cannot</u> be considered as a continuous operator for continuous densities g;

$$\int_{\Gamma\setminus\{x\}} g(y) \left(\frac{\partial}{\partial n_x} \frac{1}{|x-y|}\right) d\sigma(y)$$

does **not** define a bounded linear mapping in $C^0(\Gamma)$ any more because of (3.3) |26|.

4. THE ORDINARY PANEL METHOD AND ITS CONVERGENCE

For the approximate solution of (1.9) let $\Gamma = \bigcup_{l=1}^{L} \Gamma_l$, where every Γ_l is a segment of a Lyapounov surface. Each Γ_l can be considered to be the image of a $C^{1+\alpha}$-application $x(t)$ defined on a polygonal parameter domain $U_l \subset \mathbb{R}^2$, $\Gamma_l = x(U_l)$. Then a family of regular triangular (or quadrangular) partitions with triangles T_j defines a corresponding family of partitions on every Γ_l and, hence, on Γ. We set $F_j = x(T_j) \subset \Gamma$. By $p_j \in F_j$ let us denote the image of the center of gravity of T_j. To the collection

$$\bigcup_{j=1}^{N} F_j = \Gamma \qquad \text{and} \qquad (p_j)_{j=1}^{N}$$

we associate the one point Gaussian quadrature formula

$$\int_\Gamma f \, d\sigma \simeq \sum_{j=1}^{N} f(p_j) g_j \quad \text{where} \quad g_j = \int_{F_j} d\sigma \quad .$$

Correspondingly, we approximate the function u in (1.9) by a piecewise constant trial function

$$u_h(x) = u_h(p_j) \quad \text{for all } x \in F_j \ , \quad j = 1,\ldots,N.$$

Here and in the following $h > 0$ denotes the parameter of meshwidth, $h = \max_{j=1,\ldots,N} \operatorname{diam}(F_j)$. The classical panel method then corresponds to the collocation method: Find $u_h \in H_h$ = {space of all piecewise constant functions on $\{F_j\}_{j=1}^{N}$} such that

$$u_h(p_k) - \sum_{j=1}^{N} \left(\frac{1}{2\pi} \int_{y \in F_j} \left(\frac{\partial}{\partial n_y} \frac{1}{|p_k - y|}\right) d\sigma\right) u_h(p_j)$$

$$= -2V\psi(p_k), \qquad k=1,\ldots,N. \qquad (4.1)$$

Note that with our construction the edge and corner points γ will correspond to boundary parts of the polygon U_l and, hence, $\gamma \subset \bigcup_{j=1}^{N} \partial F_j$. Therefore $p_j \notin \gamma$. Of course, a corresponding modification of the weights according to (1.8) with grid points on γ can be defined easily but this would be more appropiate for piecewise linear elements defined in U_1 where the collocation points correspond to the corners of ∂F_j.

The entries of the influence matrix in (4.1) are given by the solid angles

$$a_{kj} = \frac{1}{2\pi} \int_{y\in F_j} \frac{\partial}{\partial n_y} \frac{1}{|p_k - y|} d\sigma \tag{4.2}$$

which for practical computations are further approximated, for instance by

$$a_{kj} \simeq \frac{1}{2\pi} \frac{\vec{n}(p_j)\cdot(\vec{p}_j - \vec{p}_k)}{|p_k - p_j|^3} g_j \text{ for } |p_k - p_j| \geq d \text{ or } |p_j - \gamma| \geq d$$

where d>0 is a fixed constant. (Further approximations see in |11|, |23|, |30|, |31|). By

$$P_h f(x) = f(p_k) \quad \text{for all } x\in F_k \tag{4.3}$$

we define a projection operator $P_h : C^0(\Gamma) \longrightarrow H_h$. The step functions as well as $C^0(\Gamma)$ can be imbedded into the space of the Baire functions of the first class which we equip with the extension of (2.1), the weighted norm

$$\|u\|_{B_w} = \sup_{x\in\Gamma} |u(x)w(x)|.$$

With respect to this norm the Baire functions form a Banach space B_w |18, p.442|. Also P_h in (4.3) is well defined in B_w having the operator norm

$$\|P_h\|_{B_w, B_w} = 1.$$

The equations (4.1) can be written as the equation

$$u_h - P_h K P_h u_h = -2P_h V\psi$$

in B_w since (1.8) extends naturally to step functions. With the same arguments as in |30| and |11| for $B_w(\Gamma)$ as for $B_1(\Gamma)$ one obtains:

Theorem 6 |2|, |30|, |31|: Let $\omega(K)<1$. Then there exists $h_o>0$ such that for all $h\in(0,h_o]$ the equations (4.1) are uniquely solvable and stable satisfying

$$\| |I-P_hKP_h|^{-1} \|_{B_w,B_w} \leq c \qquad (4.4)$$

with a constant c which is independent of h.
The proof hinges essentially on the decompositions (2.7) with

$$\| P_h K_\delta P_h \|_{B_w,B_w} \leq q < 1$$

and the collective compactness of the family $\{P_h C_\delta P_h\}$ in B_w where $C_\delta : B_w \to C^k(\mathbb{R}^3)$ for every $k \in N$.
Since the solution of (4.1) defines a projection onto H_h, Céa's lemma is a simple consequence of (4.4).

Lemma 7 |4|: For the solution of (4.1) we have the quasioptimality result

$$|u_h(x) - U(x)| \leq c \inf_{v_h \in H_h} \| U - v_h \|_{C^o(\Gamma)} \qquad (4.5)$$

For $\psi \in L_\infty(\Gamma)$ as for the nonlifting flow we easily find

$$V\psi \in C^{\tilde{\lambda}}(\mathbb{R}^3)$$

for every $\tilde{\lambda} \in (0,1)$, where $C^{\tilde{\lambda}}(\mathbb{R}^3)$ denotes the space of Hölder continuous functions with exponent $\tilde{\lambda}$. Then the solution U of (1.9) will be in some Hölder space $C^\lambda(\Gamma)$ where the exponent $\lambda \in (0,1)$ depends on the Hölder exponent of the Lyapounov pieces Γ_l and on $(1-\omega(K))$. This regularity result can be obtained under slightly stronger restrictions on Γ from results by Maz'ya |16, Theorem 7| in Hölder spaces or for Lipschitz domains with the single layer approach from results by Dahlberg and Kenig |6| providing $G \in L^p$, $p>2$. The Hölder continuity can be combined with the approximation properties of the family H_h.

Theorem 8: Let Γ be piecewise Lyapounov, $\omega(K)<1$ and $U \in C^\lambda(\Gamma)$. Then

$$|u_h(x) - U(x)| \leq c\, h^\lambda \, \| U \|_{C^\lambda(\Gamma)}$$

The proof follows from

$$\inf_{v_h \in H_h} \max_{x \in \Gamma} |U(x) - v_h(x)| \leq c h^\lambda \, \| U \|_{C^\lambda(\Gamma)} \qquad (4.6)$$

and (4.5). The estimate (4.6) follows from the corresponding result with $\lambda=1$ for smooth Γ and continuously differentiable U by interpolation of Hölder-Zygmund spaces.

The result of Theorem 6 and Lemma 7 can essentially be carried over to the adjoint equation (3.2) in $C^*(\Gamma)$, correspondingly. To this end we associate with the collocation points $(p_j)_{j=1}^N$ the finite dimensional space of Dirac measures

$$H_h^* = \{ \sum_{j=1}^N \gamma_j \delta(p_j - x) \mid \gamma_j \in \mathbb{R} \} \subset C^*(\Gamma)$$

and define by $\Pi_h: C^* \longrightarrow H_h^*$ via

$$\int_\Gamma \varphi \, d(\Pi_h G) := \sum_{j=1}^N \varphi(p_j) \int_{F_j} dG \qquad \text{for all } \varphi \in C^0(\Gamma)$$

a corresponding projection. Then the approximate equations to (3.2) are given as to find $G_h \in H_h^*$ such that

$$G_h - \Pi_h K^* \Pi_h G_h = \Pi_h \Psi .$$

With

$$G_h = \sum_{j=1}^N \gamma_j \, \delta \, (p_j - x)$$

the approximate equation is equivalent to the quadratic system of linear equations for the γ_j,

$$\gamma_j - \sum_{k=1}^N a_{kj} \gamma_k = \int_{F_j} d\Psi \quad , \qquad j=1,\ldots,N \quad . \tag{4.7}$$

The coefficient matrix in (4.7) is the adjoint to (4.1) and, hence, this projection method is stable due to Theorem 6 by duality. Hence, we again have Céa's lemma.

Lemma 9: Let $\omega(K)<1$. Then for the same $h \in (0,h_o]$ as in Theorem 6 we have unique solvability of (4.7) and the asymptotic estimate

$$\left| \int_\Gamma \varphi(dG - dG_h) \right| \leq c \inf_{\chi_h \in H_h^*} \left| \int_\Gamma \varphi (dG - dX_h) \right| \tag{4.8}$$

for any $\varphi \in C^0(\Gamma)$.

This is a rather weak convergence result. Since we require the right hand side to tend to zero one needs for the signed measure $G \in C^*(\Gamma)$ as well as for the test function φ regularity properties. For the potential $U(x)$ of the flow problem with $\psi \in L_\infty(\Gamma)$ we have Hölder continuity as was previously pointed out. Here (4.8) yields the asymptotic error estimate

$$|U(x) - \sum_{j=1}^{N} \frac{\gamma_j}{|x-p_j|}| \leq c \inf_{\xi_j \in \mathbb{R}} |U(x) - \sum_{j=1}^{N} \frac{\xi_j}{|x-p_j|}| . \qquad (4.9)$$

and the right-hand side tends to zero for fixed $x \notin \Gamma$ and $h \to 0$.

Comparing (4.9) with Lemma 7 and Theorem 8 it seems that the direct method with the classical Fredholm integral equation (1.9) in $C^0(\Gamma)$ and an approximation like (4.1) is better suited for the numerical computation of ideal flows around obstacles having corners and edges.

REFERENCES

|1| Angell, T.S., Kleinman, R.E., Kral, J.: "Double layer potentials on boundaries with corners and edges", Comment. Math. Univ. Carolinae **27** (1986).

|2| Bruhn, G., Wendland, W.: "Über die näherungsweise Lösung von Funktionalgleichungen", Internat. Schriftenreihe Numer. Math. 7 (1967), pp. 136-164.

|3| Burgago, YU.D., Maz'ya, V.G.: "Potential Theory and Function Theory for Irregular Regions", Seminar in Mathematics, V.A. Steklov Math. Inst., Leningrad Vol. 3 (1967), Consultant Bureau, Plenum Publ., New York 1969.

|4| Céa, J.: "Approximation variationelle des problèmes aux limites", Ann. Inst. Fourier (Grenoble) **14** (1964), pp. 345-444.

|5| Costabel, M.: "Boundary integral operators on Lipschitz domains: Elementary results", (Preprint 898, FB Math., Th Darmstadt 1985), to appear in SIAM J.Math.Anal.

|6| Dahlberg, B.E.J., Kenig, G.E.: "Hardy spaces and the Neumann problem in L^p for Laplace's equation in Lipschitz domains", Annals of Mathematics **125** (1987), pp. 437-465.

|7| Fredholm, I.: "Sur une nouvelle méthode pour la résolution du problème de Dirichlet", Kong. Vetenskaps-Akademiens

Förh., Stockholm, 1900.

|8| Gauss, C.F.: "Allgemeine Lehrsätze in Beziehung auf die im verkehrten Verhältnisse des Quadrats der Entfernung wirkenden Anziehungs- und Abstoßungskräfte, Werke V, 2nd ed., Göttingen (1839), pp. 194-242.

|9| Hackbusch, W., Nowak, Z.P.: "On the fast matrix multiplication in the boundary element method by panel clustering", in Boundary Elements IX, Vol.1 (eds. C.A. Brebbia, W.L. Wendland, G. Kuhn) Springer-Verlag Berlin-Heidelberg-New York-London-Paris-Tokyo (1987), pp. 463-474.

|10| Kantorowitsch, I.W., Akilow, G.P.: "Funktionalanalysis in Normierten Räumen", Akademie-Verlag, Berlin, 1964.

|11| Kleinman, R.E., Wendland, W.L.: "On Neumann's method for the exterior Neumann problem for the Helmholtz equation", J. Math. Anal. Appl. **57** (1977), pp. 107-202.

|12| Kral, J.: "The Fredholm method in potential theory", Trans. Amer.Math. Soc. **125** (1966), pp. 511-547.

|13| Kral, J.: "Integral Operators in Potential Theory", Lecture Notes Math. 823, Springer-Verlag, Berlin, Heidelberg 1980.

|14| Kral, J.: "Boundary regularity and normal derivatives of logarithmic potentials", (Preprint 159 A, Dept. Math. Sci., University of Delaware 1985), to appear.

|15| Kral, J., Wendland, W.: "Some examples concerning applicability of the Fredholm-Radon method in potential theory", Aplikace Matematiky **31** (1986), pp. 293-308.

|16| Maz'ya, V.G.: "Boundary integral equations of elasticity in domains with piecewise smooth boundaries", in Equadiff 6 Proceedings (eds. Vosmansky, J., Zlamal, M.) J.E.Purkyne University, Dept. of Mathematics, Brno (1985), pp. 235-242.

|17| McLean, W.: "Boundary integral methods for the Laplace equation", Ph.D. Thesis, Australian National University, Camberra 1986.

|18| Natanson, I.P.: " Theorie der Funktionen einer Reellen Veränderlichen", Akademie-Verlag, Berlin 1961.

|19| Netuka, I.: "Smooth surfaces with infinite cyclic

variation", Časopis pro pěst. matematiky **96** (1971), pp.86-101.

|20| Netuka,I.: "Fredholm radius of a potential theoretic operator for convex sets", Časopic pro pěst. matematiky **100** (1975) pp.374-383.

|21| Neumann,C.: "Untersuchungen über das logarithmische und Newtonsche Potential", Teubner, Leipzig 1877.

|22| Neumann,C.: "Über die Methode des arithmetischen Mittels", Hirzel, Leipzig 1887 (Part. 1), 1888 (Part.2).

|23| Nowak,Z.P.: "Panel clustering technique for lifting potential flows in three space dimensions", these proceedings (1987).

|24| Radon,J.: "Über lineare Funktionaltransformationen und Funktionalgleichungen", Sitzungsber. Akad. Wiss. Wien **128** (1919), pp. 1083-1121.

|25| Radon,J.: "Über Randwertaufgaben beim logarithmischen Potential", Sitzungsber. Akad. Wiss. Wien **128** (1919), pp. 1123-1167.

|26| Riesz,F., SZ.-Nagy,B.: "Vorlesungen über Funktionalanalysis", VEB Verlag der Wiss., Berlin 1956.

|27| Taylor,A.E.: "Functional Analysis", John Wiley & Sons, New York 1967.

|28| Verchota,G.: "Layer potentials and regularity for the Dirichlet problem for Laplace's equation in Lipschitz domains", J. Funct. Anal. **59** (1984), pp. 572-611.

|29| Wendland,W.: "Lösung der ersten und zweiten Randwertaufgaben des Innen- und Außenraumgebietes für die Potentialgleichung im $\mathbb{R}^3$ durch Randbelegungen", Doctor's Thesis TU Berlin D83, 1965.

|30| Wendland,W.: "Die Behandlung von Randwertaufgaben im $\mathbb{R}^3$ mit Hilfe von Einfach- und Doppelschichtpotentialen", Numer. Math. **11** (1968), pp. 380-404.

|31| Wendland,W.L.: "Boundary element methods and their asymptotic convergence", in Theoretical Acoustics and Numerical Treatment (ed. Filippi,P.) CISM Courses and Lectures 277, Springer-Verlag, Wien-New York (1983), pp. 135-216.

A SUBSONIC PANEL METHOD FOR DESIGN OF 3-DIMENSIONAL COMPLEX CONFIGURATIONS WITH SPECIFIED PRESSURE DISTRIBUTION

Krzysztof Kubryński
Instytut Lotnictwa. Zakład Aerodynamiki
Al. Krakowska 110/114, 02-256 Warszawa, Poland

SUMMARY

A cheap, iterative method for aerodynamic design of complex 3-dimensional configurations /such as wing-body combination/ with prescribed surface pressure distribution is presented. The method is based on the panel method with numerical optimization. In each iteration cycle the geometry modifications are modelled by surface transpiration. Some examples of application of this method for wing-body combinations are presented.

INTRODUCTION

During the last fifteen years many panel methods have been developed for the determination of the potential flowfield about real, complex aeroplane configurations. Almost all these methods have the following features:

- the possibility of accurate and detailed description of the given configuration
- versatility
- good evaluation of the velocity and pressure distribution on the surface in flow conditions in which the potential flow model is justified
- the possibility of the weak viscous-potential interaction approximation in conjunction with suitable boundary layer computational methods
- high computational efficiency and low cost

On account of these features the panel methods have became the essential tools in the aerodynamic design at sub- and supersonic Mach numbers [1-4]. Over the last years many investigators have extended this method for transonic speeds [5-8] but until now these methods have not become the true alternative of the finite--difference methods.

One of the most important goal in the aerodynamic design is such aircraft shape definition which fulfills such requirements as: high $C_{L_{max}}$ and Ma_{kr}, low C_{D_0}, appropriate boundary layer stability and stall progression etc. These requirements however depend on the appropriate surface pressure distribution. In order to avoid the design of the aircraft shape by trial and error method it is of value to possess an inverse method which computes the surface geometry for the specified pressure distribution. It is especially interesting in the case of complex configurations such as wing-body or wing-nacelle combination. The panel methods which have been originally developed for flow analysis problems can also be used to solve the design problems [9-12] . The inverse methods presented in many papers are usually suitable only for design of a wing with specified pressure distribution. The method presented in this paper has no such restrictions and can be used for design of more complex configurations.

DESCRIPTION OF THE METHOD

Direct formulation. The method is based on linearized theory of compressible flow [14]. The Prandtl-Glauert equation

$$\beta^2\varphi_{xx}+\varphi_{yy}+\varphi_{zz}=0 \;;\; \beta^2=1-Ma_\infty^2 \tag{1}$$

is assumed to govern the perturbation velocity potential φ in the flowfield. The x-axis is parallel to free-stream direction. Total velocity vector is equal to:

$$\bar{V}=\bar{V}_\infty+\bar{v}=\bar{V}_\infty+\nabla\varphi\,. \tag{2}$$

The total mass flux vector defined as

$$\bar{W}=\rho/\rho_\infty\;\bar{V} \tag{3}$$

can be linearized

$$\bar{W}=\bar{V}_\infty+\bar{w} \tag{4}$$

where linearized vector of mass flux perturbation w is equal to:

$$\bar{w}=(\beta^2 u,v,w)=(\beta^2\varphi_x,\varphi_y,\varphi_z)\,. \tag{5}$$

Equation (1) rewritten as $\nabla.(\bar{V}_\infty+\bar{w})$ express the conservation of mass to the first order in perturbation quantities.
The boundary conditions /so-called linearized mass flux boundary conditions/ can be written as:

$$\bar{W}.n=(\bar{V}_\infty+\bar{w}).\bar{n}=\dot{m}_s/\rho_\infty \tag{6}$$

and express the intensity of mass outflow through the surface. With equations (1) and (6) the second order pressure formula is closely connected:

$$Cp_2=-2(\bar{V}_\infty.\bar{v}+\bar{v}.\bar{w})/V_\infty^2 \tag{7}$$

in the sense, that forces and moments obtained by integrating this pressure over the surface satisfy the d'Alambert's paradox. The second order pressure formula agrees with the isentripic one

$$C_p = \frac{\left[1+\frac{k-1}{2} Ma_\infty^2 (1-V^2/V_\infty^2)\right]^{\frac{k}{k-1}} - 1}{\frac{k}{2} Ma_\infty^2} \qquad (8)$$

to the first order in perturbation quantities. The fundamental solution of Eq. (1) is the subsonic source:

$$\psi = -1/\left(4\pi r_{\beta QP}\right). \qquad (9)$$

Applying Greens Theorem to the flowfield and using Eq. (6) and (9) the perturbation velocity potential can be expressed as:

$$E\varphi_{(P)} = \oint\!\!\oint_{S_B} \frac{-\left(\frac{\dot{m}_s}{\varrho_\infty} - \bar{V}_\infty \cdot \bar{n}_{(Q)}\right)}{4\pi r_{\beta QP}} + \varphi_{(Q)}\beta^2 \frac{\bar{r}_{QP} \cdot \bar{n}_{(Q)}}{4\pi r_{\beta QP}^3}\, dS + \iint_{S_W} \langle\varphi\rangle_{(Q)} \beta^2 \frac{\bar{r}_{QP} \cdot \bar{n}_{(Q)}}{4\pi r_{\beta QP}^3}\, dS, \qquad (10)$$

$$E = \begin{cases} 0 & \text{if } P \in I'', \\ 1/2 & \text{if } P \in S'', \\ 1 & \text{if } P \in E''. \end{cases}$$

The jump of potential across the wake $\langle\varphi\rangle$ is determined by Kutta condition at the trailing edge: it is constant in streamline direction and equal to the trailing edge value.

Equation (10) is solved using low order panel method /plane panels with constant source and doublet distribution/ with indirect Dirichlet boundary conditions: zero perturbation potential is specified in the internal side of the surface

$$\varphi_I = 0. \qquad (11)$$

Mathematically this is equivalent to the conditions given by Eq. (6) on the external surface.

Such formulation has many advantages:

- it exhibits higher order characteristics in comparison with methods using the velocity formulation for the same order of singularity distribution [13],
- shows weak sessitivity to the panel spacing distribution,
- is cheaper than the methods with velocity formulation /less expensive computations of AIC - matrix/,
- led to better results then source methods in regions with sharp concave corners [13],
- is better numerically conditioned than source methods.

It has also some disadvantages compared with velocity formulation, especially it is less forgiving in respect to closed surface /for example the open wing tip/. This is especially true for high order formulation /with additional control points at the sides of network/ and can be usually ignored in the present, low order formulation /Fig. 1/.

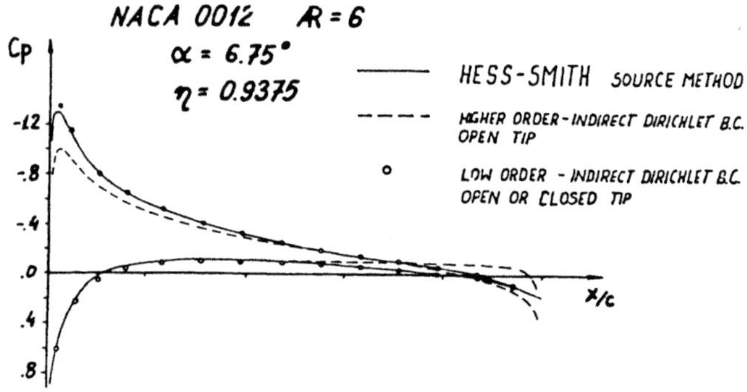

Fig. 1 Pressure distribution near the tip of the wing

The integral equation (10) is replaced by a system of linear equations of the form:

$$[A]\{\varphi\} = [B]\{\dot{m}_s/\rho_\infty - \bar{V}_\infty \cdot \bar{n}\} \tag{12}$$

which is solved to obtain the perturbation potential distribution over the surface. The velocity distribution on the surface is obtained by differentiation of the perturbation potential on the surface and adding the free-stream contribution. In the local panels coordinate system:

$$V_t = \bar{V}_\infty \cdot \bar{t} + \partial\varphi/\partial t \ , \qquad V_s = \bar{V}_\infty \cdot \bar{s} + \partial\varphi/\partial s \ . \tag{13}$$

The pressure distribution is computed by the isentropic formula (8) and aerodynamic coefficients are obtained by integrating pressure distribution using the second order formula (7).

Inverse method. In this case we want to obtain the geometry with specified pressure distribution on the surface. Because such geometry usually does not exist we must be content if we find the geometry with approximate pressure distribution. If we define the geometry in the form:

$$\text{GEOMETRY} = \text{INITIAL GEOMETRY} + \sum_{i=1}^{ND} X_i \cdot (\text{i-th BASIC SHAPE}) \tag{14}$$

the invers problem can be reduced to finding the X vector which minimizes the error in pressure distribution

$$E = \sum_{j=1}^{NP} W_j \cdot (C_{p_j}^D - C_{p_j})^2 \tag{15}$$

where: W_j - weight function on j-th panel,
$C_{p_j}^D$ - target pressure coefficient on j-th panel,
C_{p_j} - its actual value.

This is made using numerical optimization technique. The most serious problem is how to express the pressure distribution as a function of the geometry alteration. Direct utilization of the panel method leads to high cost of inverse computations.

One way to overcome this problem is [11] finding the optimal transpiration distribution and computing the new geometry /new stream surface/ which corresponds to this outflow. It is however not easy to compute such surface /especially for complex configurations/ and it can lead to nonrealistic shapes.

In the present method the basic geometry alterations are modelled by surface transpirations and after finding the optimal linear combination of this transpirations it is possible to compute directly the new geometry using Eq. (14).

The basic part of this method is the computation of transpiration which modells the basic geometry alteration. In a natural /or streamline/ coordinate system, as shown in Fig. 2 the expression of the transpiration is as follows:

$$w_{TR} = \frac{1}{\varrho_\infty}\left[\frac{\partial(\varrho U h)}{\partial \xi} + (\varrho U h)\frac{\partial \mu}{\partial \eta}\right]. \tag{16}$$

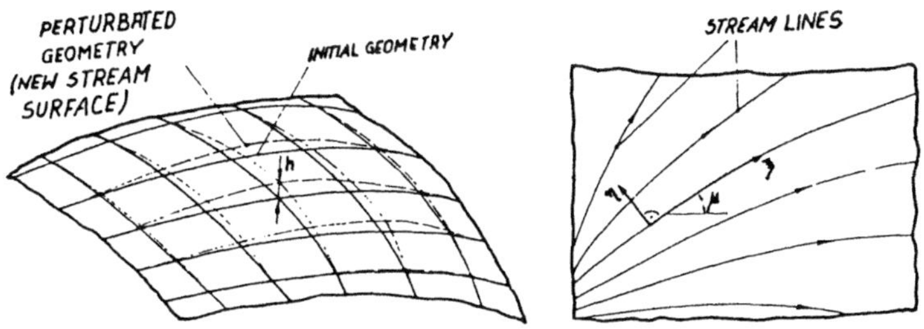

Fig. 2 Alteration of initial geometry and natural coordinate system on the surface

The mean value of the transpiration over the panel area can be obtained by mass flux balance in the volume enclosed by initial surface and the modelled stream surface, Fig. 3.

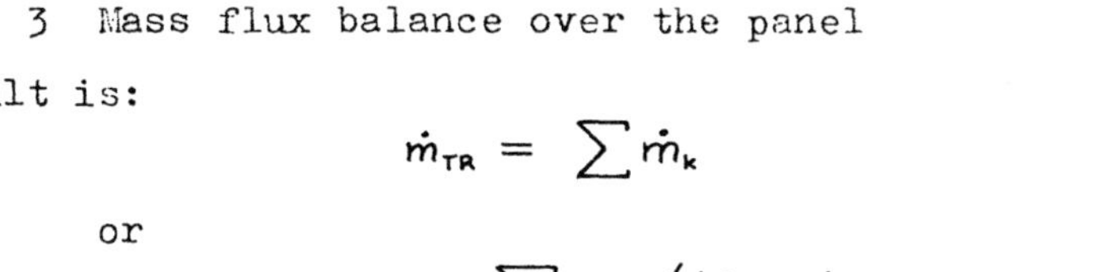

Fig. 3 Mass flux balance over the panel

The result is:

$$\dot{m}_{TR} = \sum \dot{m}_k \tag{17}$$

or

$$w_{TR_j} = \sum \dot{m}_k / (A_j \varrho_\infty) \tag{17a}$$

where: mass flux through the k-th wall of the hexahedron between j-th panel and the new stream surface

$$\dot{m}_k = \int_{L_k} (\rho U_{N_k} h)\, dl \quad , \tag{18}$$

U_{N_k} - component of tangent velocity normal to k-th panel side ,

mass flux through j-th panel surface

$$\dot{m}_{TR} = \rho V_{TR} A_{j\text{-th panel}} = \rho_\infty w_{TR} A_j \ . \tag{19}$$

The incremental potential distribution due to surface transpiration can be expressed /similar to Eq. (12) / as:

$$[\delta\varphi]_i = [A]^{-1} [B] \{w_{TR}\}_i \tag{20}$$

where $\{w_{TR}\}_i$ is the transpiration distribution associated with the i-th basic geometry alteration. The potential distribution on the surface defined by (14) can be expressed as:

$$\{\varphi\} = \{\varphi_0\} + \sum_{i=1}^{ND} X_i \cdot \{\delta\varphi\}_i \ . \tag{21}$$

The velocity on the surface is computed by formula:

$$\begin{aligned} V_t &= \bar{V}_\infty . \bar{t}' + \partial\varphi/\partial t \ , \\ V_s &= \bar{V}_\infty . \bar{s} + \partial\varphi/\partial s \end{aligned} \tag{22}$$

where $\bar{t}'$ and $\bar{s}'$ are taken from the new geometry /this partially takes into account the nonlinearity of the problem and significantly accelerate the convergence of the method/.

To find the optimal $\bar{X}$ vector the conjugate gradient method is used.

Because of the nonlinear nature of the design problem it is solved iteratively using the geometry obtained in the previous iteration as initial in the actual one.

The block diagram of the method is shown in Fig. 4.

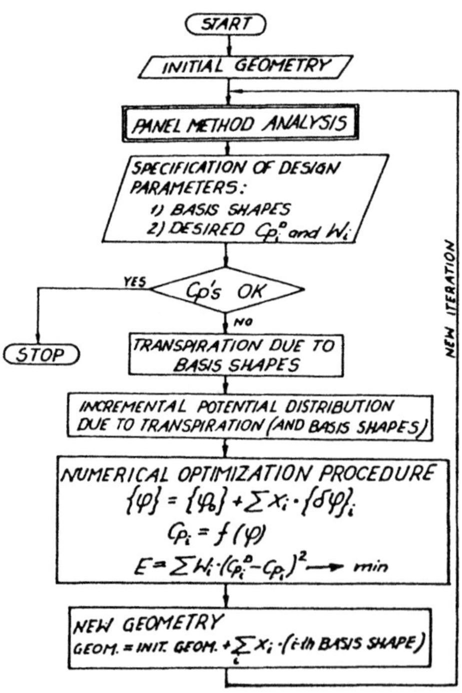

Fig. 4 Block diagram of the method

SAMPLE RESULTS

Wing with constant section. Wing is of rectangular planform having aspect ratio of 8. The target pressure distribution corresponds to pressure distribution on the wing with airfoil NACA 4415 at $\alpha = 5^\circ$. The initial geometry has the shape of profil NACA 0012. Twenty two basic geometry alterations were used, representing the movement parallel to z-axis of all panel endpoints lying on a proper wing generating line /except points lying on leading and trailing edges/. Results are shown in Fig. 5.

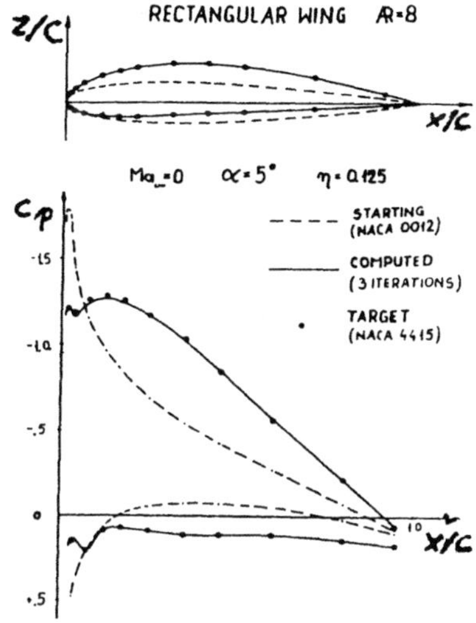

Fig. 5 Comparison of initial, target and computed pressure distribution and airfoil shape on the rectangular wing

Wing-body combination I. This configuration consists of an axisymmetrical fuselage and a swept wing with section NACA 0012 mounted below midheight of the fuselage. The target pressure distributions on the upper wing surface in three sections near the fuselage are the same as at the midspan of the wing /it was required to obtain the straight isobar lines near the fuselage/. As basic design shapes were selected changes of fuselage diameter at 21 sections. Fig. 6 shows geometry of the fuselage and the wing, the isobar patterns before and after two design iterations and pressure distributions at three sections concerned.

Wing-body combination II. This configuration has the same wing as the previous one, but has more complicated fuselage /with cockpit and not axisymmetrical/. The target pressure distribution was the same as in the previous example /taken from the midspan of the wing/. As the basic design shapes were selected 10 changes of the fuselage width over the wing. Fig. 7 shows the geometry, the isobar patterns and the pressure distributions on the upper surface of the wing. It must be mentioned that in these two examples constraints on maximum value of geometry changes /maximum diameter or width/ were used and probably it could be possible to obtain better results, especially in this example.

Wing-body combination III. This configuration has an axisymmetrical fuselage and a swept wing. The initial wing geometry

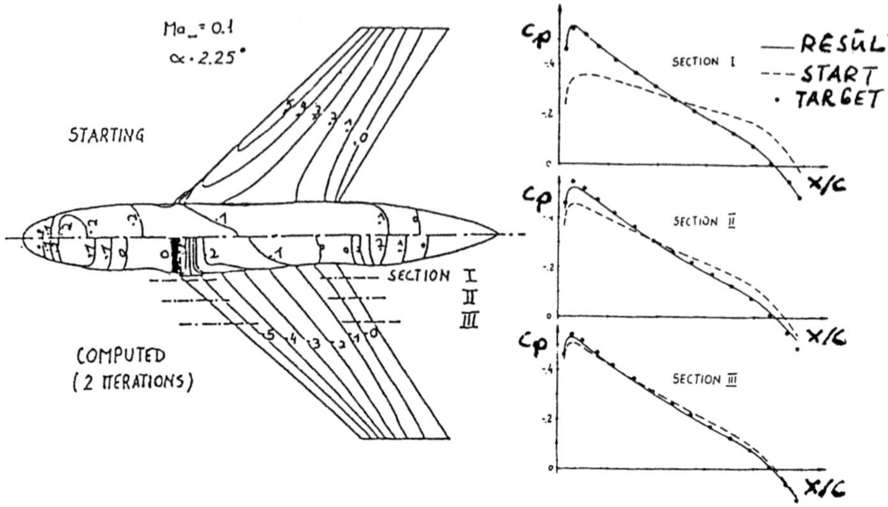

Fig. 6 Geometry of the wing-body combination I and pressure distributions on the upper surface

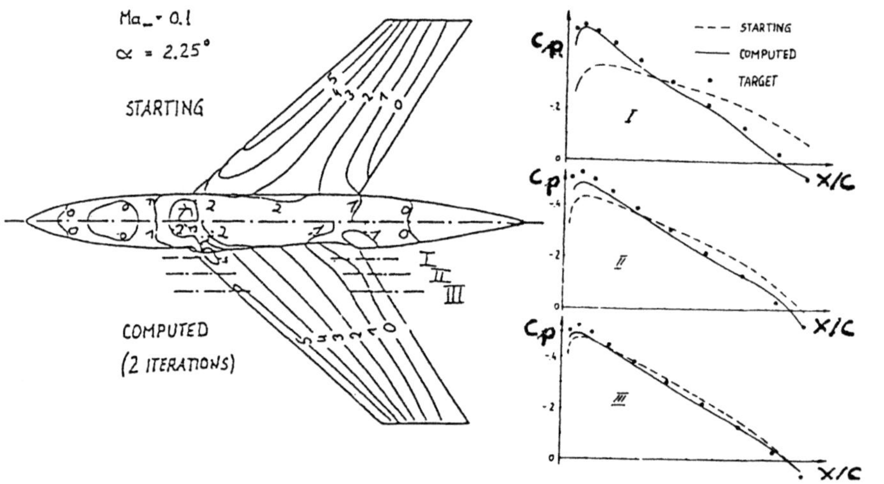

Fig. 7 Geometry of the wing-body combination II and pressure distributions on the upper surface

has constant airfoil NACA 0012 section. The target pressure distribution has been obtained by modyfing the upper surface of the wing in two sections: near the wing-fuselage intersection /upper surface of NACA 23018 airfoil/ and in the section at the break of the trailing edge /upper surface of NACA 4418 airfoil/. As basic design shapes relocations of the upper surface network points were selected in such a way, that one shape alteration corresponds to the relocation of three points.

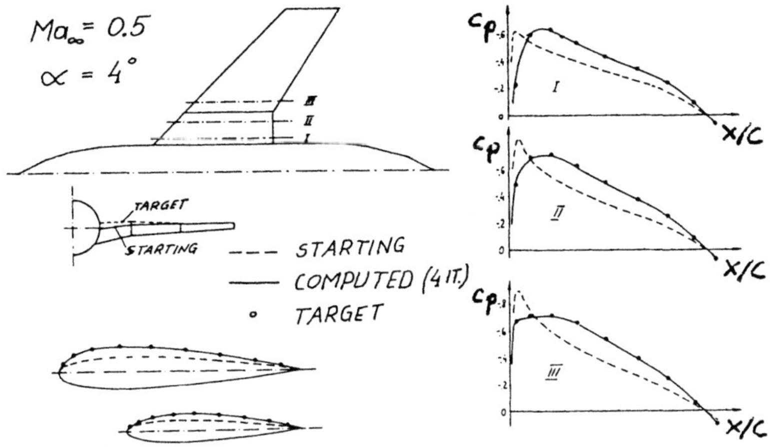

Fig. 8 Geometry of the wing-body combination III, pressure distribution on the upper surface and the initial, target and computed after 4 iterations geometry of the wing-body intersection line and the wing section at the break of the trailing edge

Wing-body-tank combination. This configuration consists of axisymmetrical fuselage, axisymmetrical tank at the tip of the wing and a wing with nearly "roof-top" pressure distribution on the upper surface at design conditions /$Ma_\infty = 0.82, \alpha = 3°$/ without tanks. In the presence of the tank there appeared high pressure peak near the leading edge at the wing tip. It was required to eliminate this by changing of the tank geometry. As basic design shapes changes of tank diameter at twenty sections were selected. Results are shown in Fig. 9.

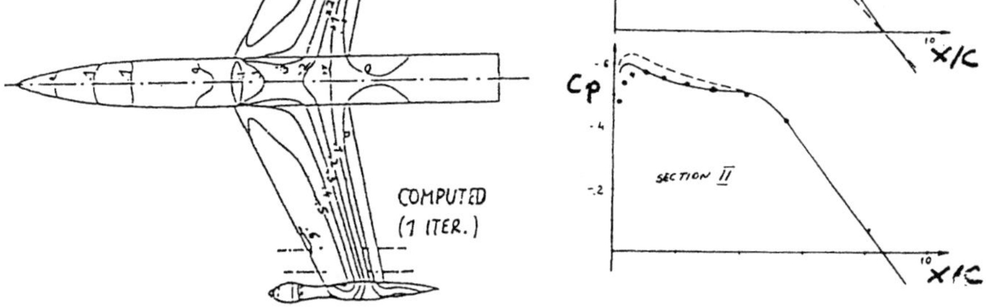

Fig. 9 Design of the wing-body-tank combination

CONCLUDING REMARKS

The method, ilustrated by the above examples is suitable to design complex configurations with prescribed pressure distribution. It allows to redefine the geometry of a single element as well as simultaneously redefine few elements /for example the wing and the body with changing wing-body intersection line/. The method is quite quickly convergent and cheap. The cost of a single design cycle /without flow analysis/ i.e. computation of transpiration and incremental potential distribution, optimization and geometry redefinition took in the present examples on the average about 70% of the flow analysis cost.

REFERENCES

1. Tinoco E. N., Johnson F. T., Freeman L. M., "Application of Higher Order Panel Method to Realistic Supersonic Configurations" - Journal of Aircraft Vol.17, No.1, 1980.
2. Maskew B., "Prediction of Subsonic Aerodynamic Characteristics: A Case for Low-Order Panel Methods" - Journal of Aircraft Vol.19, No.2, 1982.
3. Fornasier L., "Treatment of Supersonic Configurations by an Updated Low-Order Panel Method" - Journal of Aircraft Vol.21, No.4, 1984.
4. Miranda L. R., "Application of Computational Aerodynamics to Aeroplane Design" - Journal of Aircraft Vol.21, No.6, 1984.
5. Tseng K., Morino L., "Nonlinear Greens Function Method for Unsteady Transonic Flows" in "Transonic Aerodynamics" Edited by D. Nixon, AIAA Series, Progress in Aeronautics and Astronautics Vol.81, 1982.
6. Ravichandran K. S., Arora N. L., Singh R., "Transonic Full Potential Solutions by an Integral Equation Method" - AIAA Journal Vol.22, No.7, 1984.
7. Kandil O. A., Yates E. C., "Computation of Transonic Vortex Flows Past Delta Wings - Integral Equation Approach" - AIAA Pap. 85-1582.
8. Erickson L. L., Strande S. M., "A Theoretical Basis for Extending Surface-Paneling Methods to Transonic Flow" - AIAA Journal Vol.23, No.12, 1985.
9. Woodward F. A., "Analysis and Design of Wing-Body Combinations at Subsonic and Supersonic Speeds" - Journal of Aircraft Vol.5, No.6, 1968.
10. Johnson F. T., Ehlers F. E., Rubbert P. E., "A Higher Order Panel Method for General Analysis and Design Applications in Subsonic Flow" - Lecture Notes in Physics Vol.59, 1976.
11. Malone J. B., "A Subsonic Panel Method for Iterative Design of Complex Aircraft Configurations" -Journal of Aircraft Vol.19, No.10, 1982.
12. Hawk J. D., Bristow D. R., "Subsonic Surface Panel Method for Airframe analysis and Wing Design" - AIAA Pap. 83-341.
13. Bristow D. R., Grose G. G., "Modification of the Douglas Neumann Program to Improve the Efficiency of Predicting Component Interference and High Lift Characteristics" - NASA Contractor Report 3020.
14. Ward G. N., "Linearized Theory of Steady High-Speed Flow" - Cambridge University Press 1955.

Evaluation of a Higher-Order Panel Program for Subsonic Flow

Per Lötstedt
SAAB Aircraft Division, SAAB-SCANIA AB
S-581 88 Linköping, Sweden

Abstract

An efficient higher order panel program for three-dimensional subsonic flow problems based on an internal Dirichlet boundary condition has been developed for CRAY-1. The most time-comsuming part, the assembly of the system matrix, is vectorized. The system of linear equations is solved by a block SOR method. The panel program has four levels of approximation: from low order to high order. The obtained accuracy and the execution time for two different levels are compared. Conclusions are drawn concerning the efficiency of low and high order approximations. The execution speed of the program is compared with the speed of other similar panel programs of low and high order. Comparison is also made with results for a complete aircraft from wind tunnel experiments.

1. Introduction

A panel program PHOBOS (Panel program using Higher Order BOundary Singularities) has been developed for analysis of subsonic linearized inviscid flow about arbitrary three dimensional configurations. The potential of the flow is represented by the freestream potential and a distribution of sources and doublets on the surfaces of the configuration and on wake surfaces. The singularity distributions are determined by specifying the potential at the interior boundaries (see [9]). Panel methods based on this Dirichlet condition have gained popularity in recent years. In low order methods [6,8] the sources and doublets are constant on flat panels. The higher order methods [4] use linearly or quadratically varying singularities on curved panels. The panel method in [3] has flat panels and higher order approximations of the singularity distributions. The present program allows the user to choose the level of approximation for his particular application. It is also possible to compare different levels of approximation with respect to numerical accuracy and efficiency within the same program in numerical experiments. The conclusion from these experiments is that a higher order method is more cost effective. The primary purpose of the implementation of the method is to obtain an efficient program on CRAY-1 but a VAX version is also available. The structure of the code is such that the most time-consuming parts vectorize on CRAY-1.

2. Numerical formulation

The potential Φ of the flow satisfying Laplace's equation is a sum of three parts:

Φ_μ, the induced potential from the doublets,
Φ_σ, the induced potential from the sources,
Φ_∞, the freestream potential,

$$\Phi = \Phi_\infty + \Phi_\mu + \Phi_\sigma . \tag{2.1}$$

The potential in the interior Φ^i of a body with a closed surface is taken to be Φ_∞(cf. [4,8]). The source distribution is determined by Φ^i and the normal velocity at the surface which usually vanishes. Then the surface is divided into networks. A network is represented by a bicubic spline approximation as in [5]. Each network consists of a number of panels. On each panel there is an unknown doublet density which is computed by solving a system of linear equations.

The singularity distributions on a panel are approximated by the first terms in their Taylor expansions

$$\mu(\xi,\eta) = \mu_0+\mu_\xi \xi+\mu_\eta \eta +0.5\mu_{\eta\eta}\eta^2+\mu_{\xi\eta}\xi\eta+0.5\mu_{\xi\xi}\xi^2, \tag{2.2a}$$

$$\sigma(\xi,\eta) = \sigma_0+\sigma_\xi \xi+\sigma_\eta \eta +0.5\sigma_{\eta\eta}\eta^2+\sigma_{\xi\eta}\xi\eta+0.5\sigma_{\xi\xi}\xi^2, \tag{2.2b}$$

where ξ and η are local panel coordinates (cf. [5]). In the computations the derivatives in (2.2) are replaced by numerical difference formulas involving the values of the singularities associated with neighboring panels.

The integrals to be evaluated on a panel P_i are

$$\Phi_{\mu i}(x,y,z) = \frac{1}{4\pi}\int_{P_i} \mu(\xi,\eta)\hat{n}\cdot\nabla_\xi\left(\frac{1}{R}\right) dP_i ,$$
$$\Phi_{\sigma i}(x,y,z) = \frac{1}{4\pi}\int_{P_i} -\sigma(\xi,\eta)\frac{1}{R} dP_i , \tag{2.3}$$
$$R = \sqrt{(x-\xi)^2+(y-\eta)^2+(z-\zeta)^2} .$$

The potentials Φ_μ and Φ_σ in (2.1) are sums of the corresponding induced potentials (2.3) from the N panels

$$\Phi_\mu = \sum_{i=1}^{N} \Phi_{\mu i}, \quad \Phi_\sigma = \sum_{i=1}^{N} \Phi_{\sigma i} . \tag{2.4}$$

The integrals in (2.3) are split into two parts as in [5]. A flat projection P_i' of the curved panel P_i is determined. On P_i' the integrals can be computed analytically. Then a correction is added taking the curvature of the panel into account. The panel surface is assumed to vary quadratically about its center.

There is a far-field approximation of the integrals (2.3) based on a simple numerical quadrature rule. If the numerical error in the approximation is sufficiently small then the simple formula replaces the complicated analytical evaluation.

Two different Kutta conditions have been implemented. Let μ_u and μ_ℓ be the doublet distributions on the upper and lower side of the trailing edge of the wing and let μ_W be the doublet distribution on the wake strip behind the wing. Then the first Kutta condition is:

1. $\mu_W = \mu_u - \mu_\ell$.

Let V_u and V_ℓ be the speeds in the x direction on the upper and lower sides at the trailing edge of the wing. Then the second Kutta condition is:

2. $V_u = V_l$.

It is possible to distinguish four levels of approximation:

1. $\mu(\xi,\eta)=\mu_0$, $\sigma(\xi,\eta)=\sigma_0$ in (2.2), flat panels, Kutta condition 1, cf. QUADPAN[6],VSAERO[8],
2. $\mu(\xi,\eta)$ as in (2.2a), $\sigma(\xi,\eta)=\sigma_0+\sigma_\xi\xi+\sigma_\eta\eta$, flat panels, Kutta condition 2, cf. MCAERO[3],
3. $\mu(\xi,\eta)$ and $\sigma(\xi,\eta)$ as level 2, an intermediate level of curvature correction, Kutta condition 2, cf. PANAIR[4],
4. $\mu(\xi,\eta)$ and $\sigma(\xi,\eta)$ as in (2.2), full curvature compensation, Kutta condition 2.

At level 4 all second order terms in the integrands and the curvature corrections are included. The compressibility is taken into account by the Goethert rule [13]. The details of the evaluation of the integrals and the approximations are given in [7].

3. The panel program

There are three distinct parts of the PHOBOS program:

1. the discretization of the surfaces of the configuration with user specified data,

2. the assembly of the doublet and source influence matrices,

3. the solution of the system of linear equations for the doublet densities.

In the first part the computing time is proportional to the number of panels N. The amount of work needed in the first part is usually insignificant in comparison with the work in the remaining two parts.

The computing time in the second part is proportional to N^2, since N^2 elements of two matrices A and B are calculated. Often more than 90% of the total CPU-time for the CRAY implementation is spent in this part of the panel program. Hence, it is of vital importance to optimize the FORTRAN code such that the integral evaluations and the matrix assembly vectorize on CRAY-1. In the doublet matrix A, each row corresponds to a control point at the midpoint of a panel where the boundary condition is fulfilled and each column corresponds to a doublet density on a panel. The source matrix B is organized in the same manner. The data for a set of panels are collected in a number of vectors. Then the influence coefficients in A for one control point are computed using the panel data. The calculations proceed until the coefficients for all control points have been determined for all sets of panels. B is assembled in the same way. The vectors are chosen as long as possible in order to minimize the computing time on CRAY.

The system of linear equations derived from (2.1) is

$$A\underline{\mu} = -B\underline{\sigma}, \qquad (3.1)$$

where $\underline{\mu}$ is the vector of unknown doublet distributions on the panels and the vector $\underline{\sigma}$ contains the known source distributions. Three different ways of solving (3.1) are compared in [1]:

- Gaussian elimination, a direct method,

- block Successive-Over-Relaxation (block SOR), an iterative method,
- a multigrid method for integral equations, an iterative method.

The work is proportional to N^3 in the first method. Thus it is impractical for large problems. The second method is a variant of the classical SOR method applied to a blocked matrix A. The geometry of the configuration determines the blocking of A. Numerical experiments have verified that the computing time is proportional to N^2 for this method. A multigrid method for solving the linear equations was suggested in [11,12]. The method in [11] was implemented in [1] with the block SOR method in the smoothing steps.

The panel method is implemented on a CRAY-1A with a primary memory of 1 Mword. Since the whole matrix must be resident in the primary storage in a straightforward implementaion of the first method, the number of panels N is restricted to about 900. The present versions of the iterative methods can solve systems with 5500 unknowns. This limitation is caused by the lengths of the vectors in the FORTRAN program. These methods need only store a number of rows of A in the primary memory. There have been no problems with divergence in the iterative methods. Typical results from the comparison in [1] are displayed in table 1. They show that the preferred method is block SOR, since it is simpler than the multigrid method and yet faster.

Table 1. Comparison of execution times in s on CRAY-1A for three different equation solvers (* indicates an estimated time) [1].

unknown variables	block SOR with optimal parameter	best multigrid strategy	Gaussian elimination
492	0.283	0.449	1.322
1148	1.255	2.398	16.850*

Table 2 contains the CPU-time in seconds for a computation of the flow about a straight wing with a NACA0012 profile on two different machines in single precision and four levels of approximation described sect. 2. The conclusion is that the major part of the computing time is required by the integral evaluation and the computation of A and B. A substantial reduction in CPU-time is achieved with far-field approximations for this wing. Note that the scalar machine VAX11 devotes a greater share of the total CPU-time to the solution of the system of equations than CRAY-1 does.

In a comparison with other panel programs the computing time for a flow simulation with a certain number of panels or variables is one measure of the efficiency of the programs. The level of approximation of the higher

Table 2. The execution times in s for a wing and wake with 365 variables.The level of approximation is altered and the integral evaluation is either only analytical or analytical with far-field approximations.

machine	approx. level	discretization	integral evaluation		matrix assembly	equation solver
			analyt.	analyt.+far-field		
CRAY-1A	1	1.31	5.83	1.63	4.99	0.48
	2		6.85	2.30	6.93	
	3		8.14	2.83	6.92	
	4		10.06	3.76	7.75	
VAX11/782	1	30.75	698.27	-	155.78	295.13

order panel program PANAIR[4] is comparable to level 3 in PHOBOS. The version reported on in [14] uses 1305 s on CRAY 1-S for a model with 865 panels. A similar example is a wing-body combination from [7] where 813 panels were used. The total execution time in this case was about 90 s with PHOBOS. A comparison between approximation level 1 and QUADPAN[6], a low order panel method, is made in fig. 1. Fig. 1 and other experiments verify that the CPU-time is proportional to N^2.

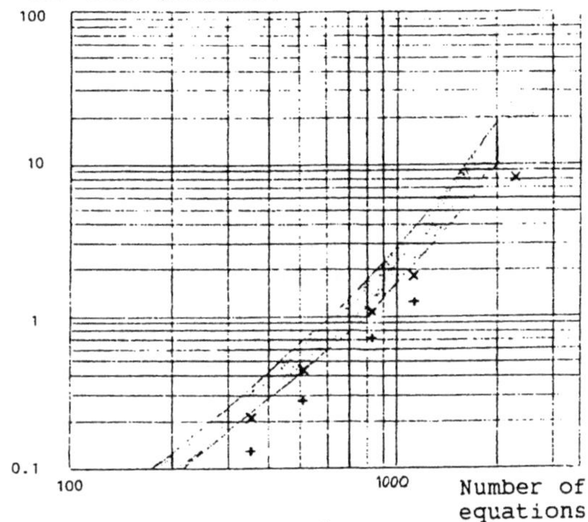

Fig. 1. The run times of PHOBOS at level 1 are compared with the run times of QUADPAN as reported in [6]. The shaded area is the estimate for QUADPAN on CRAY-1S. The CPU--times for PHOBOS using analytical evaluation (x) and mixed analytical-numerical evaluation (+) of the integrals are plotted in the same graph for comparison.

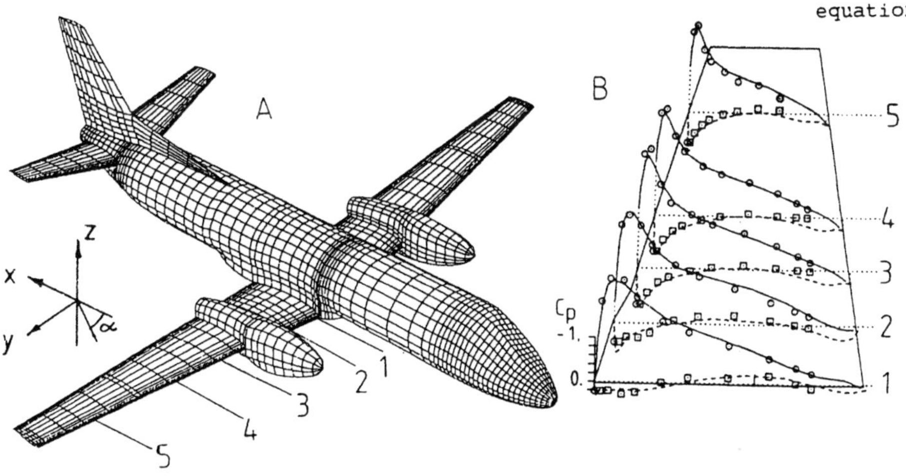

Fig. 2. A. The paneled geometry of SAAB SF 340. B. The computed pressure coefficent c_p on the wing (—upper side,---lower side) is compared with the c_p measured in wind tunnel (⊙ upper, ⊡ lower) at α=6° and M=0.15.

4. Comparison with wind tunnel results

The computational results from the panel program are compared with wind tunnel measurements in this section. The example chosen is a complete aircraft SAAB SF 340. The paneled geometry is depicted in fig. 2A. The number of variables in the system of linear equations for this model was 2332. The total CPU time on CRAY-1A with method 1 was about 490 s. On approximation level 3 the CPU time increased by about 30% as in table 2. In fig. 2B the computed pressure coefficient c_p obtained by method 1 is compared with the wind tunnel experiments at the wing stations marked in fig. 2A. The angle of attack α is 6° and the Mach number is 0.15. Comparisons with results from other panel programs and other wind tunnel experiments are made in [7].

The flow about SAAB SF 340 has been computed in an unsymmetric case with an α and a side slip angle β. The number of variables in the system of equations was 4912 and the total CPU time with method 1 was about 1400 s. Using a model consisting of 1100 panels the pressure distribution on the automobile SAAB 9000 has been determined. Engine inlets have been simulated successfully by allowing the flow to penetrate some of the panels.

5. Comparison between low and high order methods

Let us split the error in the computational results into two parts: the modeling error and the numerical error. The modeling error is due to the fact that the equation we solve is the linearized potential flow equation without viscous effects instead of the compressible Navier-Stokes equations. The numerical error is caused by the discretization of the potential flow equation. The first part can be assessed by comparisons between accurate numerical calculations with a panel method and reliable wind tunnel experiments. The second part has been studied theoretically for external problems with spline approximations of the singularities and smooth boundaries, see e.g. [2], and by comparing different panel programs in numerical experiments, see e.g. [16,17]. In [10] Oskam computed the numerical error in the solution from a high order method applied to a symmetric Karman-Trefftz airfoil, for which an exact solution is available. The same approach was taken in [15], where different levels of approximation of the singularity distributions and the Kutta condition are compared.

To be able to draw conclusions from the results in [15] for a three dimensional wing we make the following assumption: The major contribution to the error in Φ is from the flow in the x and z directions. This assumption holds true for wings with a small sweep angle and a large aspect ratio, where the character of the flow is almost twodimensional on most parts of the wing.

The wing profile in our case is an unsymmetric Karman-Trefftz profile, fig. 3A, with 15% thickness. The relative error in Φ for an airfoil is defined by

$$e = \sqrt{\sum_{i=1}^{n} (\Phi(\xi_i)-\Phi_i)^2 h_i / \sum_{i=1}^{n} \Phi(\xi_i)^2 h_i} . \qquad (5.1)$$

The number of panels on the airfoil is n, the length of the i:th panel is h_i, the exact solution at ξ_i on the airfoil is $\Phi(\xi_i)$ and the computed solution is Φ_i in (5.1).

Let j be the index running in the y direction and ξ_{ij} be a point on the wing surface. The panel width in the y direction is H_j and the length along the profile is $c_j h_i$. Moreover, suppose that the flow is two-dimensional over the wing. Since there is no flow in the spanwise direction the potential is constant on rays starting at points on the root profile and ending at points with the same distribution on the tip profile. Then the relative error E in the three-dimensional case corresponding to (5.1) is

$$E^2 = \sum_{j=1}^{m}\sum_{i=1}^{n} (\Phi(\xi_{ij})-\Phi_{ij})^2 c_j h_i H_j / \sum_{j=1}^{m}\sum_{i=1}^{n} \Phi(\xi_{ij})^2 c_j h_i H_j =$$

$$\sum_{j=1}^{m} c_j H_j \sum_{i=1}^{n} (\Phi(\xi_{ij})-\Phi_{ij})^2 h_i / \sum_{j=1}^{m} c_j H_j \sum_{j=1}^{n} \Phi(\xi_{ij})^2 h_i =$$

$$\sum_{j=1}^{m} c_j H_j e^2 \sum_{i=1}^{n} \Phi(\xi_i)^2 h_i \Big/ \sum_{j=1}^{m} c_j H_j \sum_{i=1}^{n} \Phi(\xi_i)^2 h_i = e^2. \tag{5.2}$$

We infer from (5.1), (5.2) and the assumption that the dominant part of the relative error E in Φ for a wing is e in (5.1).

The logarithm of the numerical error is displayed as a function of n in figs. 3B,C. The slope p of an asymptotic dependence

$$e = \mathcal{O}(n^p) \tag{5.3}$$

is added in the figures.

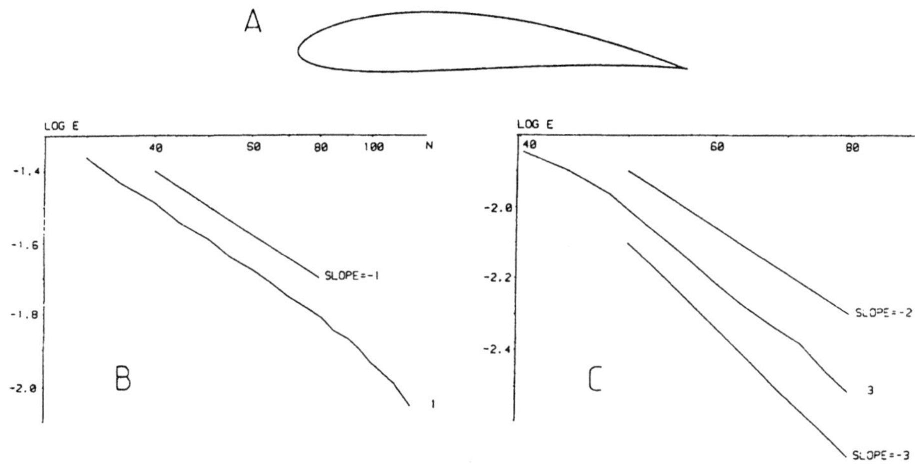

Fig. 3. A. The 15% thick Karman-Trefftz profile. B. The logarithm of the numerical error using the low order method (level 1) as a function of the number of panels n is depicted for the two-dimensional airfoil in fig. 3A. C. As above for level 3.

The potential Φ is the basic dependent variable. The velocity v is obtained by the same second order differentiation of Φ in the panel program at all levels of approximation. If the numerical error in Φ is of $\mathcal{O}(n^k)$, $k<0$, and a smoothly varying function in $\mathbb{R}^3$ then the error in v will be of $\mathcal{O}(n^l)$, $l=\max(k,-2)$. The numerical error depends not only on the number of panels but also on the wing geometry and α.

In fig. 3B from [15] the error e for the low order method 1 is plotted when $\alpha=5°$. The slope p in (5.3) is -1 as expected. The error in the higher order method 3 is shown in fig. 3C. The asymptotic branch has a slope between -2 and -3. Let n be the number of panels around the profile and m the number of panels in the spanwise direction. Then it follows from fig. 3B that the approximate behavior of the error e_1 in the first order method is an error constant over n

$$e_1 = 1.04/n. \tag{5.4}$$

From fig. 3C we obtain an approximate relation between the error e_3 of method 3 and n

$$e_3 = 22.4/n^2. \tag{5.5}$$

Let the total number of panels be

$$N = nm. \tag{5.6}$$

It follows from table 2 that the computational work W measured in CPU seconds on CRAY-1 for the two methods is

$$W_1 = 9.5\cdot 10^{-5} N^2 \text{ s}, \tag{5.7}$$

$$W_3 = 1.26\cdot 10^{-4} N^2 \text{ s}. \tag{5.8}$$

For a moderately swept wing a reasonable relation between n and m is

$$m = n/8. \tag{5.9}$$

A refinement of the paneling on a wing in the x direction will result in a refinement also in the y direction so that the error because of the spanwise variation remains small in comparison with the computed two dimensional error. Combine the equations (5.4–5.9) to obtain

$$W_1 = 1.8\cdot 10^{-6}/e_1^4, \tag{5.10}$$

$$W_3 = 9.9\cdot 10^{-4}/e_3^2. \tag{5.11}$$

For a small number of panels, n=40, we have N=200, e_1=0.026 and W_1=3.9 s with the low order method according to (5.9), (5.4) and (5.7). In order to achieve the same accuracy with the high order method, e_3=0.026, we conclude from (5.11) that the work is 1.5 s and from (5.9) and (5.6) that the number of panels is about 110. The smaller the required numerical error is, the greater the advantage is to use the higher order method. In an accurate computation spending 100s of CPU-time on CRAY-1 the estimates of the numerical errors for the low and high order methods are

$$e_1 = 1.2\cdot 10^{-2}, \quad e_3 = 3.1\cdot 10^{-3},$$

with the number of panels 1030 and 890, respectively. The breakeven point between the methods is at

$$e = 0.043, \quad W = 0.54 \text{ s},$$

which corresponds to 24 panels on the profile with method 1. This number is generally considered too small to obtain engineering accuracy and the computed errors in figs. 3B and 3C were not located on the asymptotic branches for n=24.

The extra cost in CPU-time of adding curvature corrections to flat panels in method 2 to obtain method 3 is about 8% in table 2. The slope p was found in [15] to be equal for two methods, slightly different from the methods 2 and 3, with and without compensation for curved panels, but the error constant was reduced by 1/3 with curvature corrections. The total work for the same numerical error is smaller when method 3 is employed instead of method 2, if the results from [15] carry over to our panel program. Furthermore, the method with curved panels in [15] was less sensitive to the panel distribution. Two methods similar to method 3 and 4 were also compared in [15]. The difference in the error was hardly observable.

The conclusion of the above investigation is that even for small requirements on the accuracy in the computations the use of higher order methods seems to pay in terms of computing costs. The smaller the acceptable error is in Φ, the more CPU time is saved by a higher order method. Of course, it is useless to reduce the numerical error beyond a certain limit, where the modeling error due to neglected compressibility and viscosity dominates the total error. However, higher order methods are superior also for a moderate and often used number of panels on a wing.

Acknowledgement

The author has benefitted from many discussions on the panel method with B. Arlinger. Basic subroutines for spline approximation and numerical differentiation have been taken from a panel program developed by L. E. Eriksson [5]. The results in fig. 2 were obtained in cooperation with O. Sehlén. J. Sowa computed the error curve in fig. 3C.

References

1. A. Ålund, Iterative methods to compute the singularity distribution in three dimensional panel methods, Report L-0-1 R97, SAAB-SCANIA, Linköping, Sweden, 1984.
2. D. N. Arnold, W. L. Wendland, The convergence of spline collocation for strongly elliptic equations on curves, Numer. Math., 47 (1985), 317-341.
3. D. R. Bristow, Development of panel methods for subsonic analysis and design, NASA Contractor Report 3234, 1980.
4. R. L. Carmichael, L. L. Erickson, PANAIR - A higher order panel method for predicting subsonic or supersonic linear potential flows about arbitrary configurations, AIAA paper 81-1255, 1981.
5. L. E. Eriksson, Development of a computer code for a three-dimensional higher order panel method for subsonic potential flow, FFA Report 138, The Aeronautical Research Institute of Sweden, Stockholm, Sweden, 1983.
6. C. E. Johnston, H. H. Youngren, J. S. Sikora, Applications of QUADPAN: An advanced low-order panel method, AIAA Professional Study Seminar, Aerodynamic analysis using panel methods, Colorado Springs, CO, 1985.
7. P. Lötstedt, A three-dimensional higher order panel method for subsonic flow problems - Description and applications, Report L-0-1 R100, SAAB-SCANIA, Linköping, Sweden, 1984.
8. B. Maskew, Prediction of subsonic aerodynamic characteristics: A case for low-order panel methods, J. Aircraft, 19 (1982), 157-163.
9. L. Morino, C.-C. Kuo, Subsonic potential aerodynamics for complex configurations: A general theory, AIAA J., 12 (1974), 191-197.
10. B. Oskam, Asymptotic convergence of higher-order accurate panel methods, J. Aircraft, 23 (1986), 126-130.
11. B. Oskam, J. M. J. Fray, General relaxation schemes in multigrid algorithms for higher order singularity methods, J. Comput. Phys., 48 (1982), 423-440.
12. H. Schippers, Applications of multigrid methods for integral equations to two problems from fluid dynamics, J. Comput. Phys., 48 (1982), 441-461.
13. A. H. Shapiro, The Dynamics and Thermodynamics of Compressible Fluid Flow, Vol 1, Ronald Press Co, New York, 1953.
14. L. D. Snyder, L. L. Erickson, PANAIR prediction of NASA Ames 12-foot pressure wind tunnel interference on a fighter configuration, AIAA paper 84-0219, 1984.
15. J. Sowa, A comparative study of the numerical accuracy of different panel methods, Report L-0-1 R109, SAAB-SCANIA, Linköping, 1986.
16. W. Z. Strang, C. H. Berdahl, E. L. Nutley, A. J. Murn, Evaluation of four panel aerodynamic prediction methods (MCAERO,PANAIR,QUADPAN and VSAERO), AIAA paper 85-4092, 1985.
17. H. S. Sytsma, B. L. Hewitt, P. E. Rubbert, A comparison of panel methods for subsonic flow computation, AGARDograph No 241, AGARD, Neuilly--sur-Seine, France, 1979.

CALCULATION OF FLOW CHARACTERISTICS FOR GENERAL CONFIGURATIONS USING PANEL METHODS

B. Maskew
Analytical Methods, Inc.
2133 - 152nd Avenue N.E.
Redmond, Washington, 98052, U.S.A.

SUMMARY

Extension of the panel method formulation are reviewed for modeling the effects of viscous, separated and vortex flows. Such extensions are necessary for the practical treatment of the flow about general configurations. Further developments based on an unsteady approach and also a local zonal treatment of nonlinear effects are briefly discussed. These developments should make the panel method even more attractive in the future.

INTRODUCTION

The capability to predict flow characteristics about general configurations is important to the efficient design of various transport vehicles. Such predictions are crucial to the design of aircraft but they also benefit automobile, ship and train design. Knowledge of the flow characteristics in the early design stages impacts the final performance and behavior characteristics of the vehicle and helps to avoid costly changes after manufacture.

A common feature of transport vehicles is the complex geometry that must be treated together with the effects of viscosity, flow separation and vortex flows. Although the capability to model such flows is advancing at a significant pace, due to developments in codes treating the Navier-Stokes equations and to the ever increasing power of computers, a potential flow panel method is still the only practical approach for dealing with the flow about complicated configurations. In such cases the panel method's surface integral formulation offers distinct advantages over a finite difference or finite element based method which requires a computational mesh to be constructed throughout the flow field. These advantages appear in terms of ease of use, versatility and economy in computing effort, all of which have helped to establish the panel method as an important, practical engineering tool for predicting flow characteristics about general configurations.

In recent years extensions of the panel method formulation to allow simple modeling of the nonlinear effects of viscosity, separated flow and vortex wakes has increased the attractiveness of panel methods even further. The present paper reviews the basic formulation for these extensions and uses example calculations from the VSAERO panel code to demonstrate their general applicability to a range of problems.

BASIC FORMULATION

The vehicle is assumed to be immersed in a uniform onset flow, $\mathbf{V}_\infty$. Regions of the flow field that are dominated by viscous and rotational effects are assumed to be confined to thin surface boundary layers and thin free shear layers or vortex wakes, Figure 1. The rest of the flow field is regarded as inviscid and irrotational so that the fluid velocity, $\mathbf{V}$, can be written

$$\mathbf{V} = \mathbf{V}_\infty - \nabla\phi \, , \tag{1}$$

where ϕ is the perturbation velocity potential due to the presence of the vehicle. Initially we assume the flow is incompressible so that ϕ satisfies Laplace's equation, $\nabla^2\phi = 0$.

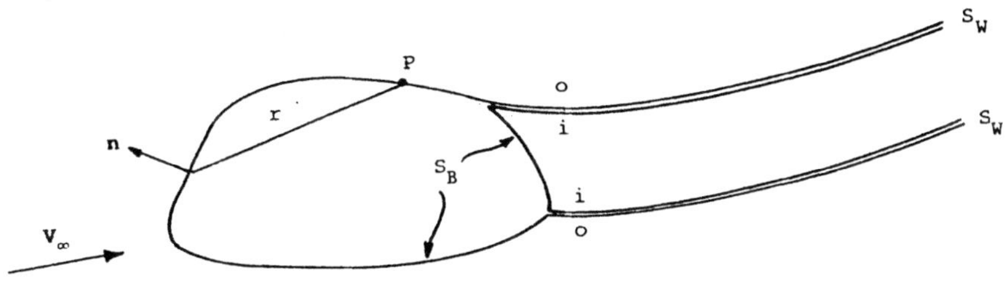

Fig. 1 Section through the Body and its Wake

Figure 1 shows a section through a body and its wake. The simplified wake model encloses the separated region with in free vortex sheets, S_W. Applying Green's theorem to the quantities, ϕ and $1/r$, for a point, P, in the flow domain but on the surface, S_B, gives [1],

$$0 = \iint_{S_{B-P}} \phi \, n \cdot \nabla(\tfrac{1}{r}) \, dS_B - 2\pi\phi_P - \iint_{S_{B-P}} \tfrac{1}{r} \, n \cdot \nabla\phi dS_B + \iint_{S_W} (\phi_o - \phi_i) \, n \cdot \nabla(\tfrac{1}{r}) \, dS_W \, . \tag{2}$$

The normal vector, $\mathbf{n}$, is directed into the flow domain: r is the distance from P to a surface element, dS. The outer and inner elements of each wake surface, S_W, with potentials, ϕ_o, ϕ_i, respectively, have been combined.

The first two terms in Eq. (1) represent the contribution from a surface doublet distribution of strength, $\mu = \phi$. The second integral expression represents the contribution from a surface source distribution of strength, $\sigma = \mathbf{n} \cdot \nabla\phi$, which can be evaluated at the outset. Thus,

$$\sigma = \mathbf{n} \cdot \mathbf{V}_\infty - \mathbf{n} \cdot \Omega \times \mathbf{R} - V_N - \partial/\partial s(U_e \delta^*), \quad (3)$$

where the second term represents the normal component of velocity due to a state of rotation, Ω, of the body-fixed frame. V_N is a specified normal velocity of the fluid relative to the body surface; V_N is zero for a solid boundary, but is nonzero when treating inlet/exhaust problems. The last term in Eq. (3) represents the displacement effect due to the rate of boundary layer growth along computed streamlines and is provided by an integral boundary layer calculation as described later.

The third integral in Eq. (2) is the contribution from a doublet distribution, $\mu_W = \phi_o - \phi_i$, on the wake surfaces. The value of μ_W at the upstream edge of S_W is the jump in potential on S_B as the separation line is crossed. This satisfies the Kutta condition of zero load there. μ_W, therefore, is a function of the (unknown) surface doublet distribution. The case of separated (closed) wake surfaces will be discussed later; in the meantime, on regular (open) wakes the value of μ_W remains constant along mean streamlines in the wake [2]. A source distribution can be included on S_W to model the effects of entrainment.

NUMERICAL PROCEDURE

Figure 2 shows the numerical solution procedure installed in the VSAERO code [2]. This is briefly outlined here in order to relate to the nonlinear treatment.

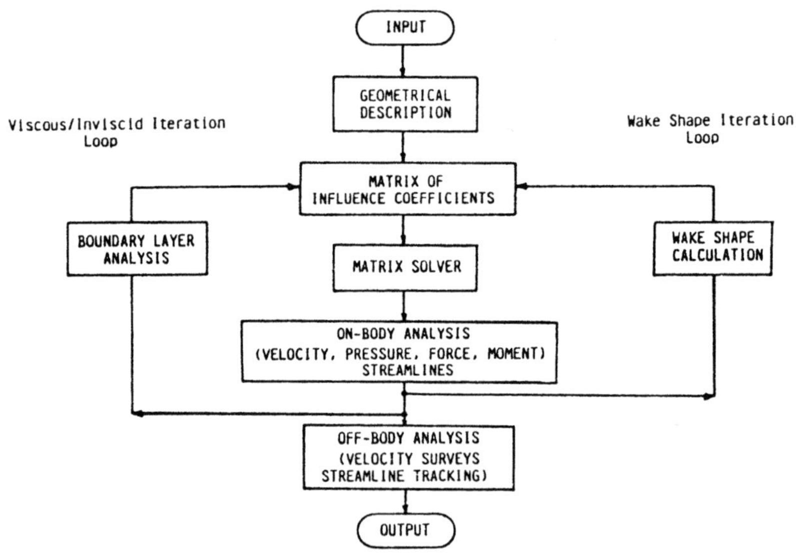

Fig. 2 Flow Diagram for the VSAERO Code

First, the surface of the vehicle is represented by a large number (N) of flat quadrilateral panels on each of which the doublet and source singularities are uniformly distributed. This allows the integrals in Eq. (2) to be evaluated piecewise over each quadrilateral for the point, P, placed at a control point at the center of each panel in turn. Thus, an N x N matrix of influence coefficients is constructed for the unknown doublet values.

Following a solution of the above equations, the on-body analysis routine evaluates the velocity and, hence, pressures on each surface panel and calculates the force and moment acting on the vehicle by summing all the surface panel contributions. Figure 2 shows the arrangement for the iteration loops for the wake shape relaxation and boundary layer routine calculations which are described below.

<u>Wake Shape Iteration</u>

With the panel singularity values known, the velocity vector can be computed at any point in the flow field using the panel velocity influence coefficient and summing over all surface and wake panels [2]. In the wake shape calculation procedure the velocity is computed at each wake panel corner point and each "streamwise" edge is aligned with the average of its two end velocities. A predictor/corrector scheme is applied at each step.

The wake shape calculation is performed in a marching procedure, starting at an upstream station and progressing through a set of vertical wake grid planes, Figure 3.

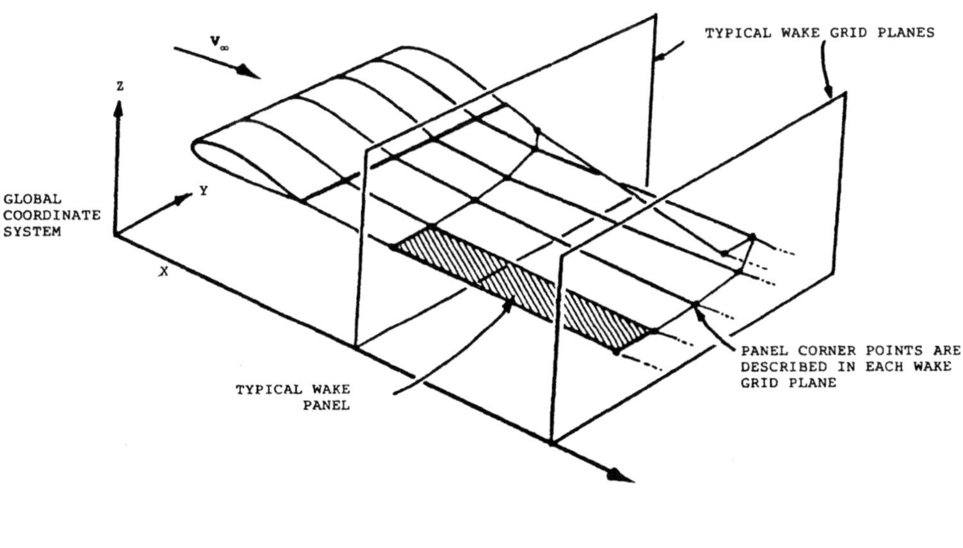

Fig. 3 Wake-Grid-Plane Scheme

For a given set of points, $\mathbf{RW}_{i,M}$; i=1, ..., NL_M, in the Mth wake grid plane, the set of points in the next plane is

$$\mathbf{RW}^{j}_{i,M+1} = \mathbf{RW}^{J}_{i,M} + \tilde{\mathbf{V}}^{j}_{i} * DX/|\tilde{V}^{j}_{xi}|;\ i=1,\ ...,\ NL_M, \qquad (4)$$

where NL_M is the number of wake lines intersecting the Mth plane; DX is the distance between the wake grid planes; J is the number of predictor/corrector cycles in use; $\tilde{\mathbf{V}}^{j}_{i}$ is the mean velocity across the interval for the ith line in the jth predictor/corrector cycle; i.e.,

$$\tilde{\mathbf{V}}^{j}_{i} = (\mathbf{V}^{J}_{i,M} + \mathbf{V}^{j}_{i,M+1})/2.$$

In each predictor/corrector cycle, the complete set of points in each wake grid plane is moved and the current movement vectors are applied to all corresponding points downstream before the next set of velocities is computed. When all the wake grid planes have been processed, the wake panel parameters are recomputed using the modified corner points. The influence coefficient matrix is then updated for the next solution, Figure 2.

This simple model has been very effective for computing various vortex/surface interactions; e.g., [3]. An example of a canard vortex/wing interaction from [4] is shown in Figure 4. This is a model of the Rockwell propulsive lift concept. The wake shape after one iteration cycle at $\alpha = 4°$ is shown in Figure 4(a) and reflects the mutual interaction between the vortex path and the surface solution. The VSAERO calculated chordwise pressure distributions to either side of the vortex passage are in close agreement with experimental data, Figure 4(b), and show the upwash/downwash effects on the outboard/inboard sides, respectively.

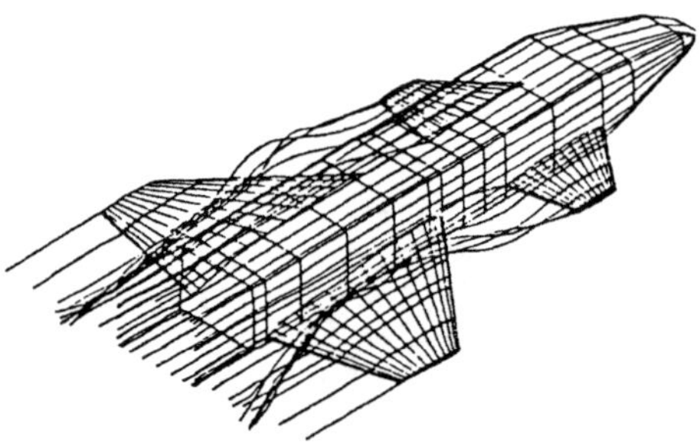

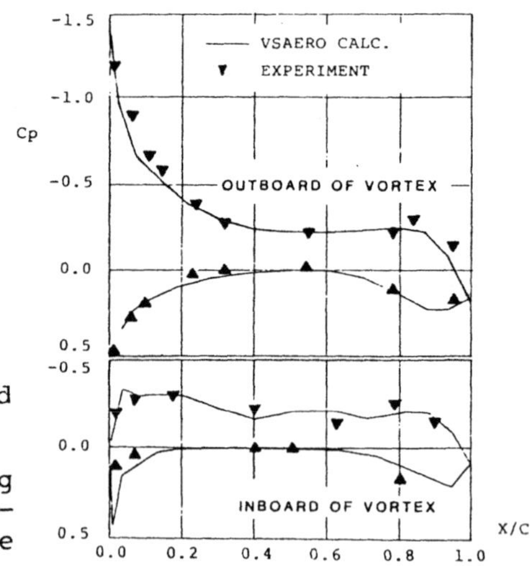

(a) VSAERO Paneling with Relaxed Wake

(b) Calculated and Measured Wing Chordwise Pressure Distributions near the Vortex Passage

Fig. 4 Rockwell Propulsive Wing Concept

Viscous Effects

Once the wake-shape iteration cycles are completed, the viscous/inviscid iteration loop is activated, Figure 2. First a family of (external) surface streamlines is computed, based on a second-order surface stream function formulation [5]. Integral boundary layer calculations [6] are then performed along each streamline using the computed (three-dimensional) velocity distributions. Pseudo axisymmetric flow conditions are assumed, based on local streamline curvature and convergence/divergence. The computed boundary layer displacement source values are redistributed over the surface panels in order to re-evaluate Eq. (3) for the next pass.

The streamline boundary layer calculation procedure is very convenient for treating complex configurations but it can be expected to break down in regions of large cross flow. Even so, the procedure has given surprisingly good results over a range of complicated problems; e.g., [7]. Such experience has justified keeping this approach rather than substituting a three-dimensional boundary layer analysis.

Separated Flow

When the boundary layer analysis predicts an extensive region of separation, a closed vortex sheet wake model is used, Figure 1. In this case, μ_W in Eq. (2) varies along the mean streamlines on S_W, with the gradient equal to the local transverse vorticity value. This model has had considerable success in two-dimensional problems [8] and some success in certain three-dimensional applications [9].

The condition of zero load on the free vortex sheet requires a jump in total pressure, ΔH, to compensate for the jump in tangential velocity across the sheet. Using $\mathbf{V}_i$, $\mathbf{V}_o$, for the inner and outer velocities, respectively, ΔH can be evaluated at the separation location. Thus,

$$\Delta H/q_\infty = -2|\mathbf{VM}_{SEP} \times \gamma_{SEP}|/V_\infty^2, \qquad (5)$$

where $\mathbf{VM}_{SEP} = (\mathbf{V}_o + \mathbf{V}_i)/2$ is the mean convection velocity at separation and $\gamma_{SEP} = \mathbf{n} \times (\mathbf{V}_o - \mathbf{V}_i)$ is the vorticity vector at separation, with $\mathbf{n}$ being the outward normal as before. q_∞ is the free stream dynamic pressure.

The jump in total pressure will be assumed constant along a mean streamline on S_W, taking its value from the conditions at the separation line. Thus, at distance, s, along the streamline, the transverse component of vorticity is:

$$\gamma(s) = \gamma_{SEP}\, f(s),$$

where $f(s) = |\mathbf{VM}_{SEP}|/|\mathbf{VM}(s)|$ is a weighting function applied to the current (unknown) γ_{SEP} value and $|\mathbf{VM}(s)|$ is the computed convection speed evaluated at the set of points down each wake line during the previous wake shape iteration.

An example calculation from [10] is shown for a Mazda RX7 automobile. Mazda kindly supplied the geometric sections defining the body surface and also experimental centerline pressure data. The VSAERO calculation used 1100 panels on one side of the vertical plane of symmetry. The full body and ground plane effects are represented by imaging. Figure 5(a) shows the panel model and the final wake shape after three viscous/potential flow iterations and three wake shape iterations. The close agreement between calculated and experimental pressures on the body centerline is shown in Figure 5(b). The separated wake model modifies the basic potential flow (wakeless) solution near the base of the body and provides good agreement on the base pressure. The case required about 10 minutes of CPU time on a CRAY XMP computer and included off-body calculations for 6 streamlines and 3200 velocity survey points.

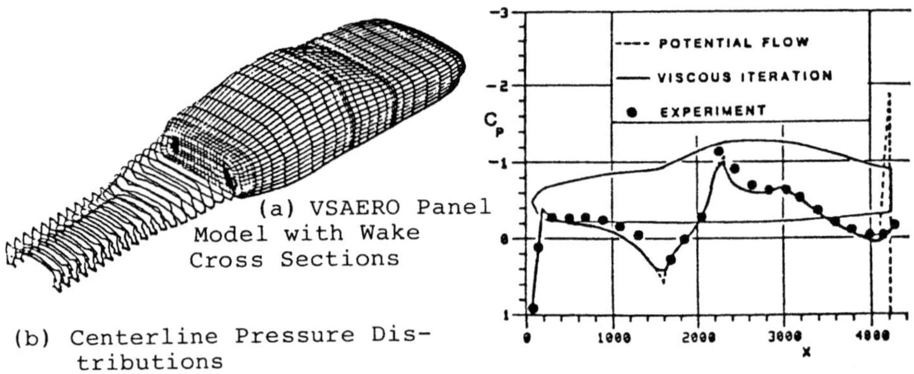

(a) VSAERO Panel Model with Wake Cross Sections

(b) Centerline Pressure Distributions

Fig. 5 Mazda RX7 Automobile

TIME-STEPPING APPROACH

Although the above vortex sheet model has been successful in predicting separated flow characteristics for a number of configurations, there have been difficulties in some applications. A more promising approach to the treatment of extensive separations appears to be offered by a time-stepping method [11] in which the wake configuration is grown by convection in time rather than being relaxed in space. The configuration is started impulsively from rest and moved forward over a number of small time steps. At each step a set of new wake panels is created along the separation line and all the previous wake panel corners are convected downstream along the computed velocity vectors using a simple Euler scheme [11]. The unsteady Kutta condition,

$$\partial\mu/\partial t + V_M(\partial\mu/\partial s) = 0$$

provides the conditions for the strength of each new wake panel as it is being formed. The pressure coefficient in the unsteady case is now

$$Cp = (VS^2 - V^2 + 2\partial\phi/\partial t)/V_\infty^2, \qquad (6)$$

where **VS** is the surface absolute velocity; **V** is the fluid velocity relative to the body-fixed frame. The $\partial\phi/\partial t$ term in Eq. (6) is directly related to the jump in total pressure, Eq. (5), required for the base separated region. In the present case, however, it provides the base pressure directly, details of variations in the wake vorticity (or doublet gradient) being accounted for in the computed convection of the wake panel corners.

The dynamic effects of changing conditions are included in the calculations. For example, Figure 6 from [11] shows the very close agreement between calculated and experimental data for lift and drag development as an airfoil pitches from 0 to 55 degrees angle of attack. In this case, coupled unsteady boundary layer calculations [11] provide a separation location which varies with time, and the effect of the ultimate leading-edge separation and ensuing vortex convection over the upper surface are captured in the detailed time histories from the calculations.

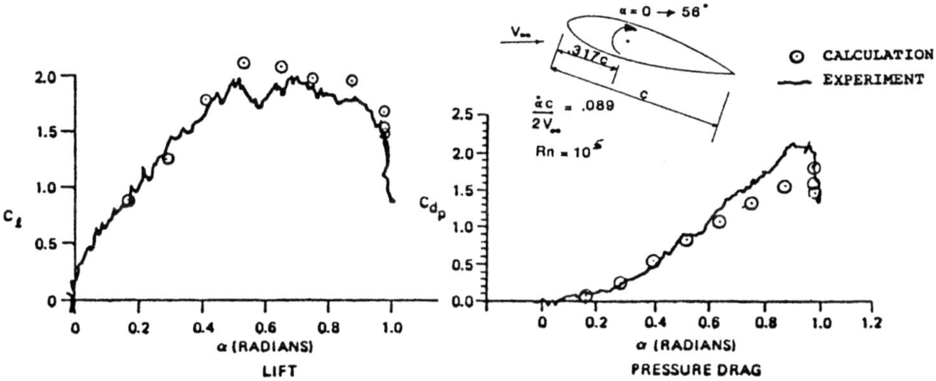

Fig. 6 Calculated and Measured Lift and Pressure Drag on a NACA 0012 Airfoil during Pitch-up

ZONAL MODELING

When a nonlinear region extends over a significant volume, e.g., jet plumes, vortex cores, etc., then a local zonal treatment is possible. The appropriate flow equations are applied in the nonlinear region using boundary conditions provided by the panel code. One area of particular interest to aircraft calculations is the treatment of local supersonic regions in a transoncic flow problem. The continuity equation in the "compressible region" becomes

$$\nabla^2\phi = 1/\rho \; \mathbf{V} \cdot \nabla\rho, \qquad (8)$$

where ρ is the density.

The nonzero value for $\nabla^2\phi$ leaves a volume integral in Green's Theorem for the compressible region, ε. Thus, Eq. (2) has an additional source term on the right-hand side; viz.,

$$\iiint_{\varepsilon} \frac{1}{r} \cdot \frac{1}{\rho} \cdot V \cdot \nabla\rho d\varepsilon \quad .$$

Piers and Slooff [12] treated the volume source integral using a field panel model. Since that time several other solutions have appeared, but by and large have not shown a computational advantage over finite difference based methods. A simplified treatment of the source term, coupled with an iterative scheme, has been investigated recently. This treatment appears to give a compressible solution without much extra effort over that for the basic incompressible panel method. A preliminary calculation is shown in Figure 7 for a NACA 0012 airfoil at $\alpha = 3.86$, $M_\infty = .75$. The first pass shows the Karman-Tsien correction applied to the incompressible solution. This is immediately converted by the first approximation to the nonlinear source term. A subsequent pass through the viscous/potential iteration gives a very encouraging agreement with measured data. This treatment of the source term is being refined and needs to be evaluated over a range of conditions.

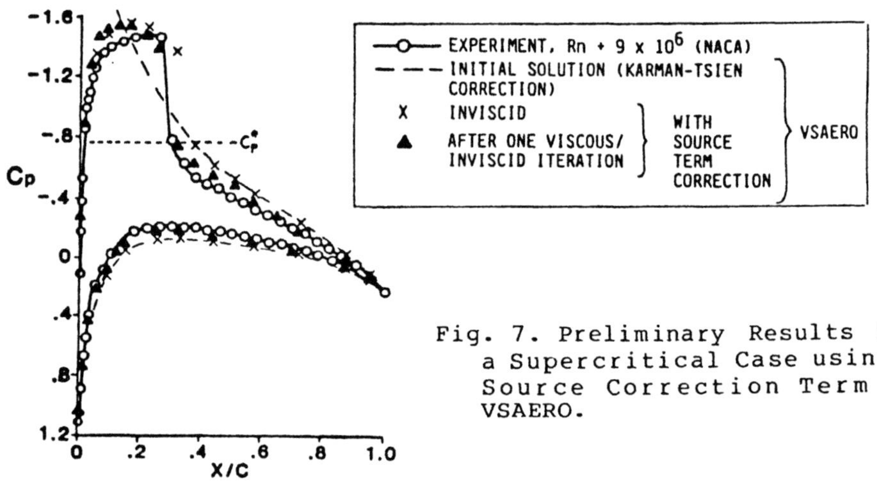

Fig. 7. Preliminary Results for a Supercritical Case using a Source Correction Term in VSAERO.

CONCLUSIONS

Panel methods are important tools for predicting vehicle aerodynamics. They are versatile, easy to use and are economical in computing effort. Coupled boundary layer calculations, separated flow modeling and free vortex sheet relaxation allow many "real flow" effects to be included. Current developments in the areas of unsteady flows and transonic flows promise to make panel methods even more attractive in the future.

REFERENCES

[1] Lamb, H.Y., _Hydrodynamics_, 6th Edition, Dover Publications, New York, N.Y., Article No. 58, 1945.

[2] Maskew, B., "Program VSAERO, A Computer Program for Calculating the Nonlinear Aerodynamic Characteristics of Arbitrary Configurations; Theory Document", _NASA CR-4023_, November 1985.

[3] Maskew, B., "Predicting the Aerodynamic Characteristics of Vortical Flows on Three-Dimensional Configurations Using a Surface Singularity Panel Method", Paper No. 13 in _AGARD CP-342_, April 1983.

[4] Nathman, J.K., "A Study of Aerodynamic Control in Stalled Flight--Leading-Edge Vortex Formation Analysis", Air Force Wright Aeronautical Laboratories, AFWAL-TR-84-3090, February 1985.

[5] Vaidyanathan, T.S., "A Flow Field Analysis Procedure Based upon Velocity Potentials", Paper Presented at AIAA Applied Aerodynamics Conference, Danvers, MS, July 1983.

[6] Dvorak, F.A., Maskew, B. and Woodward, F.A., "Investigation of Three-Dimensional Flow Separation on Fuselage Configurations", USAAMRDL-TR-77-4, 1977.

[7] Paynter, G.C., Koncsek, J.L., Turczeniuk, B., Clark, D.R. and Strash, D.J., "Application of CFD Design Technology in Development of the JVX Engine Inlet", Presented at 41st Annual Forum of the American Helicopter Society, May 1985.

[8] Maskew, B. and Dvorak, F.A., "The Prediction of CLMAX Using a Separated Flow Model", _J. Am. Hel. Soc._, April 1978.

[9] Maskew, B., Rao, B.M. and Dvorak, F.A., "Prediction of Aerodynamic Characteristics for Wings with Extensive Separations", Paper No. 31 in _AGARD CPP-291_, September 1980.

[10] Summa, J.M. and Dvorak, F.A., "Computing Automobile Aerodynamics by an Integral Method", Proc. International Conference on Supercomputer Applications in the Automotive Industry, (C. Marino, ed., Computational Mechanics Publications), Zurich, Switzerland, October 1986.

[11] Maskew, B. and Dvorak, F.A., "Predicting Dynamics Separation Characteristics of General Configurations", AIAA Paper 86-1813, Presented at 4th Applied Aerodynamics Conference, June 1986.

[12] Piers, W.J. and Slooff, J.W., "Calculation of Transonic Flow by Means of a Shock-Capturing Field Panel Method", Proc. AIAA Computational Fluid Dynamics Conference, July 1979.

PANEL CLUSTERING TECHNIQUE FOR LIFTING POTENTIAL FLOWS IN THE THREE SPACE DIMENSIONS

Z.P. Nowak

Institut für Informatik und Praktische Mathematik
2300 Kiel 1, Olshausenstraße 40

1. INTRODUCTION

We present a new type of the panel method, based on the use of an efficient self-adaptive quadrature technique for singular integrals. For this technique, a preprocessing phase is introduced, during which a set of panel clusters is defined and certain integral moments of the singularity distributions on the clusters are calculated. During the subsequent discretization and solution phases these moments are used for quadratures over the clusters, the size of which is proportional to the distance from the control points. Near a control point, the clusters are identical with panels. In such near field areas exact quadrature formulas are applied. Since the number of arithmetic operations is the same for clusters of all sizes, our technique is more efficient than the earlier ones (e.g. Johnson [5]), for which the surface is split into individual panels.

The first presentation of the panel clustering technique was made by Nowak [6]. For a wide class of integral operators and arbitrary orders of consistency it was shown by Nowak and Hackbusch [7] that the BI methods with clustering can be made more efficient than the FD or FE methods applied to the underlying boundary value problems. The present approach was found to be useful by Jakob [4], who provided an independent confirmation of its advantages for the potential calculations in the flow-field.

2. THE BI FORMULATION OF THE LIFTING POTENTIAL FLOW PROBLEM

2.1. Green's formula. Let S be the union of body surfaces and W be the union of wake surfaces, immersed in the infinite flow region R outside S+W , with the uniform velocity V_∞ far from S. Let n(p) denote the unit normal at $p \in S$ or $p \in W$, pointing to R. The normals n(p) at $p \in W$ always point to the same part of R, which lies on a chosen side of the wake surface. We use the following abbreviations for the source and doublet integrals on a part G of S+W:

$$\underset{G}{⨍} f(p) = -\frac{1}{4\pi}\int_G \frac{f(q)}{|p-q|}\, dS_q$$

$$\underset{G}{⨍}\varphi(p) = -\frac{1}{4\pi}\lim_{\bar{p}\to p}\int_G \frac{n(q)\cdot(q-\bar{p})}{|q-\bar{p}|^3}\varphi(q)\,dS_q$$

for $p \in S+W$, where $\bar{p}\to p$ denotes the approach from this part of R, which is pointed to by n(p) or by the wake normal at p, if p belongs to the trailing edge of a wing. In the sequel, $\varphi(p)$ denotes the limit of $\varphi(\bar{p})$, obtained by the same approach $\bar{p}\to p$.

The BI formulation of the potential incompressible lifting flow problem is based on Green's formula for the perturbation velocity potential φ :

$$\varphi(p) = \underset{S}{⨍} f(p) + \underset{S}{⨍}\varphi(p) + \underset{W}{⨍}\mu(p) \qquad (1)$$

at $p \in S+W$ where $f=-V_\infty \cdot n$ on S, and μ is the doublet density distribution on W. The source integral in (1) can be represented in the form

$$\underset{S}{⨍} f(p) = -V_x^\infty \underset{S}{⨍} n_x(p) - V_y^\infty \underset{S}{⨍} n_y(p) - V_z^\infty \underset{S}{⨍} n_z(p) \qquad (2)$$

2.2. Treatment of wakes. The real wake W is approximated with a fixed surface $\hat{W}$, independent of the flow conditions.

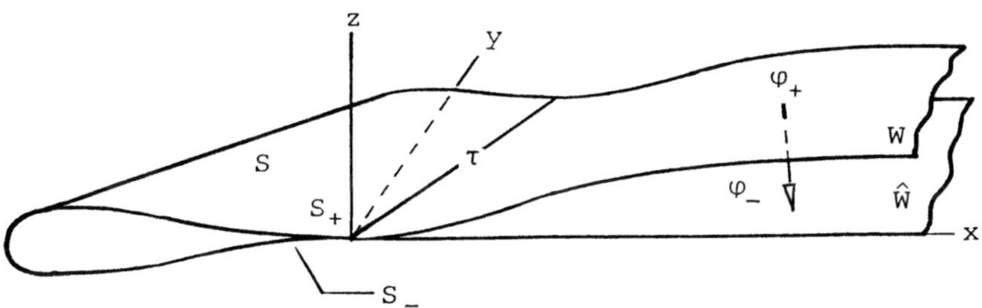

Fig. 1 A wing and the wake surface

In the sequel, $\hat{W}$ will be identified with a panel system. It will be assumed that the trailing edges are cusps. Such an assumption seems to be justified also for the wings with a non-zero trailing edge angle, due to the occurrence of the boundary layer separation near such edges. The fictitious wake $\hat{W}$ is tangent to the cusp and hence also to W at the trailing edges. Let φ_+ and φ_- denote the distributions of φ on both sides of W. In the situation illustrated in Fig. 1, the function φ_+ is smoothly continued in the direction shown by the arrow, up to $\hat{W}$. In other situations, φ_- is continued in the opposite direction. We may assume that φ_+ and φ_- are defined on both sides of $\hat{W}$. For simplicity let $\hat{W}$ be a part of the (x,y)-plane in a Cartesian coordinate system (Fig. 1) in which W is defined by an equation $z=\Delta(x,y)$ (i.e. vortex roll-ups are not allowed). The averages of the potential and velocity on both sides of $\hat{W}$ are denoted by

$$\bar{\varphi} = (\varphi_+ + \varphi_-)/2 \quad ; \qquad \bar{V} = V_\infty + \nabla\bar{\varphi} \quad .$$

Since W and $\hat{W}$ are close to each other, the maximum value of the average normal velocity

$$a = \max_{q\in\hat{W}} |\bar{V}_z(q)|$$

is small. We shall assume that $\Delta(x,y)$ and the singularity distributions on $\hat{W}$:

$$\hat{\sigma} = \partial(\varphi_+)/\partial z - \partial(\varphi_-)/\partial z \;, \; \hat{\mu} = \varphi_+ - \varphi_-$$

can be developed into the power series of a, with the leading terms $a\Delta_1(x,y)$, $a\sigma_1(x,y)$ and $\mu_o(x,y)$ respectively. Substituting these series into the pressure equality and impermeability conditions on both sides of W, we arrive at the following equations for the leading terms on $\hat{W}$:

$$\bar{V}_x \; \partial\mu_o/\partial x + \bar{V}_y \; \partial\mu_o/\partial y = 0 \quad , \qquad (3)$$

$$\bar{V}_x \; \partial(a\Delta_1)/\partial x + \bar{V}_y \; \partial(a\Delta_1)/\partial y + (a\Delta_1)\nabla^2\bar{\varphi} = \bar{V}_z \quad , \qquad (4)$$

$$a\sigma_1 = \partial\mu_o/\partial x \; \partial(a\Delta_1)/\partial x + \partial\mu_o/\partial y \; \partial(a\Delta_1)/\partial y + (a\Delta_1)\nabla^2\mu_o \quad ,$$

where $\nabla^2 = \partial^2/\partial x^2 + \partial^2/\partial y^2$. Since $\bar{V}_z$ in (4) is costly to calculate, we shall ignore the small terms $O(a)$ and assume that $\hat{\sigma}=0$ and $\hat{\mu}=\mu_o$ on $\hat{W}$. The hyperbolic equation (3) for $\hat{\mu}$ is supplemented with the following initial condition along each trailing edge τ:

$$\hat{\mu}(t) = \varphi_{s+}(t) - \varphi_{s-}(t) \quad , \qquad t\in\tau \quad , \qquad (5)$$

where φ_{s+} and φ_{s-} denote the distributions of φ on the two adjacent parts of $\bar{S}$, lying on both sides of $\hat{W}$ (Fig. 1). The condition (5) is necessary to cancel the unbounded term $O(1/|q-t|)$, which appears in the expression for the perturbation velocity $\nabla\varphi(q)$ at $q\in R$. For the cusped trailing edges the remaining unbounded term $O(\log|q-t|)$ vanishes if we impose the restriction

$$\partial\hat{\mu}(t)/\partial x = \partial(\varphi_{s+}(t))/\partial x - \partial(\varphi_{s-}(t))/\partial x \qquad (6)$$

at $t\in\tau$, where the x-direction can be chosen as in Fig. 1. The other terms in $\nabla\varphi$ tend to the bounded limits on τ.

3. BASIC PANEL METHOD

During a preliminary phase of this research, a "basic" panel method has been implemented for the non-lifting potential flows. The method has the following characteristic properties:

(i) S is approximated with a system $\hat{S}$ of plane triangular panels with the vertices on S;

(ii) φ on S is approximated with a continuous function $\hat{\varphi}$ on $\hat{S}$, linear on each panel of $\hat{S}$;

(iii) f on S is approximated with

$$\hat{f}(p) = -V_x^\infty \hat{n}_x(p) - V_y^\infty \hat{n}_y(p) - V_z^\infty \hat{n}_z(p)$$

at $p \in \hat{S}$, where $\hat{n}_x(p)$, $\hat{n}_y(p)$, $\hat{n}_z(p)$ are linear on each panel of $\hat{S}$, continuous on each part of $\hat{S}$ which approximates a smooth piece of S, and equal to $n_x(p)$, $n_y(p)$ and $n_z(p)$, respectively, when p is a panel vertex on S;

(iv) the control points are placed at the panel vertices;

(v) the exact quadrature formulas (see e.g. [5]) are used for calculating the source and doublet integrals over panels;

(vi) the discrete system is solved by the Q-R decomposition method;

(vii) the velocity on S is obtained by differentiating a bi-cubic spline fit to the values of φ at the control points.

The improved and generalized method, presented in the sequel, has the same properties (i), (ii), (iii), (iv) and (vii). The property (v) is preserved in the near fields of the control points. From now on, all functions are piecewise linear approximations and the surfaces are panel systems. For simplicity we shall write f, n_x, n_y, n_z, φ, μ, S and W instead of $\hat{f}$, $\hat{n}_x$,...,$\hat{W}$. The set of the control points on a part G of S+W will be denoted by C(G).

4. SELF-ADAPTIVE QUADRATURE TECHNIQUE

4.1. <u>Far field quadrature formulas</u>. As in Fig. 2 let p be a control point on S+W (briefly, $p \in C(S+W)$), let G be a cluster of panels and q: the variable of integration.

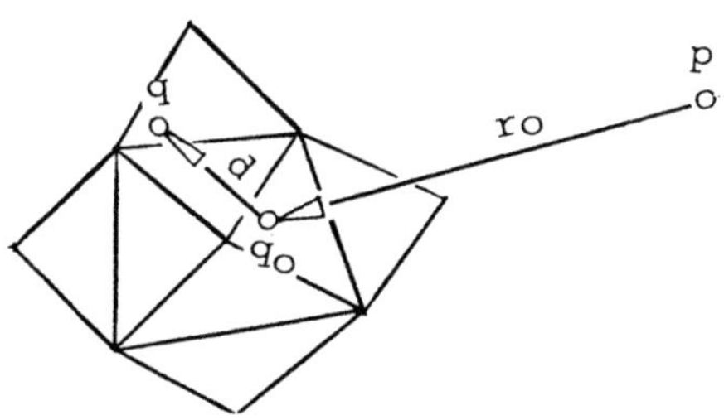

Fig. 2 A cluster of panels

We shall denote $r=q-p$, $r_o=q_o-p=(x_o,y_o,z_o)$, $d=q-q_o=(d_x,d_y,d_z)$ and

$$\alpha = r_o \cdot d/|r_o|^2 \quad , \quad \beta = |d|/|r_o| \quad .$$

The kernel of the source integral can be written as

$$|r|^{-1} = [(r_o+d)\cdot(r_o+d)]^{-1/2} = |r_o|^{-1}(1+2\alpha+\beta^2)^{-1/2}$$

$$= |r_o|^{-1}(1-\alpha-\beta^2/2+3\alpha^2/2+e(\alpha,\beta)) \quad , \qquad (7)$$

where $|e(\alpha,\beta)|=\beta^3+O(\beta^4)$. Neglecting the remainder $e(\alpha,\beta)$ is equivalent to replacing the source density f on G with a corrected function $\tilde{f}=f+\delta_f$. For sufficiently distant clusters (small β) the scaled correction can be estimated as follows

$$|\delta_f/f| = |r||e(\alpha,\beta)|/|r_o| = (1+\beta)(\beta^3+O(\beta^4))$$
$$\leqq 2\beta^3 = 2|d|^3/|r_o|^3 \quad .$$

We shall neglect $e(\alpha,\beta)$ if $|\delta_f/f|<\varepsilon$ on G. This requirement is satisfied if

$$\max_{q\in G}(|d(q)|/|r_o|) \leqq (\varepsilon/2)^{1/3} \quad . \qquad (8)$$

Then the following approximate formula for the source integral over G can be derived

$$\int_G \tilde{f}(p) = f_1/|r_o|+f_2/|r_o|^3+\ldots+f_{11}y_oz_o/|r_o|^5 \quad , \qquad (9)$$

where $f_1, f_2, \ldots, f_{11}$ are certain integral moments of f on G. We present the formulas for two of these moments

$$f_1(G;f)=\int_G f(q)dS_q, \ldots, f_{11}(G;f)=3\int_G d_y(q)d_z(q)f(q)dS_q \quad . \qquad (10)$$

Let us now proceed to the case of the doublet integral. In this case the expansion of the same order as (7) is made for the factor $|r|^{-3}$, involved in the kernel. As previously, it can be seen that neglecting the remainder of this expansion is equivalent to replacing the doublet density φ with a corrected function $\tilde{\varphi}=\varphi+\delta_\varphi$, where $|\delta\varphi/\varphi|\leqq\varepsilon$ if (8) holds. The doublet integral on G is approximated with

$$\oint_G \tilde{\varphi}(p) = \oint_1/|r_o|^3+\ldots+\oint_{23}x_oy_oz_o/|r_o|^7 \quad . \qquad (11)$$

We present the formulas for the two integral moments in (11):

$$\oint_1(G;\varphi) = \int_G \varphi(q)d(q)\cdot\gamma(q)dS_q, \ldots,$$
$$\oint_{23}(G;\varphi)=15\int_G \varphi(q)[d_x(q)d_y(q)\gamma_z(q)+d_x(q)d_z(q)\gamma_y(q) \qquad (12)$$
$$+d_y(q)d_z(q)\gamma_x(q)]dS_q \quad ,$$

where $\gamma=(\gamma_x,\gamma_y,\gamma_z)$ is the unit normal to G, constant on each panel of G. For a control point v let X_v be the continuous piecewise linear function on S+W, such that $X_v(v)=1$ and $X_v(v')=0$ at the control points $v'\neq v$. Since $\varphi(q)$, $q\in G$, is the sum of terms $\varphi(v)X_v(q)$ with $v\in C(G)$, we can represent the moments (12) in the form:

$$\oint_i (G;\varphi) = \sum_{v \in C(G)} \oint_i (G;X_v)\varphi(v), \ i=1,2,\ldots,23 \quad . \tag{13}$$

4.2. Preprocessing phase consists of two stages:
Stage I. At the beginning of the preprocessing phase a set F of panel clusters is defined. The set F is composed of subsets $F_1, F_2,\ldots,F_L$, called levels. The clusters of level F_1 are identified with all the panels of S and W. In general, a cluster of the level F_l, $l=2,3,\ldots,L$, is a union of a certain number of clusters of the level F_{l-1}. The clusters of the same level are disjoint. The total number of clusters in F is O(m), where m is the number of panels.
Stage II. All the moments $f_i(G;f)$, $i=1, 2,\ldots,11$; $G\in F$, and $\oint_i(G,X_v)$, $i=1,2,\ldots,23$; $G\in F$; $v\in C(G)$, are calculated and stored. These moments are sums of the integrals of the third or lower order polynomials over panels, obtained exactly from the Newton-Cotes quadrature formula of the third degree (Engels [1], p.280). The number of the arithmetic operations involved (the complexity) and the storage required are O(m).

4.3. Far field discretization phase consists of two stages:
Stage I. For each $p\in C(S+W)$ the following procedure is performed:

The condition (8) is checked for all the clusters of the level F_L. If (8) holds for a cluster G, then G is included in the set FF(p) of the clusters in the far field of p. Otherwise, G is split into the clusters of the level F_{L-1}. After completing this procedure for all the clusters of the level L, we obtain, as a by-product, a certain number of clusters of the level F_{L-1}. All these clusters are again split or included in FF(p), depending on the result of the test (8). Such a procedure is continued until a certain lowest level F_l has been achieved. The set of all the panels in the remaining clusters of the level F_l will be denoted by NF(p).

As a result, for each $p\in C(S+W)$ we determined a set FF(p) of the far field clusters and a set NF(p) of the near field panels.
Stage II. For each $p\in C(S+W)$ we calculate the far field parts of the three source integrals on the right-hand side of (2) by adding the contributions (9) of all the clusters $G\in FF(p)$, $G\subset S$, due to n_x, n_y and n_z.
Remark. If S+W is a symmetric surface then the calculation of the present phase can be performed only for the control points on the main part of S+W, which is transformed into the other parts by reflections in the symmetry planes. Let a control point p lie in the main part and let p' be its reflection. Then FF(p') and NF(p') are the reflections of FF(p) and NF(p), and the values of the three source integrals at p and p' may differ only in sign.

4.4. Near field discretization phase. For each control point p on S+W, or on the main part of S+W, the contributions of the panels in NF(p) are added to the three source integrals. The exact formulas are used for these quadratures. Let $S_{nf}(p)$ denote the near field of p, i.e. the union of all the panels in NF(p). The exact formulas are also used for determining the near field influence coefficients for the doublet integral operator:

$$c(p,v) = \oint_{S_{nf}(p)} X_v(p) \quad , \quad v \in C(S_{nf}(p)) \quad . \tag{14}$$

The coefficients $c(p,v)$ are stored in the form of a sequence.

4.5. <u>The complexity of discretization</u>. Let h denote a characteristic panel size. To retain the consistency $O(h^2)$ of the basic method of Section 3 we must choose $\varepsilon=O(h^2)$ for the criterion (8). For such a choice it can be seen that NF(p) contains $O(m^{2/3})$ panels for each $p\in C(S+W)$ and that FF(p) is composed of $O(m^{2/3}\log m)$ clusters. Since the number of points in $C(S+W)$ is $O(m)$, the total complexity of the two phases of discretization and the storage required are $O(m^{5/3}\log m)$. As shown in [7], the storage requirements and the complexity can be reduced to $O(m^{1+\lambda})$ for any $\lambda>0$, by choosing the appropriate length of the truncated power series expansion (7), and of the analogous expansion used for the doublet integral.

5. ITERATIVE SOLUTION PROCEDURE

5.1. <u>Iterative schemes</u>. The equation (3) for $\mu=\mu_0$ gives rise to the following iterative scheme:

$$\frac{\delta\mu_i}{\delta x}(q) = \frac{V_y^\infty+\delta\bar{\varphi}_{i-1}(q)/\delta y}{V_x^\infty+\delta\bar{\varphi}_{i-1}(q)/\delta x}\;\frac{\delta\mu_{i-1}(q)}{\delta y} \quad , \; i=1,2,\ldots \tag{15}$$

at $q\in C(W)$, where $\delta/\delta x$ and $\delta/\delta y$ are the second order accurate finite difference approximations for $\partial/\partial x$ and $\partial/\partial y$. The right-hand side of (15) will be denoted by $r_i(q)$. The i-th iterand μ_i will be split as follows

$$\mu_i = \mu_1^i + \mu_2^i \quad , \tag{16}$$

where

$$\frac{\delta\mu_1^i}{\delta x}(q)=r_i(q) \text{ at } q\in C(W) \text{ and } \mu_1^i(t)=0 \text{ at } t\in C(\tau) \quad , \tag{17}$$

$$\frac{\delta\mu_2^i}{\delta x}(q)=0 \text{ at } q\in C(W) \text{ and } \mu_2^i(t)=\varphi_{s+}^{\,i}(t)-\varphi_{s-}^{\,i}(t) \text{ at } t\in C(\tau) . \tag{18}$$

Green's formula (2) gives rise to the following iterative scheme

$$\varphi_i(p)-\oint_S^{(nf)}\varphi_i(p)-\oint_W^{(nf)}\mu_2^i(p)=\oint_S f(p)+\oint_W^{(nf)}\mu_1^i(p)+ \oint_S^{(ff)}\varphi_{i-1}(p)+\oint_W^{(ff)}(\mu_1^i+\mu_2^{i-1})(p) , \; i=1,2,\ldots, \tag{19}$$

at $p\in C(S)$, where the superscripts (nf) and (ff) denote the near field and the far field parts of the doublet integral. The solution of (18) can be represented in the form

$$\mu_2^i(q) = \varphi_{S+}{}^i(t(q)) - \varphi_{S-}{}^i(t(q)) \tag{20}$$

for $q \in C(W)$, where $t(q) \in C(\tau)$ has the same x coordinate as q. The continuous piecewise linear function $\mu_2^i(q)$, equal to (20) at $q \in C(W)$, is substituted into the second integral in (19). The resulting linear algebraic system for $\varphi_i(p)$, $p \in C(S)$, is incomplete since each $p \in C(\tau)$ carries the two values: $\varphi_{S+}^i(p)$ and $\varphi_{S-}^i(p)$. Supplementing this system with the finite difference equations for (6):

$$\frac{\delta\varphi_{S+}{}^i(p)}{\delta x} - \frac{\delta\varphi_{S-}{}^i(p)}{\delta x} = r_i(p) \quad , \quad p \in C(\tau) \quad , \tag{21}$$

we arrive at the complete system of the linear algebraic equations

$$A\varphi_i(p) = B_i(p) \quad , \quad p \in C(S) \tag{22}$$

for all the values of $\varphi_i(p)$, $p \in C(S)$.

5.2. Post-discretization phase. The influence coefficients (14) and the coefficients in the finite difference equations (21) serve to construct the sparse matrix A of the system (22); (20) is used for the construction. The predicted symmetry or anti-symmetry properties of the potential are used for reducing the size of A. The non-zero coefficients in A are stored in form of the sequence. The ILU decomposition (Hackbusch [2], p. 205) of A is made and the L, U sequences are stored for further use during the solution phase. If $\varepsilon = O(h^2)$ is chosen for (8) then the storage required for the sequences A, L and U is $O(n^{5/3})$ and the complexity of the ILU decomposition is $O(n^{7/3})$.

5.3. Solution phase. At the beginning of the solution phase the source integral with the density f is obtained at each $p \in C(S+W)$, as the combination (2) of the three source integrals calculated during the discretization phases. Next follows the iterative solution procedure. The i-th iteration consists of the three stages:

Stage I: Calculate $r_i(p)$ for each $p \in C(W)$ and solve (17).
Stage II: Calculate $B_i(p)$ for each $p \in C(S)$ and solve (22).
Stage III: Calculate μ_2^i on W from (20) and then μ_i from (16); calculate $\bar{\varphi}_i(p)$ for all $p \in C(W)$ from Green's formula

$$\bar{\varphi}_i(p) = \oint_S f(p) + \oint_S \varphi_i(p) + \oint_W \mu_i(p) - \mu_i(p)/2. \tag{23}$$

Stage II involves the calculation of the near and far field integrals on the right hand side of (19). The near field integral is the combination of the near field influence coefficients (14) and the given values $\mu_1^i(p)$, $p \in C(W)$. The calculation of the far field integrals is preceded by the calculation of the moments given by the formula (13), where we put $\varphi = \varphi_{i-1}$ if $G \subset S$, and $\varphi = \mu_1^i + \mu_2^{i-1}$ if $G \subset W$. After this preliminary calculation the sum of the two far field integrals in (19) is obtained for each $p \in C(S)$ by adding the contributions (11) of all the clusters in FF(p). The solution φ_i of (22) is determined to sufficient accuracy by one iteration of the ILU method. The doublet integrals in (23) are obtained by the same procedures as in the Stage II.

The iterative process starts with $\bar{\varphi}_o=\mu_o=\mu_o^o=0$ on C(W) and $\varphi_o=0$ on C(S). In the first iteration, the right-hand side of (19) is reduced to the source integral. If ε in (8) is $O(h^2)$ then the complexity of each following iteration is $O(m^{5/3}\log m)$. In the last iteration the calculation of (23) is not needed.

6. PRACTICAL CALCULATIONS

The present method served as a basis for a system of programs for determining the lifting flows around arbitrary configurations of the body and wake surfaces. The preprocessing program (Section 4.2) is run only once for a given geometry of the panel system. The discretization program (Sections 4.3 and 4.4) is run only once for a given accuracy ε. The post-discretization program (Section 5.2) is run only once for given symmetry properties of the flow-field. Afterwards, the calculations for the various flow cases with these symmetry properties are made by re-running only the solution program (Section 5.3).

The preliminary calculations were made for the non-lifting flow around a sphere and for the lifting flow around the RAE wing with the NACA 0012 cross section (for the wing data see Systma et al [9]), at 5° angle of attack. In both cases, the symmetry properties of the surfaces and of the flow-field were taken into account.

The control points on 1/8 of a sphere were placed at k equal increments of the two angles of the spherical coordinate system. The calculations were made for k=4,6,8,10 and 12, and for the accuracy parameter $\varepsilon=\varepsilon_1=0.037$ and $\varepsilon=\varepsilon_2=0.59/k^2=0.24h^2$, where $h=\pi/(2k)$ is the characteristic panel size. Table I presents the values of the mean square errors e_φ^b, e_V^b for φ and $V=\nabla\varphi$ on C(S), obtained by the basic method of Section 3 (basic errors), and the mean square errors e_φ^r, e_V^r, added to these functions by the present method (relative errors) as a result of the approximate far-field quadratures (9) and (11).

The relative errors are very small when compared with the basic ones, even for $\varepsilon=\varepsilon_1$. It seems that even for a large number of panels, ε may be set equal to a fixed value, independent of the panel size. For such a choice the radius of the near field is O(h) (for $\varepsilon=\varepsilon_1$ it was about 2h) and the number of panels in each NF(p) is O(1). The complexity of the near field discretization and post-discretization phases is reduced to O(m). For k=12 and $\varepsilon=\varepsilon_1$ the discretization was about 10 times faster than for the basic method. However, the far field quadratures in (19) and (23) still cost $O(m^{5/3}\log m)$ arithmetic operations in each iteration of the solution phase. Fortunately, the iterations converge very fast. For k=12 and $\varepsilon=\varepsilon_1$ the maximum corrections $\varphi_i-\varphi_{i-1}$ in the first 8 iterations are: 0.577,

$0.993_{10}-1$, $0.174_{10}-1$, $0.305_{10}-2$, $0.535_{10}-3$, $0.939_{10}-4$, $0.165_{10}-4$, and $0.289_{10}-5$. These values decrease at the rate of 0.175 per iteration. For k=12 and $\varepsilon=\varepsilon_2$ the convergence rate was about 0.06. In general, the level of the basic errors is achieved after 3 iterations, which require 2 evaluations of the far field integrals in (19) and (23).

Table I Mean square basic and relative errors

k	e_φ^b	e_φ^r		e_v^b	e_v^r	
		$\varepsilon=\varepsilon_1$	$\varepsilon=\varepsilon_2$		$\varepsilon=\varepsilon_1$	$\varepsilon=\varepsilon_2$
4	$0.184_{10}-1$	$0.442_{10}-4$	$0.442_{10}-4$	$0.264_{10}-1$	$0.863_{10}-4$	$0.863_{10}-4$
6	$0.819_{10}-2$	$0.103_{10}-3$	$0.235_{10}-4$	$0.118_{10}-1$	$0.268_{10}-3$	$0.428_{10}-4$
8	$0.458_{10}-2$	$0.120_{10}-3$	$0.104_{10}-4$	$0.661_{10}-2$	$0.391_{10}-3$	$0.174_{10}-4$
10	$0.292_{10}-2$	$0.109_{10}-3$	$0.864_{10}-5$	$0.421_{10}-2$	$0.378_{10}-3$	$0.229_{10}-4$
12	$0.202_{10}-2$	$0.154_{10}-3$	$0.590_{10}-5$	$0.291_{10}-2$	$0.523_{10}-3$	$0.109_{10}-4$

We examined also the convergence rate for the ILU method of Stage II, Section 5.3. For k=12 and $\varepsilon=\varepsilon_1$ this inner iterative process introduced the following corrections into φ_i: 0.577, $0.436_{10}-4$, $0.242_{10}-8$, ... for i=1, and $0.993_{10}-1$, $0.711_{10}-5$, $0.413_{10}-9$,... for i=2. It is clear that one ILU iteration is sufficient.

The calculations for the RAE wing were made with 23*11 control points on the wing quarter, including 23*3 points on the tip closure, and with 4*9 points on the half of the wake. The value of the accuracy parameter was set equal to $\varepsilon=0.25$, $\varepsilon=0.125$ and $\varepsilon=0.065$. In these three cases the ratio of the number of the control points in $C(S_{nf}(p))$ and in $C(S+W)$, averaged for all $p\in C(S+W)$, was equal to 0.21, 0.255 and 0.31, respectively. For more elongated wings or wing-fuselage combinations these ratios will be less. Table II shows the values of the maximum corrections $\varphi_i-\varphi_{i-1}$ of the potential on the wing surface and of the maximum scaled errors $(\mu_i-\mu_{10})/\mu_{10}$ on the wake (μ_{10} agrees with the limit μ_∞ to seven or more decimal places).

The convergence rate is about $0.9*\varepsilon$. Even for $\varepsilon=0.25$ the first three iterations are sufficient to achieve an engineering accuracy of about 1% for the doublet distribution on W, and hence for the lift distribution (20) along the trailing edge. On the Siemens 7760 computer, comparable to IBM 4341, the three iterations cost about 110, 120 and 130 seconds of the CPU time, for the respective cases $\varepsilon=0.25$, 0.125, 0.065.

As shown in Fig. 3, the converged results for the pressure coefficient c_p on the wing surface differ slightly from those given in [9], p. 22, due probably to the different forms of the Kutta condition used here and in [9],due to the different treatment of the wake, and to the fact that our wing is closed at the tip, as required by Green's formula (1).

Table II Convergence history for the case of the RAE wing

i	$\max(p \in C(S): \lvert\varphi_i(p)-\varphi_{i-1}(p)\rvert)$			$\max\limits_{p \in C(W)}(\lvert(\mu_i(p)-\mu_{10}(p))/\mu_{10}(p)\rvert)$		
	ε=0.25	ε=0.125	ε=0.065	ε=0.25	ε=0.125	ε=0.065
1	$9.44_{10}{-2}$	$1.01_{10}{-1}$	$1.07_{10}{-1}$	$2.03_{10}{-1}$	$1.19_{10}{-1}$	$7.29_{10}{-2}$
2	$1.41_{10}{-2}$	$9.54_{10}{-3}$	$6.46_{10}{-3}$	$4.33_{10}{-2}$	$1.52_{10}{-2}$	$4.90_{10}{-3}$
3	$2.49_{10}{-3}$	$1.01_{10}{-3}$	$3.78_{10}{-4}$	$9.07_{10}{-3}$	$1.93_{10}{-3}$	$2.76_{10}{-4}$
4	$5.25_{10}{-4}$	$1.27_{10}{-4}$	$2.25_{10}{-5}$	$1.88_{10}{-3}$	$2.45_{10}{-4}$	$1.65_{10}{-5}$
5	$1.10_{10}{-4}$	$1.61_{10}{-5}$	$1.34_{10}{-6}$	$3.94_{10}{-4}$	$3.11_{10}{-5}$	$9.85_{10}{-7}$
conv. rate	$2.1_{10}{-1}$	$1.3_{10}{-1}$	$6_{10}{-2}$	$2.1_{10}{-1}$	$1.3_{10}{-1}$	$6_{10}{-2}$

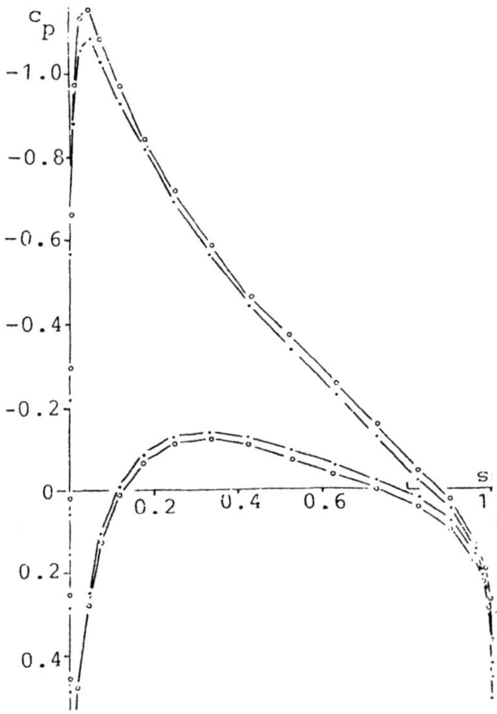

Fig. 3 c_p-distribution in the middle cross section of the wing;
·-· present results;
o-o Sytsma et al. [9].

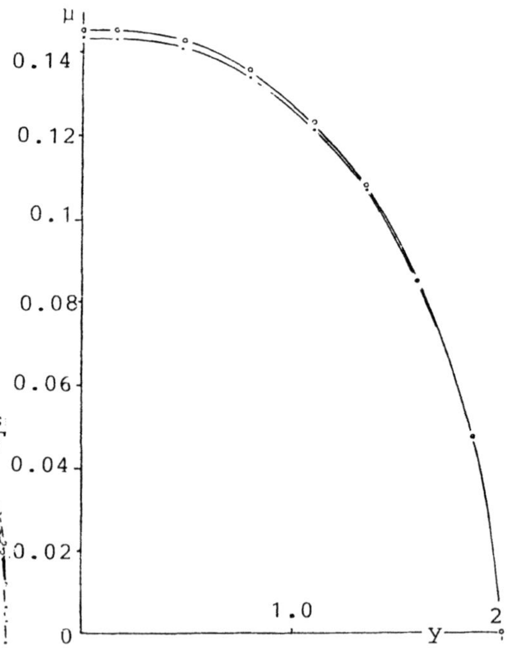

Fig.4 Lift distribution along the wing (y-axis as in Fig.1)
·-· results for ε=0.25;
o-o results for ε=0.125,0.065

The differences between our results for $\varepsilon=0.25$, 0.125 and 0.065 are too small to be visible in Fig.3. The maximum difference between the values of c_p for $\varepsilon=0.25$ and $\varepsilon=0.065$ is equal to 0.0154 in the interval $s \lesssim 0.0435$ of the steep pressure variation near the trailing edge. Outside this interval, the differences are of the order 0.005. The maximum difference in c_p between the cases $\varepsilon=0.125$ and $\varepsilon=0.065$ is 0.0016.

Fig.4 represents the distribution of the doublet strength along the trailing edge boundary of the wake, given by the formula (20). The values obtained for $\varepsilon=0.25$ and $\varepsilon=0.065$ differ by at most 1.2%. For $\varepsilon=0.125$ and $\varepsilon=0.065$ the maximum difference is 0.12%. Hence, the results are practically insensitive to the choice of ε in the interval $0 \leqq \varepsilon \lesssim 0.125$. An engineering accuracy is obtained for $\varepsilon=0.25$.

The results of these preliminary calculations allow us to conclude that the present approach may serve as a basis for a new family of efficient panel methods.

7. FINAL REMARKS

During the final phase of preparation of the present paper, it came to the author's knowledge that the clustering has been applied by Rokhlin [8] for the Laplace equation problems on a plane, discretized by means of Nyström method. In [8], the contributions of the clusters to the far field part of the doublet integral are obtained from a two-dimensional counterpart of the multipole expansion (11). The discretization phase of [8] is more complicated than that presented here in Section 4. During this phase, the local power series expansions for the far field parts of the doublet integral are determined around the nodes of the increasingly finer grids, until the finest grid is achieved. The construction involves the conversion of the multipole expansions into the truncated power series expansion. These power series are subsequently converted into another truncated series with shifted centers. As mentioned in [8], for the Laplace equation problems in the three space dimensions similar manipulations must be made on series of spherical harmonics. Contrary to the opinion expressed in [8], it seems doubtful if such a method can find serious practical applications in aerodynamics.

Our quadrature technique is simpler and more general. Instead of determining the local expansions for the integrals we calculate only what is needed: the values of these integrals at the control points. As we have seen in Section 4.1, the change from the source quadratures to the doublet quadratures is made simply by replacing formulas (9,10) with (11,12). The application of our method to other singular integrals in the two or three space dimensions is equally easy ([7]).

The complexity of the quadrature method [8] is $O(m(\log\varepsilon)^2)$, where ε is a chosen value for the quadrature error introduced by the series truncations. For $\varepsilon=O(m^{-k})$ with some $k \geqq 1$, which is the level of the truncation error of the basic Nyström method, the complexity becomes $O(m(\log m)^2)$. Recently, Hackbusch and Nowak [3] proved that the complexity of the present method can be reduced to $O(m(\log m)^4)$, where the fourth power of the logarithm is due to the higher dimensionality of our problems.

REFERENCES

[1] Engels, H., Numerical quadrature and cubature, Academic Press, London 1980.

[2] Hackbusch, W., Multigrid methods and applications, Springer, Berlin 1985.

[3] Hackbusch, W., Nowak, Z.P., On the fast matrix multiplication in the boundary element method by panel clustering (to appear).

[4] Jakob, H., Some problems for application of panel methods to complex configurations in aerodynamics, communicated during the 3rd GAMM Seminar, January 1987.

[5] Johnson, F.T., A general panel method for the analysis and design of arbitrary configurations in incompressible flow, NASA CR 3079 (1980).

[6] Nowak, Z.P., Efficient multi-level panel methods of high-order accuracy, Workshop on advanced panel methods, Universität der Bundeswehr, München, July 7th, 1986.

[7] Nowak, Z.P., Hackbusch, W., On the complexity of the panel method, Report 8608, Christian-Albrechts-Universität, Kiel, September, 1986.

[8] Rokhlin, V., Rapid solution of integral equations of classical potential theory, Journal of Comput. Physics 60, No.2 (1985), pp. 187-207.

[9] Sytsma, H.S., Hewitt, B.L., Rubbert, P.E., A comparison of panel methods for subsonic flow computation, AGARDograph 241 (1979).

SINGULARITY TREATMENT IN THE BIDIMENSIONAL TAU METHOD WITH AN APPLICATION TO PROBLEMS DEFINED ON L-SHAPED DOMAINS

Eduardo L. Ortiz,
Department of Mathematics
Imperial College of Science and Technics
Queen's Gate, London SW7 2BZ, Great Britain

SUMMARY

A recently developed multidimensional formulation of the Tau Method is applied to the numerical approximation of a model singular boundary value problem defined by Laplace´s equation in L-shaped domains. Singularity treatment in the context of the Tau Method is discussed and the singular parameters are computed with a high accuracy. A brief sketch of recent formulations of the Tau Method is given in this paper and also extensive references to the most recent literature on it.

INTRODUCTION

In a recent paper, Ortiz and Samara [26] developed an approach to the Tau Method (see Lanczos [7]), based on Ortiz´ new formulations of that method [15], which makes it easily applicable to problems in more than one dimension. Several types of partial differential equations have been treated numerically by these authors; their results show that the error function has a balancing behaviour, reminiscent of the error of the best uniform approximation by algebraic polynomials of the exact solution of the given equation. Results on the optimality of the Tau Method have been reported in recent papers of Freilich and Ortiz [4] and Onumanyi and Ortiz [14], for classes of ordinary differential equations, and by Namasivayam and Ortiz [12] for partial differential equations. Ortiz and Pun [21]-[22] have discussed the numerical approximation of nonlinear partial differential equations with this technique, and Ortiz and Pham [24]-[25] have used function theoretical and functional analytical techniques to give convergence results for the Tau Method in the context of nonlinear ordinary and partial differential equations respectively. Results of a high accuracy have been reported by Liu, Ortiz and Pun [9] for eigenvalue problems of Steklov´s type, where the spectral parameter appears in a complex form in the boundary conditions. Applications of the Tau Method to sensitive stability problems of fluid dynamics have been reported by Liu and Ortiz in [10] and [11].

Singular boundary value problems for partial differential equations have been discussed in papers of El Misiery and Ortiz [2]-[3]. However, the technique used there is the Tau-Lines Method, an hybrid technique based on the Method of Lines and the recursive formulation of the Tau Method of Ortiz [15]. The first of them is used to reduce the dimensionality of the problem and the second to find an accurate approximation of the reduced problem.

In the case of two independent variables such reduction leads, through discretization in one variable, to a system of ordinary differential equations. Therefore, the Tau Method is used in only one dimension. Experiments are being conducted with a slicing technique for higher dimensional problems.

In this paper we review a series of results recently obtained in joint work with Dr. K.-S. Pun, of British Petroleum, London, on the approximation of Laplace's equation in L-shaped domains of different proportions. The approach used here is a direct Tau Method defined in two dimensions, therefore no discretization of the given partial differential equation is required. The background to the techniques used here is given in recent papers of Ortiz [20] and Ortiz and Pun [21]-[22], to which the reader is referred. Further details on singular boundary value problems will be given in a forthcoming paper of Ortiz and Pun [23].

DIFFERENT FORMULATIONS OF THE TAU METHOD

A brief sketch of the Tau Method will be given in this section. Let us assume for the moment that **D** is the class of linear differential operators D with polynomial coefficients (or with coefficients represented by accurate polynomial approximations); that the maximum order of differentiation in D is equal to ν and that it is acting on μ-times differentiable functions of the single variable x.

(i) Recursive formulations of the Tau Method

Ortiz has shown in [15] that any linear differential operator of the class **D** is uniquely associated with a sequence of canonical polynomials $Q_n(x)$. Canonical polynomials satisfy the following functional relation:

$$D\, Q_n(x) = x^n + r_n(x),\ n \in \mathbf{N} - \mathbf{S},\ \mathbf{N} := \{0, 1, 2, \ldots\},$$

where the residual polynomials $r_n(x) \in R_S$. The latter is called the subspace of residuals of D. It has a finite dimension, equal to the number s of elements of the finite set **S**. The set **S** is defined in terms of the order ν and the height h of the differential operator D. It identifies the powers of x which cannot be generated in the image of D by applying it to a polynomial expression.

It is also shown in [15] that the sequence of canonical polynomials associated with an operator $D \in \mathbf{D}$ can be easily generated by means of a self-starting recursive relation. This is an essential point in making the Tau Method attractive from a computational point of view.

In the Tau method a given problem $D\, y(x) = f(x)$ plus supplementary conditions, is replaced by the so-called Tau Problem $D\, y_n(x) = f(x) + H_n(x)$, which has the same supplementary conditions as the original one. The Tau Problem admits an exact polynomial solution $y_n(x)$ which is called the Tau approximation of order n of y(x); the latter is regarded as a function implicitly defined by the given differential equation. The right hand side f is assumed to be a polynomial or to be expressed in terms of a sufficiently accurate polynomial approximation. Such approximation, and that of non polynomial coefficients in the equation, can be carried out automatically, by using the Tau Method itself as the data for the problem is feed into the computer program for that method. It should be pointed out that the use of polynomial approximations for mathematical functions is present, more or less explicitly than we have made it here, in any numerical method realized in a finite computer.

$H_n(x)$ is a polynomial perturbation term the norm of which satisfies prescribed minimal properties on a given interval. Chebyshev, Legendre, Hermite and other special polynomials, as well as piecewise polynomial functions, have been used depending of the specific objective of the computation (see Ortiz [16]-[17], Liu and Ortiz [8]). Namasivayam and Ortiz [13] have recently produced a detailed analysis of the approximation properties of such alternative perturbation terms.

Instead of canonical polynomials which are mapped by D into $\mathbf{X} := \{x^n\}$, $n \in \mathbf{N}$, other forms, which represent the image of D in terms of polynomial basis $\mathbf{V} := \{v_n(x)\}$, $n \in \mathbf{N}$, have also been used in the formulation of the Recursive Tau Method (see Ortiz [17]-[16]). Associated canonical polynomials are such that the operator D acting on them generates $v_n(x)$ in the image space:

$$D\, Q_n(V,x) := v_n(x) + r_n(x),$$

the residual polynomials are defined as before.

If $H_n(x)$ is a linear combination of elements $v_n(x)$ of a given basis $\mathbf{V}$, it is clear that its representation in the solution $y_n(x)$ of the Tau problem $D\, y_n(x) = H_n(x) + f(x)$ is simply given by a linear combination with the same coefficients, but with each $v_n(x)$ replaced by the corresponding $Q_n(V,x)$. This alternative form of expressing the Tau solution is relevant to the accuracy of the numerical representation of $y_n(x)$.

We wish to remark that the system of linear algebraic equations to be solved to construct the Tau approximation $y_n(x)$, that is to satisfy exactly the supplementary conditions of the given problem and to deal with the possible existence of a non-trivial subspace of residuals R_S, is one of order only equal to $\nu + s$ (as before, s is the number of elements in the finite set S). Therefore the construction of the Tau approximation $y_n(x)$ is extremely economical from a computational point of view.

The solution of such system fixes the coefficients of $H_n(x)$ which in the Tau Method terminology are called the tau-parameters. Different techniques have been developed to simplify further the construction of the sequence Q in a step by step formulation of the Tau Method (see Ortiz [18]; Onumanyi and Ortiz [14] and Hosseini Ali Abadi and Ortiz [5]).

Most of what we have said carries over to the case when $D \in \mathbf{D}$, is a operator depending on more than one variable. However, in this case the algebraic kernel of D plays a far more important role than before. Details on this approach and its applications to the numerical solution of partial differential equations are discussed in a forthcoming paper of Hosseini Ali Abadi and Ortiz [6].

(ii) The Operational formulation of the Tau Method

The objective of this approach is to reduce the Tau approximation problem to one of (linear) algebra. Hence its name.

In this formulation of the Tau Method the differential operator D is replaced by an algebraic one which, operated on an approximate polynomial expansion of the exact solution y(x), is capable of producing the same effect on its coefficients as D does. Ortiz and Samara [26] showed that such algebraic operator can easily be determined in terms of a couple of elementary matrices of a very simple structure:

$$\mu := (\mu_{ij}) : \mu(ij) := \delta_{i+1\ j}$$

and

$$\eta := (\eta_{ij}) : \eta_{ij} := (j+1)\ \delta_{i\ j+1}$$

with $\delta_{ij} := 1$, for $i = j$ and zero otherwise. They have only one line, parallel to the main diagonal, with non-zero elements. These matrices are related to the effect of differentiation and multiplication by the variable on the definition of the coefficients of a given polynomial expression. Ortiz and Samara have proved that if $a(x) := \mathbf{a}^T \mathbf{x}$, where $\mathbf{a} := (a_0, a_1, a_2, \ldots)^T$ and $\mathbf{x} := (x^0, x^1, x^2, \ldots)^T$, then

$$x^i\ [d^j/dx^j]\ a(x) := \mathbf{a}^T \eta^j \mu^i \mathbf{x}.$$

It should be remarked that, but for a multiplier, the latter is the elementary expression of any operator D ∈ D.

The observation that the supplementary conditions can also be interpreted as the result of another operator (a point evaluation functional in the case of ordinary differential equations, a differential operator of a lower dimension, defined on the boundary, in the case of partial differential equations) acting simultaneously on the solution, allowed these authors to find an extremely simplified formulation of the Tau Method in which these two parts of the problem, which have different dimensionality and represent in general interior and boundary of the problem, are combined in a matrix formulation which implicitly defines the coefficients of the Tau approximation.

A similar argument can be used for partial differential equations, details are given in the above mentioned paper of Ortiz and Samara [26] and in more recent papers of Ortiz and Pun [21]-[22] in which a numerical technique for the tretament of nonlinear problems is developed. For a theoretical discussion of the convergence of such schemes for nonlinear ordinary and partial differential equations see Ortiz and Pham Ngoc Dinh [24] and [25] respectively.

The equivalence between the latter and the Recursive formulation of the Tau Method has been discussed in Ortiz and Samara [27].

SINGULARITY BEHAVIOUR AND THE TAU METHOD

The problem to be considered in this paper is defined by

$$u(x,y) = 0 \text{ in } G,$$

and

$$u_n|_{OA} = 0,$$
$$u|_{AB} = 0,$$
$$u_n|_{BC} = 0,$$
$$u|_{CD} = 1,$$
$$u_n|_{DE} = 0,$$
$$u_n|_{EO} = 0;$$

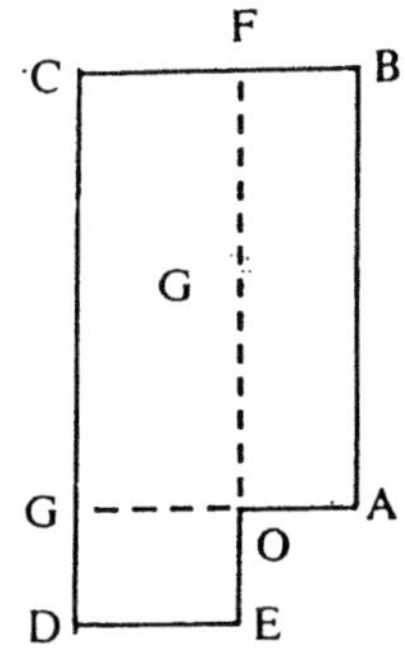

Figure 1

(1)

where G is the L-shaped domain OABCDE and u_n stands for the normal derivative of u(x,y) on given sections of the boundary of G. Results reported in this paper have been obtained by using the Operational formulation of the Tau Method.

We shall segment G into three non overlaping subdomains $G^{(i)}$, i = 1, 2, 3, with common sides OF and OG, which are parallel to the sides AB and DE respectively. The x-axis is taken on the direction OA and the y-axis on that of OF. A Tau Method bivariate approximation satisfying the given boundary conditions can be constructed easily by using the Tau-Elements approach (see Ortiz and Pun [22] and Ortiz [20]) on each of the subdomains $G^{(i)}$. The Tau solution is required to satisfy implicit matching conditions for the function and its normal derivative across the sides OF and OG of the segmented domain. The matrix corresponding to this problem has an almost block diagonal structure; it was inverted by using De Boor and Weiss´ subroutine SOLVEBLOK (see [1]). In all the calculations presented in this paper the perturbation terms are defined in a product basis (see Ortiz [19]).

The exact solution of problem (1) admits an expansion which, in polar coordinates centered at the point O, is given by

$$u(r,\theta) := a_0 + a_1 r^{2/3} \cos [2\theta/3] + a_2 r^{4/3} \cos [4\theta/3] + \dots \tag{2}$$

where $r := [x^2 + y^2]^{1/2}$ and $\theta := \arctan [y/x]$. The known form of the singularity expansion will be taken into account in the formulation of a Tau Method approximate solution.

Let us express a Tau bivariate approximation of degree N in x and M in y as:

$$u_{NMk}(x,y) := a_1 r^{2/3} \cos [2\theta/3] + a_2 r^{4/3} \cos [4\theta/3] + \dots + v_{NM}(x,y),$$

where k indicates the number of terms of the singular expansion retained for u_{NMk} and v_{NM} is a bivariate polynomial term of degrees N and M in x and y respectively. We call $u_{NMk}(x,y)$ a <u>Singular Tau Method</u> approximation of u(x,y) on G. The constant term a_0 is absorbed into the expression of $v_{NM}(x,y)$.

In the case of k = 2, conditions to be satisfied on the boundary of G (see (1)) become

$$[v_{NM}]_n|_{OA} = 0,$$
$$a_1\, r^{2/3} \cos [2\theta/3] + a_2\, r^{4/3} \cos [4\theta/3] + v_{NM}|_{AB} = 0,$$
$$a_1\, [2/3]\, r^{-1/3} \sin [\theta/3] - a_2\, [4/3]\, r^{1/3} \sin [\theta/3] + [v_{NM}]_n|_{BC} = 0,$$
$$a_1\, r^{2/3} \cos [2\theta/3] + a_2\, r^{4/3} \cos [4\theta/3] + v_{NM}|_{CD} = 1,$$
$$a_1\, [2/3]\, r^{-1/3} \sin [\theta/3] - a_2\, [4/3]\, r^{1/3} \sin [\theta/3] + [v_{NM}]_n|_{DE} = 0,$$
$$[v_{NM}]_n|_{EO} = 0. \qquad (3)$$

Having divided G into disjoint subdomains or elements $G^{(i)}$, let $u^{(i)}_{NMk}(x,y)$ be the Singular Tau Method approximation corresponding to the subdomain $G^{(i)}$. The set of $u^{(i)}_{NMk}(x,y)$, for all values of the index i, will be called a bidimensional Singular Tau-Elements Method approximation of the given problem in G. The two extra parameters a_1 and a_2 require two extra conditions. We could either use boundary perturbation (see Ortiz and Pun [23]) or impose the following two extra conditions:

$$u^{(2)}_{NM2}(0,0) := u^{(3)}_{NM2}(0,0)$$

and

$$(\partial/\partial y)\, u^{(1)}_{NM2}(x,y)|_{x=0,\ y=0} := 0,$$

we shall use this last procedure.

Let us consider the following irregular L-shaped domain G which can be inscribed in a rectangle R from which the "missing" corner is small relative to the dimensions of R.

Let

AB = 8; BC = 5; CD = 10; DE = 3 and EO = OA = 2.

A comparison between results obtained with the Singular Tau-Elements Method and those reported by Symm [28] are given in the following Table. More detailed results for different configurations of the L-shaped domain can be found in Ortiz and Pun [23]. In the last column of the table we have indicated the CPS time in seconds taken by the CDC Cyber 174 Computer System of Imperial College to produce these results.

For N = M = 6, and taking a Chebyshev product basis, the estimates of the singular coefficients a_1 and a_2 agree up to $1\mathrm{x}10^{-4}$ and $6\mathrm{x}10^{-4}$ respectively; the maximum absolute error in any of the three elements $G^{(i)}$ is below $4\mathrm{x}10^{-4}$. The computation was continued up to N = M = 9; numerical results are not reported in the Table because they remain stationary, indicating that for that number of singular terms the accuracy attainable with the Singular Tau-Elements Method had been saturated. Computer times required to solve this problem are close to those found for regular L-shaped problems, suggesting than the regularity of the L-shaped domain is not a dominant factor in this approximation.

Table: Numerical solution of Laplace´s equation on an irregular L-shaped domain with the bidimensional Singular Tau - Elements Method.

M=N	a_1	a_2	Maximum Absolute Error $G^{(1)}$	$G^{(2)}$	$G^{(3)}$	CPS Time (sec)
3	-0.2489	-0.0688	$5.84 x 10^{-3}$	$5.42 x 10^{-3}$	$2.49 x 10^{-3}$	0.4
4	-0.2569	-0.0637	$4.87 x 10^{-4}$	$8.93 x 10^{-4}$	$3.94 x 10^{-4}$	0.9
5	-0.2572	-0.0658	$2.77 x 10^{-4}$	$2.77 x 10^{-4}$	$2.58 x 10^{-4}$	2.0
6	-0.2572	-0.0656	$1.35 x 10^{-4}$	$1.21 x 10^{-4}$	$3.11 x 10^{-4}$	4.4
Reference values given by Symm [28]: a_1 = -0.2573; a_2 = -0.0662.						

The same computations were repeated taken for the perturbation terms a Legendre product basis instead of a Chebyshev one. For each value of N and M estimates of a_1 and a_2, maximum absolute errors in each element $G^{(i)}$ and CPS times were only slightly different to those reported in the previous tables.

Our computations suggest that the values reported by Symm [28] for the singular coefficients of this problem may be accurate to three decimal places. Slightly more accurate estimates, which follow from our calculations are:

a_1 := -0.25723 and a_2 := -0.0656.

Papamichael and Whiteman have reported numerical results for the same problem in [29]; they used a refined technique based on the Conformal Transformation Method. Their results agree with our estimates, constructed with low degrees of approximation (N = M = 5; CPS time 0.938 seconds) and also with the results reported by Symm, although with a loss of accuracy near the singular point O.

FINAL REMARKS

The Singular Tau-Elements Method seems to be a technique suitable for the analysis of singular boundary value problems for partial differential equations which deserves further research. A more detailed treatment of the problems discussed here will be found in Ortiz and Pun [23], where techniques for boundary perturbation are discussed in some detail. Numerical results for accurate solutions of regular and irregular L-shaped problems are given there. They are probably the most accurate estimates of stress intensity factors and related singular coefficients in the technical literature on these model problems.

REFERENCES

1.- C. de Boor and R. Weiss, "SOLVEBLOK: a package for solving almost block diagonal linear systems", ACM TOMS, 6 (1980) pp. 80-87.

2.- A.E.M. El Misiery and E.L. Ortiz, "Numerical solution of regular and singular biharmonic problems with the Tau-Lines Method", Comm. in Appl. Numer. Methods, 1 (1985) pp. 281-285.

3.- A.E.M. El Misiery and E.L. Ortiz, "Tau-Lines: a new hybrid approach to the numerical treatment of crack problems based on the Tau Method", Comp. Meth. in Appl. Mech. and Engng., 56 (1986) pp. 265-282.

4.- J.H. Freilich and E.L. Ortiz, "Numerical solution of systems of differential equations with the Tau Method: an error analysis", Maths. Comp. 39 (1984) pp. 189-203.

5.- M. Hosseini Ali Abadi and E.L.Ortiz, "Numerical solution of feedback control systems", Letters in Applied Mathematics (1987) in press.

6.- M. Hosseini Ali Abadi and E.L.Ortiz, "Algebraic kernels of differential operators and the numerical solution of the partial differential equations" (to appear elsewhere).

7.- C. Lanczos, "Trigonometric interpolation of empirical and analytical functions", J. Math. Phys, 17 (1938) pp.123-199.

8.- Liu, K. M. and Ortiz, E. L. , "Approximation of eigenvalues defined by ordinary differential equations with the Tau Method", in "Matrix Pencils", B. Kagstrom and A. Ruhe, eds., Springer-Verlag, Berlin (1983) pp. 90-102.

9.- K.M. Liu, E.L. Ortiz and K.-S. Pun, "Numerical solution of Steklov´s partial differential eigenvalue problem", in "Boundary and Interior Layers", J.J.H. Miller, ed., Boole Press (1984) pp. 244-249.

10.- Liu, K. M. and Ortiz, E. L., "Tau Method approximation of eigenvalue problems where the spectral parameter enters nonlinearly", J. Computational Physics, 76 (1987) in press.

11.- K.M. Liu and E.L. Ortiz, "Tau Method approximate solution of high order differential eigenvalue problems defined in the complex plane, with an application to Orr-Sommerfeld stability equation", Comm. Appl. Numer. Methods, 2, (1987) in press.

12.- S. Namasivayam and E.L. Ortiz, "Best approximation and the numerical solution of partial differential equations", Portugaliae Mathematica, 40 (1985) pp. 97-119.

13.- Namasivayam, S. and Ortiz, E. L., "Dependence of the local truncation error on the choice of perturbation term in the step by step Tau Method for systems of differential equations" (to appear elsewhere).

14.- P. Onumanyi and E.L. Ortiz, "Numerical solution of stiff and singularly perturbed boundary value problems with a segmented-adaptive formulation of the Tau Method", Math. Comput., 43 (1984) pp. 189-203.

15.- E.L. Ortiz, "The Tau Method, SIAM J. Numer. Analysis", 6 (1969) pp. 480-491.

16.- Ortiz, E. L. , "A recursive method for the approximate expansion of functions in a series of polynomials", Computer Physics Communications, IV, 2 (1972) pp. 151-156.

17.- Ortiz, E. L. , "Canonical polynomials in the Lanczos´ Tau Method", in "Studies in Numerical Analysis, Lanczos´ Festschrift", B.P.K.Scaife, ed., Academic Press, New York (1974) pp.73-93.

18.- Ortiz, E. L. , "Step by step Tau Method: Piecewise polynomial approximations", Comp. and Maths. with Appli., 1 (1975) pp. 381-392.

19.- E.L. Ortiz, "Polynomial condensation in one and several variables with applications", in "Topics in Numerical Analysis III", J.J.H. Miller, ed., The Royal Irish Academy, Academic Press, New York (1977) pp. 327-360.

20.- E.L. Ortiz, "Recent Progress in the Numerical Treatment of Singular Problems for Partial Differential Equations with Techniques Based on the Tau Method", in "Numerical Approximation of Partial Differential Equations", E.L. Ortiz, Ed., North Holland, Amsterdam (1987) pp. 83-98.

21.- E.L. Ortiz and K.-S.Pun, "Numerical solution of nonlinear partial differential equations with the Tau method", J. Comp. and Appl. Math., 12 & 13 (1985) pp. 511-516.

22.- E.L. Ortiz and K.-S.Pun, "A bi-dimensional Tau-Elements Method for the numerical solution of nonlinear partial differential equations with an application to Burgers´ equation", Computers and Maths. with Appli., 12 (1986) pp. 1225-1240.

23.- E.L. Ortiz and K.-S.Pun, "Numerical solution of Laplace´s equation in L-shaped domains with a bidimensional formulation of the Tau Method" (to appear elsewhere).

24.- Ortiz, E. L. and Pham, A., "On the convergence of the Tau Method for nonlinear differential equations of Riccati´s type", Nonlinear Analysis, 9 (1985) pp. 53-60.

25.- E.L. Ortiz and A. Pham Ngoc Dinh, "Linear recursive schemes associated with some nonlinear partial differential equations and their numerical solution with the Tau Method", SIAM J. Math. Anal., 18 (1987) pp. 452-464.

26.- E.L. Ortiz and H. Samara, "Numerical solution of partial differential equations with variable coefficients with an operational approach to the Tau Method", Comp. and Maths. with Appli., 31 (1984) pp. 95-103.

27.- E.L. Ortiz and H. Samara, "Some equivalence results concerning a class of polynomial methods for the numerical solution of differential equations", Imperial College, Res. Rep. NAS-78, (1978) pp. 1-36.

28.- G.T. Symm, Treatment of singularities in the numerical solution of Laplace´s equation by an integral equation method, Res. Rep. NAC 31.01.1973, Nat. Phys. Lab. (U.K.) (1973).

29.- J.R. Whiteman and N. Papamichael, "Numerical solution of two dimensional harmonic boundary value problems containing singularities by Conformal Transformation Methods", Res. Rep. No. TR/2, Mathematics Department, Brunel University (1971).

A FOURIER BOUNDARY CONDITION FOR PANEL METHOD

J. Ryan, T.H. Lê, Y. Morchoisne
Office National d'Etudes et de Recherches Aérospatiales
29, Avenue de la Division Leclerc, Châtillon, France

INTRODUCTION

The flow of an incompressible, inviscid fluid past three-dimensional bodies can be calculated with an integral representation of the potential, using an internal Dirichlet boundary condition deduced from the external Neumann boundary condition [1]. A low order panel discretisation of the resulting equation gives rise to a set of linear equations. The matrix of which, in the case of thick bodies is usally well conditioned. Applying this method to a thin wing produces a matrix with a condition number which increases greatly with the thinness of the body and the number of panels used in the discretisation. The calculated potential becomes very inaccurate in the thin areas (i.e. near the trailing edge of wings, canards, flaps, spoilers, etc...).

Two ways could be seen to solve this problem. The first was to keep the Dirichlet formulation and to increase the discretisation order [2]. The second one was to change the internal boundary condition and to keep a low order panel discretisation. For an infinitely thin body, the continuous problem is ill-defined. Though an increase of the panel method order can slightly improve the discretised problem, the essential problem would not be solved.

The second way presented here is to replace the internal Dirichlet boundary condition by an equivalent Fourier condition. This approach is shown to be very effective in the case of a thin lift augmenting wing, barely increasing the computational cost and not disturbing the potential values obtained on thick bodies with the Dirichlet method.

The plan of this article is the following :

We first analyse the reason why the Dirichlet (and also the Neumann) formulation linked to a low order panel method fails when applied to thin bodies ; from this analysis, it is seem that a linear combination of Dirichlet and Neumann boundary conditions (known as a Fourier condition*) forbids the matrix degeneracy.

The integral representation based on this Fourier formulation is then developed.

First applications are made on flattened spheres of varying thicknesses (where analytical solutions are known) which show the precision of this new formulation.

Then the case of a thin high lift wing where the Dirichlet method breaks down is successfully computed by the Fourier method.

---

*N.B : In anglo-saxon literature, this condition is also known as a Robin condition.

PROBLEM FORMULATION

The flow potential Φ round an aerodynamical body with boundary Γ is governed by the following equations :

$$\Delta\Phi = 0 \qquad \text{in } \Omega' \text{ the exterior fluid domain ,} \tag{1}$$

$$\left.\frac{\partial\Phi}{\partial n}\right|_{\Gamma} = 0 \qquad \text{where n is the external unit surface normal.}$$

This full potential Φ can be seen as $\Phi = \phi + \phi_\infty + \phi_W$ where

ϕ_∞ is the free stream velocity potential ($V_\infty = \nabla\phi_\infty$),

ϕ_W the wake velocity potential ($V_W = \nabla\phi_W$)

and ϕ the perturbation potential so that (1) becomes (2) :

$$\Delta\phi = 0 \qquad \text{in } \Omega' ,$$

$$\left.\frac{\partial\phi}{\partial n}\right|_{\Gamma} = -\,(V_\infty + V_W).n \, , \tag{2}$$

$$\phi(x) \to \phi_\infty(x) ,$$

$$x \to +\infty .$$

The solution of this problem is unique and can be written by means of an integral representation (see for example [3]) :

$$\forall\, x \in R^3 - \Gamma ,\ \phi(x) = -\frac{1}{4\pi}\int_\Gamma \mu(y)\,\frac{\partial}{\partial n_y}\frac{1}{r}\,d\gamma(y) + \frac{1}{4\pi}\int_\Gamma \sigma(y)\,\frac{1}{r}\,d\gamma \, , \tag{3}$$

$$\forall\, x \in \Gamma ,\ \frac{\phi^+(x) + \phi^-(x)}{2} = -\frac{1}{4\pi}\int_\gamma \mu(y)\,\frac{\partial}{\partial n_y}\frac{1}{r}\,d\gamma(y) + \frac{1}{4\pi}\int_\Gamma \sigma(y)\,\frac{1}{r}\,d\gamma \tag{4}$$

where $r = |x - y|$ is the distance between the two points x and y of $\mathbf{R}^3$, ϕ^+ is the external limit of the potential on the boundary, ϕ^- the internal limit, the doublet $\mu = \phi^- - \phi^+$ is the potential jump across the boundary and the source $\sigma = \frac{\partial\phi^-}{\partial n} - \frac{\partial\phi^+}{\partial n}$ the normal derivative jump across the boundary.

Applying an internal Dirichlet boundary condition $\phi^- = 0$ in equation (2) gives the following equation in μ :

$$\forall\, x \in \Gamma ,\ 2\pi\,\mu(x) - \int_\Gamma \mu(y)\,\frac{\partial}{\partial n}\frac{1}{r}\,d\gamma(y)$$
$$= -\int_\Gamma \frac{(V_\infty + V_W)(y).n(y)}{r}\,d\gamma(y) \tag{5}$$

which has a unique solution.

Applying the gradient operator to equation (4), and using an inner Neumann boundary condition gives :

$$\forall x \in \Gamma,\ 2\pi(V_\infty + V_W).n(x) - \int_\Gamma \mu(y) \frac{\partial}{\partial n_x} \frac{\partial}{\partial n_y} (\frac{1}{r})\, d\gamma(y)$$

$$= \int_\Gamma (V_W + V_\infty) n_y \frac{\partial}{\partial n_x} (\frac{1}{r})\, d\gamma(y)\ . \qquad (6)$$

The doublet distributions (μ) given by equations (5) and (6) determine the same potential outside the body.

Both equations (5) and (6) are discretised by use of a low order panel scheme, the doublet and source density μ and $(V_\infty + V_W).n$ are set to be constant per panel and the integration on a panel P with edge ∂P of $\frac{\partial}{\partial n_x} \frac{\partial}{\partial n_y} \frac{1}{r}$ is transformed by the Biot-Savard Law into $\int_{\partial P} \frac{ds \times R}{R^3}$ where ds is the differential vector arc length along ∂P. The potential is computed at a finite set of points $(x_i)_{i \in I}$ of Γ, called collocation points.

Thus, the discretisation of (5) and (6) gives two finite sets of linear equations denoted symbolically by :

- Dirichlet condition (5)
  A M = a,
- Neumann condition (6)
  B m = b.

where M is the vector of unknown singularities

a and b are the vectors of specified boundary conditions and A and B are the matrices of "influence coefficients".

For thick bodies, theses two linear systems have a low condition number, and are easily solved.

In the case of thin bodies, two collocation points can be very closed to each other while belonging to panels with opposite normal vectors.

This implies for matrix A two nearly identical lines, and for matrix B, two nearly opposite lines. Matrices A and B have a high condition number and solving these systems can become quite problematic.

More precisely, the two matrices look like this :

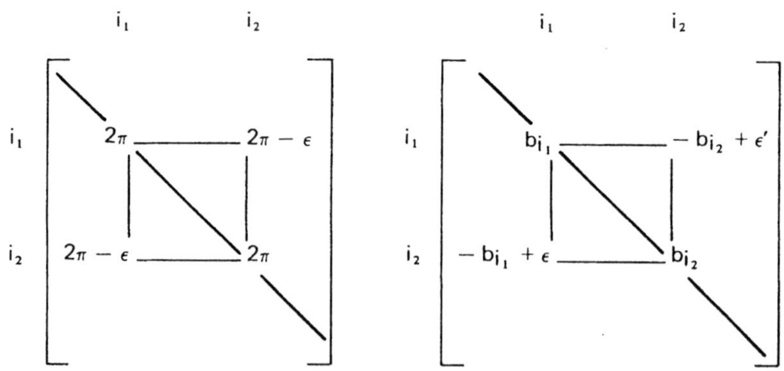

A B

- For matrix A, all diagonal terms are equal to 2π, large non diagonal terms, influence coefficients of opposite panels, are close to 2π, and the remaining terms are small and decrease with the thickness.

- For matrix B, all diagonal terms are positive, and are $\theta(1/l)$ where l is a characteristic length of a panel, large non diagonal terms in matrix A, are similar to the diagonal term but with opposite sign, and the other terms remain bounded when the thickness decreases.

From these remarks, it can be seen that the linear combination $A + \beta B$, with $\beta > 0$, $\beta \backsim \frac{2\pi}{b}$ defines the following strong diagonal dominant matrix :

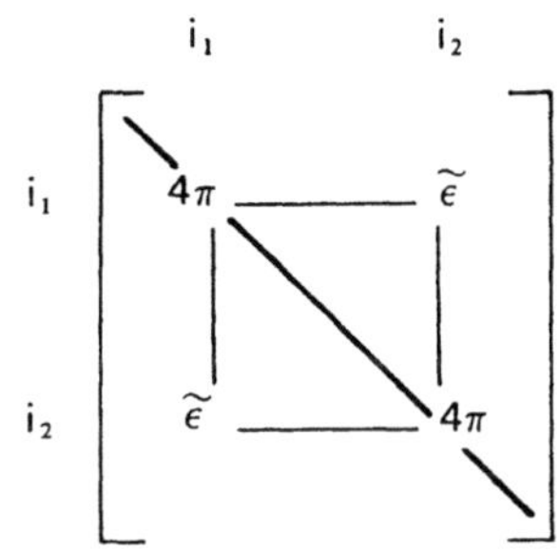

Such a linear combination is nothing but the matrix version of a Fourier condition $\phi^- + \beta \frac{\partial \phi^-}{\partial n} = 0$, linear combination of a Dirichlet and a Neumann boundary condition.

Therefore, one way of treating thin bodies, within panel methods, is to replace the usual Dirichlet inner boundary condition by a Fourier condition. This condition, for thick bodies and the same normal velocity jump, is formally equivalent to the Dirichlet condition and yields the following integral equation :

$$2\pi\, \mu(x) - \int_\Gamma \mu(y) \frac{\partial}{\partial n_y} \frac{1}{r}\, d\gamma(y) - \beta(x) \int_\Gamma \mu(y) \frac{\partial}{\partial n_x} \frac{\partial}{\partial n_y} \left(\frac{1}{r}\right) d\gamma(y)$$

$$= - \int_\Gamma \frac{(V_W + V_\infty).n(y)}{r}\, d\gamma(y) - \beta(x)\; 2\pi (V_W + V_\infty).n(x) \tag{7}$$

$$+ \int_\Gamma (V_W + V_\infty).n(y) \frac{\partial}{\partial n_x} \frac{1}{r}\, d\gamma(y) .$$

As when discretised, the Dirichlet and Neumann diagonal terms are both positive, and large non diagonal terms have opposite signs, the optimum choice for the function β, to obtain a strong diagonal matrix, is to be positive and related to the characteristic length of the panel.

With this choice, the matrix will easily be inverted (see for example [4]). For convex bodies, the diagonal dominance is in inverse ratio to the body thickness.

NUMERICAL RESULTS

To test the precision of the Fourier condition, first computations were done for flattened spheres, where the exact solution is known.

The general equation of these ellipsoïds is $\frac{x^2}{\varepsilon^2} + y^2 + z^2 = 1$ for values of ε ranging from 1 to 10^{-6}.

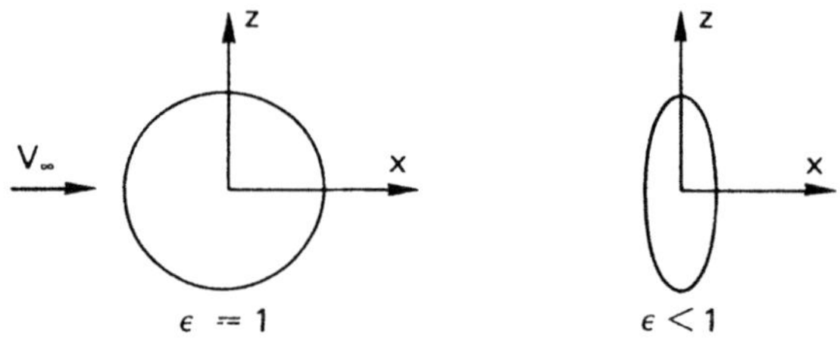

Half the sphere is meshed with 512 panels.

A Dirichlet computation takes 19s on a CRAY1S while a Fourier computation takes 22s, representing a 15% increase in cost.

The error norm used is the relative l_∞ norm.

$$E_C = \frac{\mathrm{Sup}_{i \in I} \mid V_{ex}(x_i) - V_{comp}(xi) \mid}{\mathrm{Sup}_{i \in I} \mid V_{ex}(x_i) \mid} ,$$

x_i are the barycentres of panels Γ_i indexed by I,

V_{ex} is the exact velocity and

V_{comp} the velocity computed with a low order panel method.

Results obtained are shown in the table below :

ε	Dirichlet	Fourier
1	0.7 %	0.1 %
10^{-1}	0.75%	0.15%
10^{-2}	0.85%	0.3 %
10^{-3}	1.5 %	1 %
10^{-4}	the matrix can no longer be numerically inverted	1.5 %
10^{-5}		1.5 %
10^{-6}		1.5 %

As can be seen from this table, the Fourier condition is more precise on thick ellipsoïds, and remains good when the body becomes very thin.

Another application of the Fourier method was computing the flow past a rectangular wing with a Karman Trefftz (K.T.) airfoil section (see fig. 1).

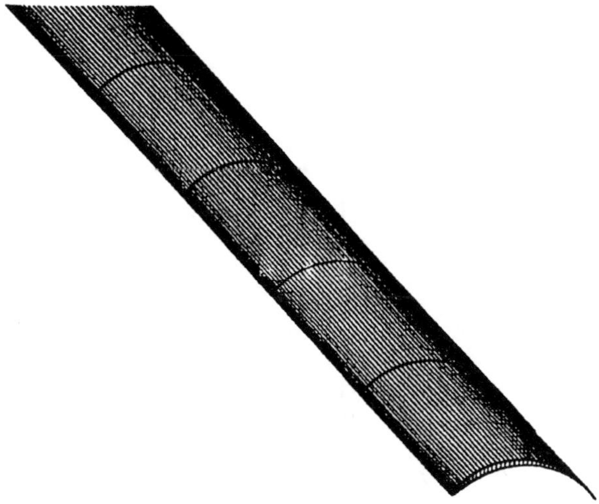

Fig. 1 — Panel distribution

Figures 2 to 4 show velocity distribution for different meshes of the K.T. wing.

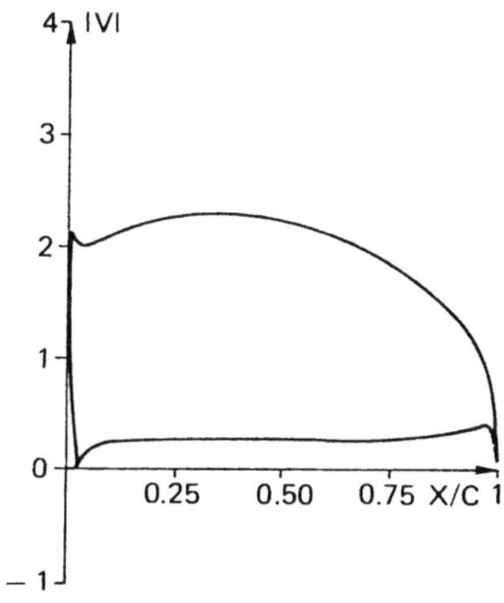

Fig. 2 — Coarse mesh Dirichlet-Fourier solution

On figure 2 is shown the common velocity distribution for a coarse mesh (50 x 10) along an airfoil section close to the plane of symmetry, the Fourier method and the Dirichlet method producing the same results.

On figure 3, is shown the velocity distribution for a finer mesh (100 x 10) along the same section as above and given by the Dirichlet method ; on figure 4, the velocity distribution computed by the Fourier method.

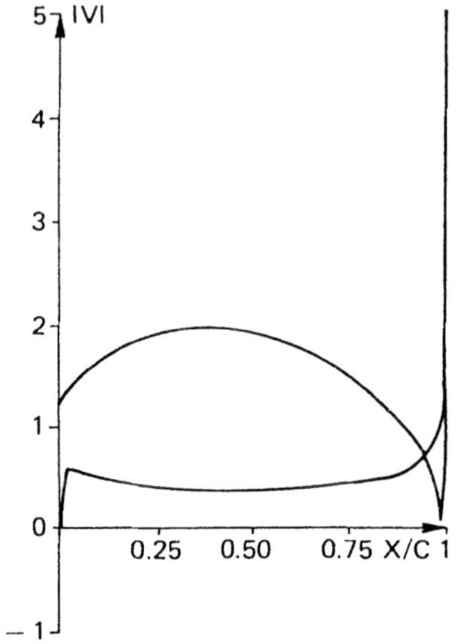

Fig. 3 – Fine mesh Dirichlet solution

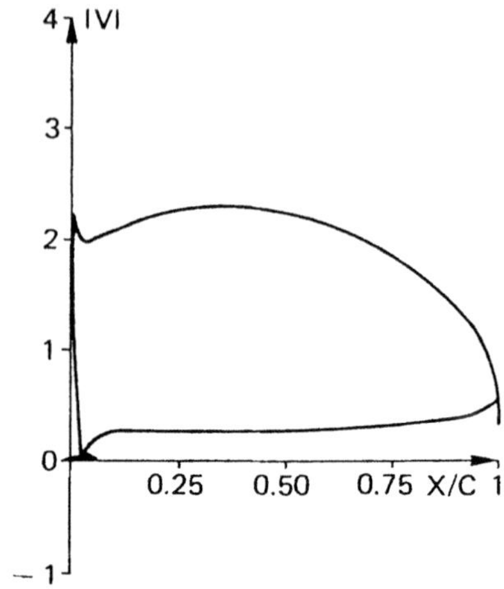

Fig. 4 – Fine mesh Fourier solution

Mesh refining for Fourier method (fig. 2 and 4) improved the smoothness of the velocity curve near the stagnation point and at the trailing edge.

While the Fourier method always produces satisfactory results, the Dirichlet method is shown to become unreliable for thicknesses dependent on the number of panels in the discretisation. This is easily understood as a greater amount of panels gives a sharper definition of the thin trailing edge where the problem lies.

CONCLUSION

In view of the above results, it can be concluded that this new approach with an internal Fourier boundary condition added to a low order panel discretisation is a reliable (and cheap) method for the study of flows past thin bodies. This will allow the analysis of very complex systems with thin and thick elements such as wing + flaps + spoilers + winglets + ... and with wakes delimiting thin dead-air regions.

REFERENCES

[1] T.H. Lê, Y. Morchoisne, J. Ryan : Techniques numériques nouvelles dans les méthodes de singularités pour l'application à des configurations tridimensionnelles complexes. Congrès AGARD, Aix en Provence, Avril 1986.

[2] D.R. Bristow and G.G. Grose : Modification of the Douglas Neuman Program. To improve the effeciency of predicting component interference and high lift characteristics. NASA Contractor Report 3020 (1978).

[3] J.C. Nédelec : Approximation des équations intégrales en mécanique et en physique. Cours à l'Ecole Polytechnique. Juin 1977.

[4] H. Boillot, T.H. Lê : CRAY1S asynchronous I/O technics and block management on boundary element code. First Word Congress on Computational Mechanics. Austin, Texas, 1986.

ON THE EFFECT OF NUMERICAL QUADRATURES IN SOLVING BOUNDARY INTEGRAL EQUATIONS

J. Saranen

Section of Mathematics, Faculty of Technology,
University of Oulu, Linnanmaa, SF-90570 Oulu, Finland

SUMMARY

In this article we derive new error estimates for basic numerical methods for boundary integral equations taking also the effect of numerical quadratures into account. The proof of these estimates is based on new stability estimates for discretized equations. Practical examples show that the new results describe the effect of quadratures more precisely than previous estimates.

INTRODUCTION

We consider solution of boundary integral equations by the main discretization methods such as the Galerkin method, collocation method and the least-squares method. For practical solution of boundary integral equations one has to utilize numerical integration for determination of the stiffness matrix and of the right hand side. In this process it is valuable to know beforehand the minimal degree of accuracy in numerical integration which preserves essential convergence properties of the proposed discretization method. The effect of numerical quadratures has been studied extensively in the case of differential equations [5]. Also some special integral equations, especially Fredholm equations of second kind, have been analysed [3], [4] . For boundary integral equations very general results are achieved by Wendland [13], [14]. The effect of numerical integration is expressed by giving asymptotic estimates for the corresponding approximate solution. The estimates obtained in [13], [14] are in a good accordance with numerical experiments but still too pessimistic if solution of potential problems is considered.

Here we present a new analysis of the numerical error. We are able to derive asymptotic error estimates improving those derived by Wendland. Numerical experiments [8] have shown that these new results describe the effect of the numerical error more precisely than the existing results.

As earlier studies our argumentation is based on consistency of the numerical approximation and on stability properties of the discretized equation. But instead of the conventional L_2-stability of the matrix equation we use an other notation of stability. This stability notationdepends on the order of the corresponding integral operator.

The analysis presented here is largely independent on the dimension of the boundary and on the character of trial functions used for the approximation. However we don't aim at the most generality and therefore discuss explicitly only the case where the boundary is a curve and where the trial functions are smoothest splines. The operators are considered in the frame of strongly elliptic pseudodifferential operators of the real order 2α. Our results coincides with those derived by Wendland only in the case of zero order $\alpha = 0$. If $|\alpha|$ increases the results differ significantly from the earlier ones. Practical consequences of this effect for the choice of quadrature rules are discussed in [8].

1. ASYMPTOTIC ERROR ESTIMATES

For simplicity we consider the equation

$$Au = f \tag{1.1}$$

where $A: H^{s+\alpha} \to H^{s-\alpha}$ is an isomorphism for all Sobolev spaces $H^s = H^s(\Gamma)$, $s \in \mathbb{R}$. Here Γ is a smooth closed curve of the complex plane. The operator A is assumed to be a strongly elliptic pseudodifferential operator on the curve Γ. Generalizations to the case where the operator A is not an isomorphism and to the case of elliptic systems are obvious [1], [13] and therefore omitted here.

We consider the space $S_h^d = S_h^d(\Delta)$ of the smoothest splines of the degree d with respect to the mesh $\Delta = \{x_k\}_1^N$. It is assumed that the mesh is quasiuniform. In the following $||\cdot||_t$ is the Sobolev norm and $(\cdot|\cdot)$ denotes the L_2-inner product.

We recall the asymptotic error estimates for the basic numerical methods. First we consider the Galerkin approximation $u_h \in S_h^d$ defined by

$$(Au_h|v) = (Au|v) , \quad v \in S_h^d . \tag{1.2}$$

We assume that $S_h^d \subset H^\alpha$ i.e. that it holds $\alpha < d + \frac{1}{2}$. The following result is true:

Theorem 1.1: *Assume that* $u \in H^s$ *for some* s *such that* $2\alpha - d - \frac{1}{2} < s \leq d + 1$. *Then the equations* (1.2) *have a unique solution for small* h *and the asymptotic error estimate*

$$||u - u_h||_t \leq ch^{s-t}||u||_s \tag{1.3}$$

is valid for $t \leq s$, $2\alpha - d - 1 \leq t < d + \frac{1}{2}$.

This result appeares in [7] under the assumption that u belongs to the "energy space" H^α. The extension to the case $2\alpha - d - \frac{1}{2} < s \leq \alpha$ was carried out by Ruotsalainen and Saranen in [9]. In particular we observe by (1.3) that Galerkin method is stable with respect to the Sobolev-norms corresponding to the values $2\alpha - d - \frac{1}{2} < s < d + \frac{1}{2}$, i.e. it holds

$$||u_h||_s \leq c\, ||u||_s \tag{1.4}$$

for those values of s.

The second example is the <u>collocation method</u> [1], [2], [6], [11], [12] . Here we assume that it holds $d > 2\alpha$ if d is an odd integer and that $d > 2\alpha - \frac{1}{2}$ if d is an even integer. If d is an odd integer we consider collocation

$$(Au_h)(t_k) = (Au)(t_k) \quad k = 1,\dots,N \tag{1.5}$$

at the nodes $t_k = x_k$ of the mesh and for even integers d we require the collocation equations at the midpoints t_k of the mesh. In this case the mesh is assumed to be smoothly graded in the sense of [2]. We have

<u>Theorem 1.2:</u> *Assume that* $u \in H^s$, $2\alpha + \frac{1}{2} < s \leq d + 1$. *For sufficiently small* h *there exists a unique collocation solution* $u_h \in S_h^d$ *with the asymptotic error estimate*

$$||u - u_h||_t \leq ch^{s-t}||u||_s \tag{1.6}$$

for $2\alpha \leq t < d + \frac{1}{2}$, $t \leq s$.

Thus the stability is valid for values $2\alpha + \frac{1}{2} < s < d + \frac{1}{2}$. The general result of Theorem 1.2 is not enough to describe the convergence of the approximate potential, if we apply mid-point collocation with trial functions of even degree to classical potential problems. In this case the pseudodifferentialoperator A has the special property that the principal symbol is an even function of the Fourier transformed variable and we have by [10]

$$||u - u_h||_{2\alpha-\tau} \leq ch^{d+1+\tau-2\alpha}||u||_{d+1+\tau} \tag{1.6'}$$

if $u \in H^{d+1+\tau}$, $0 \leq \tau \leq 1$.

Finally we consider the <u>least-squares approximation</u> $u_h \in S_h^d$ which is determined by the requirement, $S_h^d \subset H^{2\alpha}$ and

$$||Au_h - Au||_o = \min_{v \in S_h^d} ||Av - Au||_o \tag{1.7}$$

or equivalently

$$(Au_h|Av) = (Au|Av) \quad , \quad v \in S_h^d . \tag{1.8}$$

Since (1.8) is Galerkin method for the operator A*A we have by Theorem 1.1.

<u>Theorem 1.3:</u> *Assume that* $u \in H^s$, $4\alpha - d - \frac{1}{2} < s \leq d + 1$. *For sufficiently small* h *there exists a unique least-squares solution* u_h *of* (1.7) *and the asymptotic error estimate*

$$||u - u_h||_t \leq ch^{s-t}||u||_s \tag{1.9}$$

is valid for $t \leq s$, $4\alpha - d - 1 \leq t < d + \frac{1}{2}$.

By (1.9) least-squares method is stable for the values s such that $4\alpha - d - \frac{1}{2} < s < d + \frac{1}{2}$.

2. NUMERICAL INTEGRATION

Let $\{\mu_j\}_1^N$ be the functions of the standard basis in the spline space S_h^d. For the practical implementation of the boundary element methods one has to compute numerically the enries a_{ij} of the stiffness matrix and the values f_i in the right hand side. For the Galerkin method we have

$$a_{ij} = (A\mu_j|\mu_i) \ , \quad f_i = (f|\mu_i) \ .$$

For the collocation method holds

$$a_{ij} = (A\mu_j)(t_i) \ , \quad f_i = f(t_i) \ ,$$

and for the least-squares method we have

$$a_{ij} = (A\mu_j|A\mu_i) \ , \quad f_i = (f|A\mu_i) \ .$$

Let $\tilde{a}_{ij}$ and $\tilde{f}_i$ be the corresponding numerical approximations. We assume that these numerical approximations have the following asymptotic accuracies

$$|a_{ij} - \tilde{a}_{ij}| \leq ch^{\sigma} \ , \quad |f_i - \tilde{f}_i| \leq c_f \, h^{\sigma-1} . \tag{2.1}$$

Here $h = N^{-1}$ is the discretization parameter. Furthermore let $\tilde{u}_h \in S_h^d$ be the corresponding approximation such that

$$\tilde{u}_h = \sum_{j=1}^{N} \tilde{\alpha}_j \mu_j \ .$$

The function $\tilde{u}_h$ satisfies the equations

$$\sum_{j=1}^{N} \tilde{a}_{ij} \, \tilde{\alpha}_j = \tilde{f}_i \ , \quad i = 1,\dots,N, \tag{2.2}$$

whereas the theoretical approximation $u_h = \sum_{j=1}^{N} \alpha_j \mu_j$ is determined by the system

$$\sum_{j=1}^{N} a_{ij} \alpha_j = f_i \ , \quad i = 1,\dots,N \ . \tag{2.3}$$

For further discussion it is enough to consider only Galerkin and collocation methods since the least-squares method reduces to a Galerkin method.

To analyse the effect of using numerical quadratures we introduce linear operators A_h and $\tilde{A}_h : S_h^d \to S_h^d$ as well as the functions f_h and $\tilde{f}_h \in S_h^d$ such that

$$(A_h \mu_j | \varphi_i) = a_{ij} \ , \quad (\tilde{A}_h \mu_j | \varphi_i) = \tilde{a}_{ij}$$

and

$$(f_h|\varphi_i) = f_i \ , \quad (\tilde{f}_h|\varphi_i) = \tilde{f}_i$$

where $\varphi_i = \mu_i$ for the Galerkin method and where $\varphi_i = \delta_i = \delta_{t_i}$

(Dirac's distribution) for the collocation. By means of these notations equations (2.2) and (2.3) read

$$\tilde{A}_h \tilde{u}_h = \tilde{f}_h \tag{2.4}$$

and

$$A_h u_h = f_h \,. \tag{2.5}$$

The following consistency relations have already been used in [13], [14] :

<u>Lemma 2.1:</u> *For the Galerkin method we have*

$$||(A_h - \tilde{A}_h)v||_0 \leq ch^{\sigma-2}||v||_0 \,, \quad v \in S_h^d \tag{2.6}$$

and

$$||f_h - \tilde{f}_h||_0 \leq c_f h^{\sigma-2} \,. \tag{2.7}$$

In the case of the collocation method we have

$$||(A_h - \tilde{A}_h)v||_0 \leq ch^{\sigma-1}||v||_0 \,, \quad v \in S_h^d \tag{2.8}$$

and

$$||f_h - \tilde{f}_h||_0 \leq c_f h^{\sigma-1} \,. \tag{2.9}$$

We discuss first the Galerkin method. For the analysis the following stability result is crucial.

<u>Lemma 2.2:</u> *(Galerkin method). There are two different cases: If* $2\alpha < d+\frac{1}{2}$ *, then it holds*

$$||v||_{2\alpha} \leq c||A_h v||_0 \,, \quad v \in S_h^d \,. \tag{2.10}$$

If $2\alpha > d+\frac{1}{2}$ *, then for all* $t_0 < d+\frac{1}{2}$ *there exists a constant* $c = c(t_0)$ *such that*

$$||v||_{t_0} \leq c||A_h v||_0 \,, \quad v \in S_h^d \,. \tag{2.11}$$

<u>Proof</u>. Define w such that

$$Aw = A_h v$$

for $v \in S_h^d$. Then v is the Galerkin approximation of w. The function w has the regularity $w \in H^{d+\frac{1}{2}+2\alpha-\varepsilon}$ for all $\varepsilon > 0$. If $2\alpha < d+\frac{1}{2}$ we can apply the stability (1.4) by choosing $s = 2\alpha$. This yields

$$||v||_{2\alpha} \leq c||w||_{2\alpha} \leq c||A_h v||_0$$

proving (2.10). In the second case we apply (1.4) with $t_0 < d+\frac{1}{2}$.

This yields by $t_0 < 2\alpha$

$$||v||_{t_0} \leq ||w||_{t_0} \leq c\ ||A_h v||_{t_0-2\alpha} \leq c\ ||A_h v||_0$$

proving assertion (2.11). □

In the case of the Galerkin method we need the condition

$$r := \sigma - 2 + \min(0,2\alpha) > 0\ . \qquad (2.12)$$

Lemma 2.3: *(Galerkin method). Assume that (2.12) holds. For small h we have two different cases:*

If $2\alpha < d+\frac{1}{2}$ *, then it holds*

$$||v||_{2\alpha} \leq c\ ||\tilde{A}_h v||_0\ , \quad v \in S_h^d\ . \qquad (2.13)$$

If $2\alpha \geq d+\frac{1}{2}$ *, then for all* $t_0 < d+\frac{1}{2}$ *there exists a constant* $c = c(t_0)$ *such that*

$$||v||_{t_0} \leq c\ ||\tilde{A}_h v||_0\ , \quad v \in S_h^d\ . \qquad (2.14)$$

Proof. Assume that $2\alpha < d+\frac{1}{2}$. Then we obtain by (2.10), (2.6)

$$||v||_{2\alpha} \leq c(||\tilde{A}_h v||_0 + ||(\tilde{A}_h - A_h)v||_0)$$

$$\leq c(||\tilde{A}_h v||_0 + h^{\sigma-2}||v||_0)$$

$$\leq c(||\tilde{A}_h v||_0 + h^{\sigma-2+\min(0,2\alpha)}||v||_{2\alpha})\ .$$

This yields (2.13) if h is small enough. If $2\alpha \geq d+\frac{1}{2}$ we may assume for (2.14) that $0 \leq t_0 < d+\frac{1}{2}$. Then (2.11), (2.6) similarly yield

$$||v||_{t_0} \leq c\ (||\tilde{A}_h v||_0 + h^{\sigma-2}||v||_{t_0})$$

which implies the assertion since we have $\sigma - 2 > 0$ by (2.12). □

By the previous lemma we also conclude that the operator $\tilde{A}_h$ is an isomorphism in S_h^d for small h and accordingly there exists a unique approximate solution $\tilde{u}_h \in S_h^d$ satisfying (2.4).

Next we estimate the difference $u_h - \tilde{u}_h$.

Theorem 2.1: *(Galerkin method). Assume that condition (2.12) is satisfied. Then for sufficiently small h holds the estimates:*

If $2\alpha < d+\frac{1}{2}$ *, then*

$$||u_h - \tilde{u}_h||_{2\alpha} \leq ch^{\sigma-2}(c_f + ||u_h||_0)\ . \qquad (2.15)$$

If $2\alpha \leq d+\frac{1}{2}$ *, then for all* $t_0 < d+\frac{1}{2}$ *there exists* $c = c(t_0)$ *such that*

$$\|u_h - \tilde{u}_h\|_{t_o} \le ch^{\sigma-2}(c_f + \|u_h\|_o) \quad . \tag{2.16}$$

Proof. If $2\alpha < d+\frac{1}{2}$ we have by (2.13), (2.6) and (2.7)

$$\|u_h - \tilde{u}_h\|_{2\alpha} \le c\ \|\tilde{A}_h(u_h - \tilde{u}_h)\|_o$$

$$\le c\ (\|(\tilde{A}_h - A_h)u_h\|_o + \|A_h u_h - \tilde{A}_h\tilde{u}_h\|_o)$$

$$\le c\ (h^{\sigma-2}\|u_h\|_o + c_f h^{\sigma-2}).$$

Similarly (2.16) follows by using (2.14), (2.6) and (2.7) □

Finally, we can formulate the following result for asymptotic error

Theorem 2.2: *(Galerkin method). Assume that condition (2.12) is is satisfied. For all values* $2\alpha - d - 1 \le t < d+\frac{1}{2}$ *,* $2\alpha - d - \frac{1}{2} < s \le d + 1$ *we have the asymptotic error estimate*

$$\|u - \tilde{u}_h\|_t \le c\, h^{s-t}\|u\|_s \tag{2.17}$$

$$+ c\, h^{\sigma-2+\min(0,2\alpha-t)}(\|u_h\|_o + c_f).$$

Proof. Assume first that $2\alpha < d+\frac{1}{2}$ holds. Then (2.17) follows from (1.3), (2.15) by using inverse estimates for values $2\alpha < t < d+\frac{1}{2}$. In the case $2\alpha \ge d+\frac{1}{2}$ assertion (2.17) reads

$$\|u - \tilde{u}_h\|_t \le c\ [h^{s-t}\|u\|_s + h^{\sigma-2}(\|u_h\|_o + c_f)] \tag{2.18}$$

since always is $t < d+\frac{1}{2} \le 2\alpha$. But (2.18) follows from (1.3) and (2.15) by choosing $t_o = t$. □

Now we discuss the collocation method. We have

Lemma 2.4: *(Collocation method). The operator* A_h *have the stability*

$$\|v\|_{2\alpha} \le c\ \|A_h v\|_o \ , \ v \in S_h^d \ . \tag{2.19}$$

Proof. We choose an integer $\tilde{d} \ge d$ such that $\tilde{d}+\frac{1}{2} > -\alpha + \frac{1}{2}(d + 1)$. Let $I^{\tilde{d}}$ be the interpolation operator $I^{\tilde{d}} : H^s \to S_h^{\tilde{d}}$ corresponding collocation at the points t_k in (1.5). For $v \in S_h^d$ we define w such that $Aw = I^{\tilde{d}}A_h v$. Then v is a collocation solution of w where w belongs to the space $H^{\alpha+\frac{1}{2}(d+1)}$. Convergence result (1.6) gives

$$\begin{aligned} \|v\|_{2\alpha} &\le \|w\|_{2\alpha} + \|w - v\|_{2\alpha} \\ &\le c\, \|I^{\tilde{d}} A_h v\|_0 + c h^{-\alpha + \frac{1}{2}(d+1)} \|w\|_{\alpha + \frac{1}{2}(d+1)} \\ &\le c\, \|I^{\tilde{d}} A_h v\|_0 + c h^{-\alpha + \frac{1}{2}(d+1)} \|I^{\tilde{d}} A_h v\|_{-\alpha + \frac{1}{2}(d+1)} \\ &\le c\, \|I^{\tilde{d}} A_h v\|_0 \le c\, \|A_h v\|_0 \quad \square \end{aligned}$$

For the collocation we need the condition

$$r := \sigma - 1 + \min(0, 2\alpha) > 0 \,. \tag{2.20}$$

If this condition is satisfied, we obtain similarly as for the Galerkin method the following:

Lemma 2.5: *(Collocation method). Assume that (2.20) is valid. Then for small h holds the stability*

$$\|v\|_{2\alpha} \le c\, \|\tilde{A}_h v\|_0 \quad , \quad v \in S_h^d \,. \tag{2.21}$$

Accordingly the numerical solution $\tilde{u}_h$ exists and we have

Theorem 2.3: *(Collocation method). Assume that (2.20) is valid. Then for sufficiently small h holds the estimate*

$$\|u_h - \tilde{u}_h\|_{2\alpha} \le c\, h^{\sigma-1} (c_f + \|u_h\|_0) \,. \tag{2.22}$$

Finally, we state that

Theorem 2.4: *(Collocation method). Assume that (2.20) is valid. We have the asymptotic error estimate*

$$\begin{aligned} \|u - \tilde{u}_h\|_t &\le c\, h^{s-t} \|u\|_s \\ &\quad + c\, h^{\sigma - 1 + \min(0, 2\alpha - t)} (\|u_h\|_0 + c_f) \end{aligned} \tag{2.23}$$

if $2\alpha \le t \le s \le d + 1 \quad , \quad t < d + \frac{1}{2} \quad , \quad s > 2\alpha + \frac{1}{2}$.

Proofs are similar as for the Galerkin method.

Finally we consider the least-squares approximation (1.7). Besides of the assumption $S_h^d \subset H^{2\alpha}$ i.e. $2\alpha < d + \frac{1}{2}$ we need that the condition

$$r := \sigma - 2 + \min(0, 4\alpha) > 0 \tag{2.24}$$

is satisfied. By Theorem 2.2 we immediately obtain

Theorem 2.5: *(Least-squares method). Assume that (2.24) is valid. Then we have the asymptotic error estimate*

$$||u - \tilde{u}_h||_t \leq c\, h^{s-t} ||u||_s \qquad (2.25)$$

$$+ c\, h^{\sigma-2+\min(0,4\alpha-t)} (||u_h||_o + c_f)$$

for $4\alpha - d - 1 \leq t \leq s \leq d + 1$, $t < d + \frac{1}{2}$, $s > 4\alpha - d - \frac{1}{2}$.

3. DISCUSSION. EXAMPLES

3.1. Comparison with known results. Here we compare our results with those obtained by Wendland in [13], [14]. Since least-squares method reduces to a Galerkin method, we consider only Galerkin and collocation methods.

By using the notation of $r > 0$ according to (2.12) and (2.20) we have derived for both methods an estimate of the form

$$||u - \tilde{u}_h||_t \leq c\, h^{s-t} ||u||_s \qquad (3.1)$$

$$+ c\, h^{r-\min(0,2\alpha)+\min(0,2\alpha-t)} (||u_h||_o + c_f) .$$

We compare this result with the error estimate

$$||u - \tilde{u}_h||_t \leq c\, h^{s-t} ||u||_s \qquad (3.2)$$

$$+ c\, h^{r+\min(0,-t)} (||u_h||_o + c_f)$$

which has been obtained by Wendland using L_2-stability. Let $n(t)$ and $m(t)$ be the exponents in (3.1) and (3.2) which give a bound for the effect of using numerical quadratures.

We discuss only the case $\alpha \leq 0$ which is typical for integral equations. Then we have

$$n(t) = r-\min(0,2\alpha)+\min(0,2\alpha-t) = \begin{cases} r-2\alpha, & t \leq 2\alpha, \\ r-t, & t > 2\alpha \end{cases} \qquad (3.3)$$

and

$$m(t) = r+\min(0,-t) = \begin{cases} r & t \leq 0, \\ r-t & t > 0 \end{cases} . \qquad (3.4)$$

We observe that always is $n(t) \geq m(t)$ and equality holds only for nonnegative values of t unless is $\alpha = 0$. This means that for negative values of t the use of numerical quadratures has a smaller effect on the asymptotic accuracy of the solution $\tilde{u}_h$ than it can be concluded from the result (3.2). The greatest value of the difference $n(t)-m(t)$ obtained already with $t = 2\alpha$ is

$$n(2\alpha)-m(2\alpha) = -2\alpha .$$

This difference can clearly be observed in numerical computations as we shall see in later examples concerning the single layer potential where $2\alpha = -1$ and the biharmonic single layer potential where $2\alpha = -3$.

The graphs of the functions $n(t)$ and $m(t)$ are illustrated in Figure 1 below

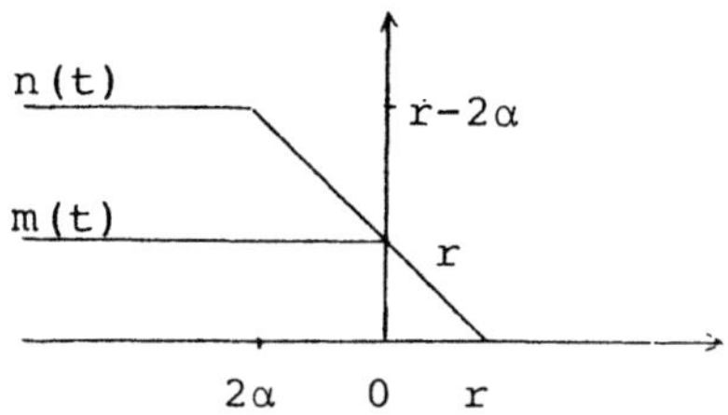

Fig.1 Graphs of the functions m(t) and n(t) with nonpositive α.

3.2. Choice of the quadrature rules. From the estimate (3.1) one can find a sufficient accuracy of the numerical quadrature rule which preserves the theoretical order of the convergence.

We compare the results (3.1) and (3.2) by giving the minimal accuracy of the numerical integration which is enough to recover the highest order of the theoretical convergence. The highest order is obtained by choosing the maximum value of s and the minimum value of t. For potential type problems this highest order of the convergence also gives the pointwise convergence rate of approximate potential.

Maximal orders of the convergence are by (1.3) and (1.6)

$$O(h^{2(d+1-\alpha)}) \quad \text{for Galerkin,} \tag{3.5}$$

$$O(h^{d+1-2\alpha}) \quad \text{for collocation.} \tag{3.6}$$

For classical potential problems using even degree splines with mid-point collocation the maximal order of the convergence by (1.6)' is

$$O(h^{d+2-2\alpha}) \ . \qquad (3.6)'$$

By our result (3.1) these orders are not affected if the numerical integration satisfies the condition $\sigma \geq \sigma_0$ with

$$\sigma_0 = 2d+4-2\alpha \quad \text{for Galerkin} \tag{3.7}$$

and

$$\sigma_0 = d+2-2\alpha \quad \text{for collocation,} \tag{3.8}$$

or corresponding to the case (3.6)' with even degree splines

$$\sigma_0 = d+3-2\alpha \ . \qquad (3.8)'$$

For comparison the previously known result (3.2) yields corresponding all these cases the condition $\sigma \geq \sigma_1$ where

$$\sigma_1 = \sigma_0 - 2\alpha \ . \tag{3.9}$$

Thus if $|\alpha|$ increases, then our result differs considerably from the previously known results.

Next we illustrate how the above results can be used in choosing particular quadrature rules. We consider the case where the operator A decomposes as $A = A_0 + A_1$ such that the part $(A_0\mu_j|\varphi_i)$ of stiffness elements can be calculated exactly and that A_1 is an integral operator with smooth kernel. The part $(A_1\mu_j|\varphi_i)$ of stiffness element is calculated numerically. We assume first that the <u>spline-weighted grid-point</u> rule [13], [14] is employed for this smooth part. In this case there are p (p is odd) node points in the quadrature rule for one dimensional integration and if product integration for the Galerkin method is employed, then we have

$$|a_{ij} - \tilde{a}_{ij}| \leq c\ h^{p+3} \tag{3.10}$$

for element of the stiffness matrix and

$$|f_i - \tilde{f}_i| \leq c\ h^{p+2} \tag{3.11}$$

for the right-hand side. Consequently we have $\sigma = p+3$ for the Galerkin method. For collocation the spline - weighted grid-point rule correspondingly gives the asymptotic accuracy (2.1) with $\sigma = p+2$. Thus in order to retain the maximal order of the convergence it suffices that $p \geq p_0$, where p_0 is the smallest positive odd integer satisfying

$$p_0 \geq \begin{cases} 2d+1-2\alpha & \text{, for Galerkin,} \\ d+1-2\alpha & \text{, for collocation} \end{cases} \tag{3.12}$$

and $p_0 \geq d+2-2\alpha$ for the mid-point collocation with even degree d corresponding to the result (3.6)'. If we instead use the previously known error estimate (3.2), then we obtain the lower bound $p \geq p_1$, where p_1 is the smallest positive odd integer satisfying

$$p_1 \geq \begin{cases} 2d+1-4\alpha & \text{, for Galerkin,} \\ d+1-4\alpha & \text{, for collocation} \end{cases} \tag{3.13}$$

and $p_1 \geq d+2-4\alpha$ for the mid-point collocation corresponding to the result (3.6)'. Since p_0 and p_1 take only odd values the requirements (3.12) and (3.13) are generally different only if $\alpha < -\frac{1}{2}$.

As the second example we mention the <u>spline-weighted Gaussian rule</u>. In this rule, if p denotes the number of abscissas for one-dimensional integration, we obtain the accuracy $\sigma = 2p+2$ for the Galerkin method and $\sigma = 2p+1$ for collocation [8]. This rule has all advantages of Gaussian type quadratures: it is easy to implement and the highest possible degree of accuracy is achieved with minimal number of abscissas. Moreover, all weights are positive implying the convergence for every smooth function.

3.3. Examples. We consider approximation of the single layer potential

$$\Phi(x) = -\frac{1}{2\pi}\int u(y)\ln|x-y|ds_y \;, \quad x \in \Omega \tag{3.14}$$

and the biharmonic single layer potential

$$\Phi(x) = -\frac{1}{8\pi}\int_\Gamma u(y)|x-y|^2\ln|x-y|ds_y \;, \quad x \in \Omega \tag{3.15}$$

in the smooth bounded domain $\Omega \subset \mathbb{R}^2$. Assume that density u is defined by the uniquely solvable boundary equation Au = f, where A = S is Symms operator for (3.14) and where A = B is the biharmonic single layer operator for (3.15). The first example appears in determation of the potential function with given Dirichlet data f. Our previous results are applicable with $2\alpha = -1$ for S and with $2\alpha = -3$ for B.

Having found the numerical approximation $\tilde{u}_h \in S_h^d$ we have the approximate potentials

$$\tilde{\Phi}_h(x) = -\frac{1}{2\pi}\int_\Gamma \tilde{u}_h(y)\ln|x-y|ds_y \;, \quad x \in \Omega \tag{3.14'}$$

and

$$\tilde{\Phi}_h(x) = -\frac{1}{8\pi}\int_\Gamma \tilde{u}_h(y)|x-y|^2\ln|x-y|ds_y, \quad x \in \Omega. \tag{3.15'}$$

For simplicity we consider the case of given smooth function f. Then inequality (3.1) yields the pointwise estimate

$$|\Phi(x)-\tilde{\Phi}_h(x)| \le c(h^{2d+2-2\alpha}+ h^{r-2\alpha}) \tag{3.16}$$

for the Galerkin method and for collocation

$$|\Phi(x)-\tilde{\Phi}_h(x)| \le c \begin{cases} h^{d+1-2\alpha}+h^{r-2\alpha}, & d \text{ odd,} \\ h^{d+2-2\alpha}+h^{r-2\alpha}, & d \text{ even.} \end{cases} \tag{3.17}$$

If the accuracy of the quadratures is too poor, then the second term in (3.16) and in (3.17) is dominating. Various testexamples calculated in [8] indicate that in such a case the rate of the pointwise convergence is of the order $h^{r-2\alpha}$ instead of the more pessimistic order h^r predicted by (3.2). Here we present numerical results with the choice $\Omega = \{x \,|\, |x| < 2\}$. In the first example with A = S we have the boundary data $f(\varphi) = 2\cos\varphi$ and in the second example with A = B we have used $f(\varphi) = 32\cos 2\pi$, $0 \le \varphi \le 2\pi$. The potential is calculated at the point x = (0.5,0). In both cases we have used the spline-weighted Gaussian rule for numerical integration solving the boundary equations by the Galerkin method with d = 1 for A = S and d = 0 for A = B. We have used the accuracies σ = 4,6,8 and 10 for numerical integration. In Table 1 and Table 2 below we have listed the experimental convergence rates by using the meshes with $h = 2^{-\nu}$, $3 \le \nu \le 8$. The results are in excellent agreement with the theoretical rates κ_0 predicted by (3.16). The only exception is the choice $\sigma = 4$ in the case A = B where the approximations converge but we have no proof.

The experimental convergence rate κ with the parameter value h is determined by

$$\kappa = \frac{\ln(|\Phi-\tilde{\Phi}_{h'}|/|\Phi-\tilde{\Phi}_h|)}{\ln(h'/h)}$$

with $h' = 2h$.

Table 1 Single layer. Experimental and theoretical convergence rates. Galerkin method. Degree d = 1.

h	$\sigma = 4$	$\sigma = 6$	$\sigma = 8$	$\sigma = 10$
1/16	2.016	4.122	5.024	5.053
1/32	2.005	4.049	5.030	5.045
1/64	2.001	4.021	4.999	5.007
1/128	2.000	4.010	4.996	4.999
1/256	2.000	4.005	4.997	5.001
κ_0	2.000	4.000	5.000	5.000

Table 2 Biharmonic single layer. Experimental and theoretical convergence rates. Galerkin method. Degree d = 0.

h	$\sigma = 4$	$\sigma = 6$	$\sigma = 8$	$\sigma = 10$
1/16	2.011	5.087	5.904	5.904
1/32	2.013	4.391	5.238	5.239
1/64	2.004	4.180	5.060	5.061
1/128	2.001	4.091	5.015	5.015
1/256	2.000	4.047	5.004	5.004
κ_0	-	4.000	5.000	5.000

The computations we carried out by Mr I. Lusikka using IBM 3083 EXO VM/SP -computer at the University of Oulu. For more examples covering also the collocation method and more general Petrov-Galerkin methods we refer to [8].

REFERENCES

[1] ARNOLD, D.N., WENDLAND, W.L.: "On the asymptotic convergence of collocation methods". Math. Comp., 41, (1983), pp. 349-381.

[2] ARNOLD, D.N., WENDLAND, W.L.: "The convergence of spline collocation for strongly elliptic equations on curves". Num. Math., 47, (3)(1985), pp. 317-343.

[3] ATKINSON, K.E.: "A Survey of Numerical Methods for the Solution of Fredholm Integral Equations of Second Kind". SIAM, Philadelphia, Penns, 1976.

[4] BAKER, C.T.H.:"The Numerical Treatment of Integral Equations". Clarendon Press, Oxford, 1978.

[5] CIARLET, P.G.: "The Finite Element Method for Elliptic Problems". North-Holland, Amsterdam-New York-Oxford, 1978.

[6] ELSCHNER, J., PRÖSSDORF, S., RATHSFELD, A., SCHMIDT, G.: "Spline approximation of singular integral equations". Demonstratio Mathematica, 28 (3), 1985, pp. 661-672.

[7] HSIAO, G.C., WENDLAND, W.L.: "The Aubin-Nitsche lemma for integral equations". Journal of Integral Equations. 3, (1981), pp. 299-315.

[8] LUSIKKA, I., SARANEN; J.: "Effect of quadratures in solving boundary integral equations by some Petrov-Galerkin methods". In preparation.

[9] RUOTSALAINEN, K., SARANEN, J.: "On the convergence of the Galerkin method for nonsmooth solutions of integral equations". Manuscript.

[10] SARANEN, J.: "The convergence of even degree spline collocation solution for potential problems in smooth domains of the plane". To appear.

[11] SARANEN, J., WENDLAND: "On the asymptotic convergence of collocation methods with spline functions of even degree". Math. Comp., 45, (1985), pp. 91-108.

[12] SHMIDT, G.: "On spline collocation methods for boundary integral equations in the plane". Math. Meth. Appl. Sci., 7, (1), (1985), pp. 74-89.

[13] WENDLAND, W.L.: "Asymptotic accuracy and convergence". pp. 289-313 of C.A. Brebbia (Ed.), Progress in Boundary Element Methods, Pentech, Press, London (1981).

[14] WENDLAND, W.L.: "On some mathematical aspects of boundary element methods for elliptic problems". In J. Whiteman, editor, Mathematics of Finite Elements and Applications V, pp. 193-227, Acad. Press, London, 1985.

ON THE EVALUATION OF AERODYNAMIC INFLUENCE COEFFICIENTS

H. Schippers
National Aerospace Laboratory NLR
Anthony Fokkerweg 2
1059 CM Amsterdam, The Netherlands

ABSTRACT

In this paper numerical schemes are discussed to reduce the computational work involved with the evaluation of the aerodynamic influence coefficients of panel methods. The applicability of these schemes is investigated to the calculation of potential flow around an airfoil. The interest concerns the decomposition of the boundary integral operator by splitting the contour integration such that the singular behaviour of the kernelfunction near the trailing edge is taken into account. The decomposition results in a compact operator and a contractive operator. The compact operator contains a smooth kernelfunction, which is approximated by piecewise linear interpolation. The interpolation provides error terms of the same order as the panel method. For a fixed number of panels the accuracy, however, has been decreased significantly.

INTRODUCTION

Panel methods for aerodynamic applications are in widespread use in the aerospace community since the 1960's. The main computational costs are related to the computation of the aerodynamic influence coefficients, which require upto 80 % of the total computational effort. Methods to reduce these costs have been studied at NLR since 1980, mostly in the form of simple model problems involving integrals with thin airfoil theory kernels. A review of the several methods that have been studied can be found in [6]. The basic concept underlying these methods is that the matrix of influence coefficients can be decomposed into a sparse matrix containing only near-field information and a dense matrix containing only far-field information. The reduction of computational costs is obtained by approximating the far-field information in a cheap way. To this end the following methods have been considered:

1. Clustering through multi-pole expansion - In this method, studied by Piers [5], the influences of distant panels on a fine grid point are grouped together in a coarse grid panel which can be interpreted as a cluster of fine grid panels. The influence of the coarse grid panel on a fine grid point is approximated by means of multi-pole expansions, i.e. the integral expression representing such influence is expanded in powers of $1/r$, r being the distance between the fine grid point and the center of the coarse grid panel.
2. Clustering through prolongation of coarse grid approximations. In this method, proposed by Slooff in [6], the far-field influence is first determined in coarse grid points utilizing the coarse grid influence coefficients and, subsequently, transposed to fine grid points by prolongation.

The purpose of the present paper is to investigate the applicability of the second approach to a first order panel method. The total velocity potential φ is assumed to be the superposition of the potential φ_∞, due to a uniform onset flow and a perturbation potential φ_ν, due to a vortex distribution at the airfoil. The boundary condition of vanishing normal velocity is replaced by the internal condition that the tangential component of $\nabla\varphi$ is prescribed. This approach, studied by Martensen [3], leads to a Fredholm integral equation of the second kind. The boundary integral operator can be decomposed into two parts such that the singular behaviour of the kernelfunction near the trailing edge is taken into account. This decomposition results in a contractive operator and a compact operator. When the integral equation is solved by a panel method the contractive operator yields a sparse matrix containing only the influence coefficients due to the panels near the trailing edge. These influence coefficients cannot be obtained through prolongation of coarse grid approximations without introducing anomalous errors. This is caused by the singular behaviour of the kernelfunction near the trailing edge. On the other hand, however, the compact operator yields a dense matrix (mostly containing far-field information) of which the coefficients behave smoothly. Therefore, it seems to be allowed to apply the above clustering approach to this dense matrix. The additional errors involved with this approach will be discussed.

CALCULATION OF POTENTIAL FLOW AROUND AN AIRFOIL

The incompressible potential flow is obtained as the superposition of an undisturbed onset flow and a flow generated by a vortex distribution at the airfoil. The boundary condition of vanishing normal velocity is replaced by the internal condition that the tangential component of the velocity is prescribed. This approach, studied by Martensen [3], leads to the following Fredholm integral equation of the second kind

$$\nu(x) - \frac{1}{\pi} \int_S \nu(y) \frac{\cos(n(x),\ x-y)}{|x-y|} \, dS_y = 2\ U.t(x),\ x \in S, \qquad (1)$$

where n(x) is the outward normal to the boundary at the point x, t(x) is the unit tangential vector at x, and U is the velocity vector of the undisturbed flow. Since this equation is not uniquely solvable (the homogenous equation has a nontrivial solution) the following side condition is added

$$\int_S \nu(y)\, dS_y = \Gamma, \qquad (2)$$

where Γ denotes the circulation. The value of Γ follows from the Kutta-condition

$$\nu(x_{te}^+) + \nu(x_{te}^-) = 0. \qquad (3)$$

Here x_{te}^+ and x_{te}^- denote the upper and lower side of the trailing edge, respectively.

For numerical reasons we add to the left-hand side of (1) a term $\lambda \log |x-x^*|$ with x^* a point inside the airfoil and λ unknown. Then the equations (1)-(2) can be shortly written as

$$(I-K)\nu + \lambda w = g, \tag{4}$$

$$\Lambda\nu \qquad = \Gamma. \tag{5}$$

The function $w(x) = \log |x-x^*|$ has been chosen because it does not belong to the null-space of $I-K^*$ with K^* the adjoint operator of K. As a consequence the augmented system (4)-(5) is uniquely solvable. Furthermore, the constant λ appears to be zero. This follows from applying the functional Λ to equation (4). It is known that $\Lambda(I-K)\nu = 0$, for all ν, and $\Lambda g = 0$. Hence, $\lambda \, \Lambda \, w = 0$. It follows that $\lambda = 0$, because $\Lambda w \neq 0$. As a consequense the solution of the augmented system (4)-(5) equals the solution of equations (1)-(2). The above concept of including the term λw was inspired by the work of Hsiao and Wendland [1, 2].
The numerical approximation of (4) is facilitated (see e.g. Ref. [7]) by decomposing the integral operator K into two parts K^1 and K^2 such that K^1 accounts for the singular behaviour of the kernelfunction near the trailing edge. To this end we introduce a cut-off function with compact support. This cut-off function has the value one in the neighbourhood of the trailing edge and decreases continuously in a monotonic way to zero (see Fig. 1). Then, the operator K^1 is defined as

$$K^1 \, \nu(x) = \frac{1}{\pi} \int_S X(y) \, \frac{\cos(n(x),\, x-y)}{|x-y|} \, \nu(y) \, dS_y \tag{6}$$

and

$$K^2 = K - K^1. \tag{7}$$

The operator K^1 defines a contraction on the Sobolev space H^0. Furthermore, it turns out that the operator K^2 is a bounded mapping from H^0 into H^α with $\alpha > 0$. The value of α depends on the smoothness of the airfoil and the size of the cut-off region.

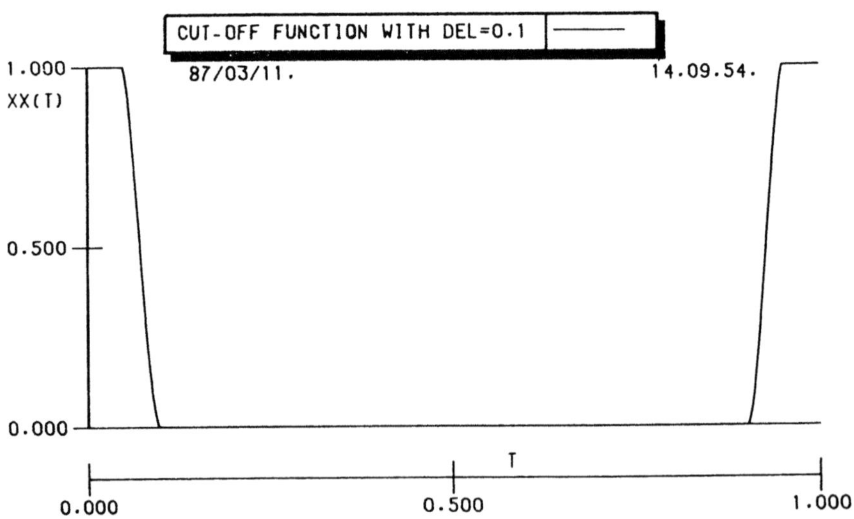

Fig. 1 Cut-off function X(t)

The integral equation (4) is solved by a first-order panel method, which is concisely described below. The airfoil is assumed to be given by some parametrization $x(t)$, $t \in [0,1]$ with $x(0) = x_{te}^-$ and $x(1) = x_{te}^+$. The interval $[0,1]$ is divided into N equidistant segments with length $h = 1/N$. Then, the begin- and end-point of the i-th panel are given by $x((i-1)h)$ and $x(ih)$, respectively. On this grid ν is approximated by a piecewise constant function ν_N and the resulting equation is solved by a collocation method. The collocation points x_i, $i = 1, 2, \ldots, N$, are given by $x_i = x((i-\frac{1}{2})h)$. Let T_N be the projection operator defined by piecewise constant interpolation at the collocation points. Then, we obtain the following approximate equations

$$(I - T_N K^1 - T_N K^2)\, \nu_N + \lambda_N\, T_N w = T_N g, \tag{8}$$

$$\Lambda \nu_N = 0. \tag{9}$$

Let the kernelfunction of (1) be shortly denoted by $k(x,y)$. Then, the (i,j)th aerodynamic influence coefficient of the first term of (8) is given by

$$a_{ij} = \delta_{ij} - \int_{S_j} X(y)\, k(x_i,y)\, dS_y - \int_{S_j} (1-X(y))\, k(x_i,y)\, dS_y. \tag{10}$$

The second term of (10) only gives a contribution for panels near the trailing edge. The kernelfunction $k(x,y)$ behaves singularly if x and y are located on opposite sides of the trailing edge. Therefore these terms are evaluated by analytical formulae replacing S_j by a linear panel. The last term of (10) can be evaluated in the same way. It is cheaper, however, to evaluate this term by applying the midpoint rule, which gives a sufficiently accurate result because the integrand is regular (i.e. if $x(t) \in C^3(0,1)$ then $(1 - X(x(t))\, k(.,x(t))$ is at least twice continuously differentiable). Hence, the last term of (10) is evaluated as $- h(1-X(x_j))\, k(x_i,x_j)\, |\dot{x}_j|$. If the collocation point x_i and the integration point x_j coincide, then the kernelfunction is evaluated by means of a small curvature expansion.

The concept of applying different approximations to evaluate the influence coefficients is also used in existing panel methods. However, the underlying criteria for selecting the appropriate approximation are different. They are based on the distance between the collocation point and the center of the panel, which results in near field, intermediate field and far field approximations. The present criteria of selection only depend on the size of the support of the cut-off function X.

The performance of the present panel method is illustrated for the calculation of potential flow around a Karman-Trefftz airfoil of which the exact velocity distribution is known. The calculations have been carried out for a symmetric airfoil with a relative thickness of 5 % and an interior trailing edge angle of $\pi/10$.

Figure 2 shows the number of correct digits (i.e. $-\log|\text{error}|$) of the numerical solutions as obtained by applying analytical formulae or midpoint rule to the evaluation of the influence coefficients of K^2. It appears that the midpoint rule yields more accurate results away from the trailing edge than the analytical formulae in which the panels are approximated by flat segments. This can be explained by observing that the integrand of $K^2\nu$,

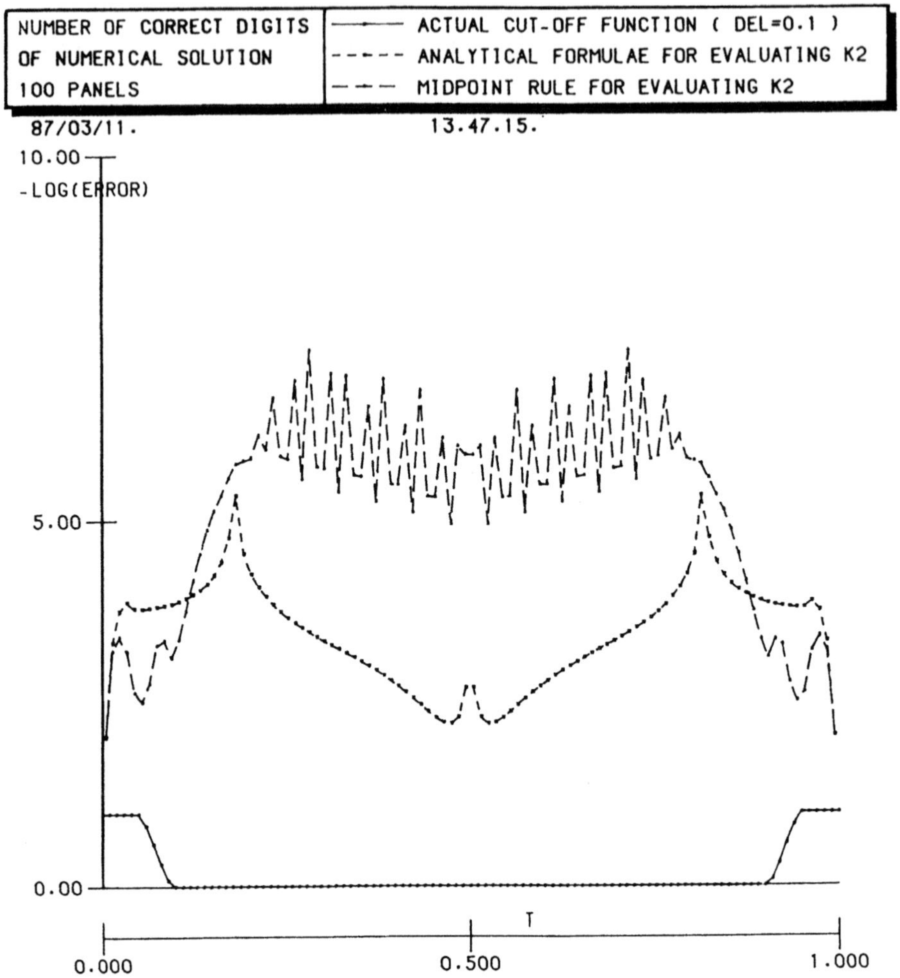

Fig. 2 Comparison of calculated error distributions of a symmetric Karman-Trefftz airfoil (thickness 5%, trailing edge angle $\pi/10$, angle of attack 0°)

given by $(1-X(x(t))\ k\ (.,\ x(t))\ \nu(x(t))$, is 1-periodic with respect to t. The most accurate rule for the integration of such a function is the trapezoidal rule, which agrees with the repeated midpoint rule in this case because the begin- and end-point of integration may be shifted half a meshwidth. A further inspection of figure 2 reveals that near the trailing edge the error increases, which is caused by the change of integration formulae and the singular behaviour of ν. This error can be decreased slightly by increasing the support of the cut-off function, so that the analytical formulae are applied in a larger region near the trailing edge.

The convergence of the above panel method is given in table 1. The error is measured by means of the discrete L_2-norm according to

$$\ell_2 = \sum_{i=1}^{N} (h(\nu_{N,i} - \nu(x_i)))^2, \tag{11}$$

where $\nu_{N,i}$ are the coefficients of ν_N, and by means of a modified norm $\ell_{2,\delta}$ that excludes the points within a small radius of the trailing edge disregarding the singular behaviour of ν. This radius is 1 % of the airfoil chord. The results of table 1 suggest a global accuracy of $O(h^2)$ in the modified norm for the case that the midpoint rule is used to evaluate the influence coefficients of $K^2\nu$. When the analytical formulae are applied to evaluate these coefficients, the global accuracy tends to decrease to $O(h)$.

Table 1 - L_2-errors of the numerical solutions for a symmetric Karman-Trefftz airfoil (thickness = 5 %, trailing edge angle = $\pi/10$, angle of attack = 0).
- Influence coefficients of K^2 are evaluated by the midpoint rule or by analytical formulae.
- Length of support of cut-off function is given by $\delta = 0.1$.

	N	L_2-norm	modified L_2-norm
mid-point rule	50	.47(-2)	.42(-2)
	100	.41(-2)	.60(-3)
	200	.10(-2)	.15(-3)
analytical formulae	50	.35(-2)	.29(-2)
	100	.23(-2)	.19(-2)
	200	.15(-2)	.11(-2)

So far nothing has been said about the computational costs. They are still dominated by the evaluation of the N^2 influence coefficients. In the next section we will investigate some numerical schemes to reduce these costs.

CLUSTERING THROUGH PROLONGATION

In this section the applicability of clustering schemes is investigated to reduce the computational costs involved with the evaluation of $T_N K^2 \nu_N$. Recall that K^2 was obtained by splitting off the singular part near the trailing edge (see eqs. 6-7). In terms of the parameter t the corresponding kernelfunction reads

$$k^2\ (\tau,t) := \frac{1}{\pi}\ (1-X(x(t))\ \frac{\cos(n(x(\tau),\ x(\tau)-x(t))}{|x(\tau)-x(t)|}\ |\frac{dx}{dt}\ (t)|, \tag{12}$$

which can proven to be twice continuously differentiable in τ and t, provided the parameter representation $x(t) \in C^3(0,1)$. This smoothness of k^2 forms the basis of the present clustering schemes, where k^2 is approximated by piecewise-linear interpolation on the midpoints of a coarse grid. This coarse grid is obtained by clustering two adjacent fine grid panels, i.e. the course grid panel $\tilde{S}_i$ consists of the collection of two grid panels $S_{2i-1} \cup S_{2i}$. Hence, the midpoints of the clusters are given by

$$\tilde{t}_i := (2i-1)\,h, \qquad i=1,\ 2,\ \ldots,\ N/2.$$

When the following reference function is used

$$\mu(z) := \begin{cases} 1+z, & \text{for} \quad -1 \leqq z < 0, \\ 1-z, & \text{for} \quad 0 \leqq z < 1, \\ 0\ , & \text{elsewhere}, \end{cases}$$

then the piecewise linear interpolation of some function g(t) on the course grid is given by

$$\tilde{g}(t) = \sum_{i=1}^{\frac{1}{2}N} g(\tilde{t}_i)\ \mu_i(t), \tag{13}$$

where the basis functions μ_i are defined by

$$\mu_i(t) := \mu((t-\tilde{t}_{i-1})/2h), \qquad 0 \leqq t < 1,$$

and its 1-periodic extension.

Next, the following approximations to k^2 are introduced.

1) Clustering through column interpolation
The kernelfunction k^2 is approximated with respect to the first argument. It is replaced by

$$\tilde{k}^2(\tau,t) := \sum_{i=1}^{\frac{1}{2}N} \mu_i(\tau)\ k^2(\tilde{t}_i,t). \tag{14}$$

The interpolation error is estimated by

$$|\tilde{k}^2(\tau,t) - k^2(\tau,t)| \leqq c\ h^2 \sup_{0<\tau<1} \left|\frac{\partial^2\ k^2}{\partial\tau^2}(\tau,t)\right|. \tag{15}$$

2) Clustering through row interpolation
This approximation follows from applying (13) to k^2 with respect to the second argument. For the interpolation error a similar estimate holds as for (15).

3) Clustering through row and column interpolation
This approximation is obtained by applying both row and column interpolation. Hence, k^2 is replaced by

$$\bar{k}^2(\tau,t) = \sum_{i=1}^{\frac{1}{2}N} \sum_{j=1}^{\frac{1}{2}N} k^2(\tilde{t}_i,\tilde{t}_j)\ \mu_i(\tau)\ \mu_j(t) \tag{16}$$

and the interpolation error is estimated by

$$|\bar{k}^2(\tau,t) - k^2(\tau,t)| \leq C\, h^2 \sup_{0<\tau,t<1} \sum_{\alpha=0}^{2} \left|\frac{\partial^2}{\partial\tau^\alpha \partial t^{1-\alpha}} k^2(\tau,t)\right|. \tag{17}$$

The influence coefficients of the fine grid are now obtained by evaluating the interpolated kernelfunction at the midpoints $t_i = (i-\frac{1}{2})\,h$, $i=1, \ldots, N$. When the evaluations have been carried out, the above clustering schemes can be written in matrix notation. Let a matrix with N columns and M rows be denoted by $K_{M,N}$. Then, the clustering schemes result in the following matrix multiplications:

- Column interpolation $P_{N,\frac{1}{2}N}\, K_{\frac{1}{2}N,N}$,
- Row interpolation $K_{N,\frac{1}{2}N}\, R_{\frac{1}{2}N,N}$,
- Row and column interpolation $P_{N,\frac{1}{2}N}\, K_{\frac{1}{2}N,\frac{1}{2}N}\, R_{\frac{1}{2}N,N}$.

Here P and R are prolongation and restriction matrices, respectively, which are sparse.

The number of evaluations of the kernelfunction k^2 needed by the respective schemes is given below:

- column interpolation $\frac{1}{2}N^2$,
- row interpolation $\frac{1}{2}N^2$,
- row and column interpolation $\frac{1}{4}N^2$.

Hence, the computational costs would have been reduced by a factor $\frac{1}{2}$ (or even a factor $\frac{1}{4}$ by applying row and column interpolation), if the accuracy is not affected. Note however that the clustering schemes provide interpolation errors of the second order, which is the same order as has been observed in the numerical experiments of the previous section.

Table 2 Comparison of modified L_2-norms of numerical solutions for a symmetric Karman-Trefftz airfoil (angle of attack = 0°)

	N = 50	N = 100	N = 200
No clustering	.42(-2)	.60(-3)	.15(-3)
Row interpolation	.27(-1)	.44(-2)	.24(-3)
Column interpolation	.18(-1)	.41(-2)	.11(-2)
Row and column interpolation	.25(-1)	.31(-2)	.12(-2)

Table 3 Comparison of modified L_2-norms of numerical solutions for a symmetric Karman-Trefftz airfoil (angle of attack = 10°)

	N = 50	N = 100	N = 200
No clustering	.54(-2)	.65(-3)	.20(-3)
Row interpolation	.27(-1)	.46(-2)	.26(-3)
Column interpolation	.19(-1)	.44(-2)	.21(-2)
Row and column interpolation	.70(-1)	.44(-2)	.22(-2)

In tables 2 and 3 numerical results are presented for a symmetric Karman-Trefftz airfoil (thickness 5 %, trailing edge angle $\pi/10$). The errors are measured by means of the modified L_2-norm disregarding the vicinity of the trailing edge. In table 2 the angle of attack is taken to be $\alpha = 0°$; in table 3 α is equal to 10°. From these tables it is observed that the error increases significantly if one of the clustering schemes is applied, in particular if clustering through column interpolation is used. The increase of the error may be explained by error estimate (15), which relates the interpolation error to the magnitude of the second derivative of k^2. It appears that the second derivative is rather large for $t + \tau = 1$ (i.e. when the collocation point and the panel segment of integration are located on opposite sides of the airfoil). For N = 200 the clustering scheme through row interpolation gives comparable results as the panel method without clustering. In this case, the costs have only been reduced by a factor ½ to $\frac{1}{2}N^2$. At these costs a similar result, however, can be obtained by applying the panel method without clustering using $\frac{1}{2}\sqrt{2}N$ panels. When N is taken to be 140, the error is .32(-3) for $\alpha = 0°$. Therefore, it is concluded that the clustering schemes discussed in this section do not reduce the computational costs involved with the evaluation of the influence coefficients of the integral operator K^2.
The convergence of the clustering schemes has been estimated by (17) to be of second order. For the scheme that uses column interpolation this order is only observed in table 2. In table 3 the convergence tends to decrease to O(h), which may be caused by the first order implementation of the Kutta-condition (3). The scheme that uses row interpolation appears to converge somewhat faster than second order.

CONCLUSIONS

In this paper a panel method has been discussed to solve the boundary integral equation for the vorticity distribution at the airfoil. This method is based on a decomposition of the boundary integral operator, which results in a contractive operator and a compact operator. The influence coefficients of the contractive operator, which yield a sparse matrix, are evaluated by analytical formulae using linear panels for the approximation of the airfoil geometry. The influence coefficients of the compact operator, supplying a dense matrix, can be evaluated by the midpoint rule. This approach yields a rather efficient panel method. The numerical experiments suggest a global accuracy of $O(h^2)$ in a modified L_2-norm, where the vicinity of the trailing edge is disregarded. In the previous section clustering schemes through prolongation of coarse grid approximations have been discussed to reduce the costs involved with the evaluation of these coefficients further. Although the interpolation errors are also of second order, the error of the numerical solution increases significantly so that no reduction of costs has been obtained. So far we have not investigated clustering schemes through multipole expansion for the fast evaluation of K^2. These schemes, as studied by Piers [5] to solve the thin airfoil integral equation, seem to be more promising. Recently, it was shown by Nowak [4] that such a panel clustering scheme can be applied successfully to three dimensional potential flow problems. For the lifting flow around a wing the costs were reduced by a factor 0.1. Therefore, it is recommendable to study these schemes further in developing more efficient panel methods for complex aerodynamic configurations.

REFERENCES

[1] HSIAO, G.C., MACCAMY, R.C.: "Solution of boundary value problems by integral equations of the first kind", SIAM Review 15 (1973), pp. 687-705.

[2] HSIAO, G.C., KOPP, P., WENDLAND, W.L.: "Some applications of a Galerkin collocation method for integral equations of the first kind", Math. Methods in the Applied Sciences 6 (1984), pp. 280-325.

[3] MARTENSEN, E.: "Berechnung der Druckverteilung an Gitterprofilen in ebener Potentialströmung mit einer Fredholmsehen Integralgleichung", Arch. Rat. Mech. and Anal. 3 (1959), pp. 235-270.

[4] NOWAK, Z.P.: Panel clustering technique for lifting potential flows in three space dimensions", Proceedings GAMM-Seminar on Panel Methods in Mechanics, Kiel, January 16-18, 1987, Vieweg Verlag, Braunschweig.

[5] PIERS, W.J.: "Beschrijving van een multi-rooster techniek voor een panelenmethode, deel II: een cluster-methode voor het vlakke plaat geval", unpublished work at NLR.

[6] SLOOFF, J.W.: "Requirements and developments shaping a next generation of integral methods", Proceedings of IMA Conference on Numerical Methods in Aeronautical Fluid Dynamics, Reading, March 30 - April 1, 1981, (also NLR MP 81007 U).

[7] WENDLAND, W.L.: "Boundary element methods and their asymptotic convergence" in : Theoretical Acoustics and Numerical Techniques (P. Filippi, ed.), CISM courses and lectures 277, Wien, New York: Springer-Verlag (1983), pp. 135-216.

A DOUBLET POINT METHOD FOR THE CALCULATION OF UNSTEADY PROPELLER AERODYNAMICS

J. Schöne

DFVLR, Institut für Entwurfsaerodynamik,

Flughafen, D-3300 Braunschweig, F.R. Germany

SUMMARY

A lifting surface theory for unsteady flow fields of propellers in incompressible medium is presented. The derivation of the governing integral equation, which relates the pressure loadings and induced effects, is outlined. In the solution method, the lifting surface is divided into panels on which one doublet and one upwash point are located. The time dependent strength of the singularity is developed in a Fourier series. This leads, together with the tangential flow condition in every upwash point, to a set of linear algebraic equations. Results of parametric study regarding the influence of discretization are shown. Comparisons between calculated and measured data of steady and unsteady propeller aerodynamics conclude the paper.

1. INTRODUCTION

One important problem of modern propellers is the appearance of unsteady flow phenomena for special configurations and operating conditions. From the view point of aerodynamic applications, three cases are of special interest for propellers having axial onflow. First there may be a nonuniform or time dependent onflow, for example the flow field behind a strut. The second case is that of unsteady motion, like flutter of blades. The third type of unsteady phenomenon comes from interference between rotating stages, when counter rotating propellers are used.

Due to the high computational expense, most of the numerical methods concerning unsteady propeller or rotor flow fields are based on linearized potential theory: Tsakonas et al [1] and Hanson [2,3] developed numerical methods for harmonically varying free stream vectors at every section of a blade. Ref. [3] treats counter rotating propellers neglecting the influence of the rear propeller on the front. Runyan and Tai [4] described a theory for a helicopter rotor in forward flight and Lesieutre and Sullivan [5] coupled a quasi steady vortex lattice method with an unsteady 2-dimensional analysis for counter rotating propellers.

The present paper describes a lifting surface theory for

unsteady, incompressible aerodynamics of propellers having axial onflow. In order to reduce the amount of numerical effort, a simple discretization scheme - compared with higher order panel methods - is used here, where only one doublet is located on a lifting element. The method is able to treat different configurations and flow conditions. An unsteady formulation is used and the interfering lifting surfaces in the case of counter rotating propellers are treated directly without iteration. The derivation of the governing integral equation, for which a numerical solution method is developed, is described by Das [6], who has formulated a very general integral equation for propeller aerodynamics, and by Dat [7], who gives a relation for a line distribution of doublets in an arbitrary motion.

2. GOVERNING EQUATIONS

Consider a space fixed reference frame with the medium at rest at infinity. For an inviscid, irrotational, and adiabatic flow field the velocity vector $\vec{V}$ can be expressed as a gradient of a velocity potential Φ

$$\vec{V} = \text{grad}\ \Phi\ . \tag{1}$$

In incompressible flow this potential fullfills the Laplace equation, $\Delta\Phi = 0$, where Φ is a function of location and time. The Neumann boundary condition requires that the relative normal velocity on the body surface vanishes. The velocities tend to zero at infinity. The solution for this problem can be found with the use of Green's theorem.

In unsteady aerodynamics it is common to introduce at this point Prandtl's acceleration potential in place of Φ. This approach does not require a prescription of the unsteady potential wake. This wake is inherent and is equivalent to a rigid wake model. In the following one has to deduce first the acceleration potential. Then its relation to the velocity potential is established because the boundary condition is formulated in terms of derivatives of Φ. Finally an integral equation is derived.

Using the equation of motion for incompressible fluids

$$\frac{D}{Dt}\vec{V} = -\frac{1}{\rho_\infty}\text{grad}\ p \tag{2}$$

an acceleration potential Ψ is introduced

$$\vec{b} = \frac{D}{Dt}\vec{V} = \text{grad}\ \Psi\ , \tag{3}$$

where D/Dt is the substantial derivation, ρ_∞ the density of the undisturbed medium, p the pressure, $\vec{b}$ the acceleration vector. Such a potential does exist, because the assumption of an irrotational velocity field implies that the acceleration vector field is also irrotational. Combining eq.(2) and (3) leads, after an integration, to the following relation:

$$\Psi = -\frac{p - p_\infty}{\rho_\infty} + F(t), \tag{4}$$

where F is a function of time only. Thus the interrelation between the velocity and acceleration potential follows from eq.(1) and (3), yielding:

$$\frac{D}{Dt}\,\mathrm{grad}\,\Phi = \mathrm{grad}\,\Psi. \tag{5}$$

The operators D/Dt and grad are exchangeable here. The operator D/Dt reduces in a fixed coordinate system, with the assumption of small velocity disturbances (in comparison with the velocity of motion), to $\partial/\partial t$. Then eq.(5) becomes

$$\frac{\partial}{\partial t}\,\Phi = \Psi + G(t). \tag{6}$$

G is a function of time only. Integrating eq.(6) with respect to time leads to the relation:

$$\Phi - \Phi_{t=-\infty} = \int_{-\infty}^{t} (\Psi + G(t))\, dt_o\,. \tag{7}$$

The induced velocity in the normal direction is obtained by taking the directional derivative ($\partial/\partial n = \vec{n} \cdot \mathrm{grad}$)

$$v = \frac{\partial}{\partial n}\,\Phi = \int_{-\infty}^{t} \frac{\partial}{\partial n}\,\Psi\, dt_o\,. \tag{8}$$

It can be seen from eq.(7), that Ψ satisfies

$$\Delta\Psi = 0 \tag{9}$$

and also the condition at infinity $\mathrm{grad}\,\Psi \to 0$. Solutions of this problem, i.e. sources and doublets, are well known. In ref. [7] it is shown with eq.(4), that a doublet distribution on a surface is related to a lift distribution acting on that surface. In the present method a propeller blade is considered as a lifting surface without thickness. Therefore the only fundamental solution taken into account are doublets. These form a sheet located on a regular helix. The normal velocity condition is determined from the real camber surface.

A solution of eq.(9), which describes a doublet distribution over a surface S_o producing a local and time dependent nondimensional pressure difference Δc_p, has the form [7]

$$\Psi_o = \frac{1}{8\pi} \iint_{S_o} W^2 \, \Delta c_p \left\{ \frac{\partial}{\partial n_o} \frac{1}{D} \right\} dS.$$

D is the distance between an observation point and a doublet, $\vec{n}_o$ the normal to the surface at the doublet and W the magnitude of the velocity of it. With eq.(8) one gets an integral equation

$$v = \frac{1}{8\pi} \iint_{S_o} \int_{-\infty}^{t} W^2 \, \Delta c_p \frac{\partial}{\partial n} \left\{ \frac{\partial}{\partial n_o} \frac{1}{D} \right\} dt_o \, dS. \qquad (10)$$

From the general integral equation deduced by Das [6] one obtains eq.(10) after introducing the simplifications described above. Eq.(10) relates the induced velocity (a function of location and time in the fixed reference frame) to the pressure distribution on an arbitrarily moving lifting surface. The unknown function is Δc_p which is a function of location on the lifting surface and of time. The induced velocity, v, is supplied by the boundary condition of tangential flow on the blade surface, which is known directly in a moving reference frame.

3. SOLUTION METHOD

The numerical solution of eq.(10) for unsteady propeller aerodynamics is the central problem of the present method. Here discrete lifting elements are used in connection with a simple arrangement of the singularities on the element.

The first step is the division of the lifting surface into N 'panels'. Instead of a continuous distribution of singularities over the panel area, here only one doublet j is located in the quarter point and one upwash point i is located in the three quarter point (Fig. 1). This so-called doublet point scheme was introduced by Ueda and Dowell [8] for unsteady wing aerodynamics.

One doublet produces a single force and therefore the lifting surface is simulated now by a system of time dependent forces. These forces acting on the panel surfaces ΔS lead to the unknown pressure differences Δc_p. Here only periodic cases are treated and therefore the pressure differences is developed into a Fourier series:

$$\Delta c_{pj} = c_{oj} + \sum_{m=1}^{n} c_{mj} \cdot \cos(m \cdot \omega \cdot t_o) + s_{mj} \cdot \sin(m \cdot \omega \cdot t_o). \qquad (11)$$

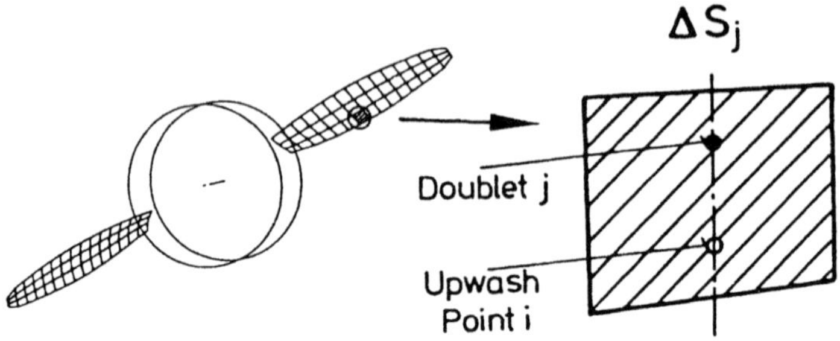

Fig. 1 Division into panels and doublet point arrangement

The unknowns are now the Fourier coefficients $c_o, c_1, \ldots c_n, s_1, \ldots s_n$ for every panel j. Δc_p varies periodically with frequency ω. For the case of counter rotating propellers with N_B blades on each of the front and rear propellers, rotating with constant amounts of rotational velocity Ω, the frequency has the value $\omega = 2 \cdot \Omega \cdot N_B$.

These steps lead to a system of linear algebraic equations (eq.(12)). The parenthesis in the matrix contains the influence of one panel j on an upwash point i at a time k. The unknowns are the Fourier coefficients and the right hand side is supplied by the boundary condition at the upwash point at different times. The equations are formulated for different times within one period. In the case of interfering lifting

$$\begin{array}{c} \\ \\ \\ k=1 \\ \\ \\ k=2 \\ \\ 2n+1 \end{array}
\left[\begin{array}{cccc}
(a_o a_1^c \ldots a_n^s)_{11} & (\;)_{12} & \cdots & (\;)_{1N} \\
(\;)_{21} & (\;)_{22} & \cdots & (\;)_{2N} \\
\vdots & & \cdots(\;)_{ij}\cdots & \\
(\;)_{N1} & (\;)_{N2} & \cdots & (\;)_{NN} \\
\hline
(\;)_{11} & \cdots & & \\
\vdots & & & \\
 & & & (\;)_{NN} \\
\hline
\vdots & & & \vdots \\
\hline
\end{array}\right]
*
\left[\begin{array}{c}
\begin{pmatrix} c_o \\ c_1 \\ \vdots \\ s_n \end{pmatrix}_1 \\
(\;)_2 \\
\vdots \\
(\;)_j \\
\vdots \\
(\;)_N
\end{array}\right]
=
\left[\begin{array}{c}
v_1 \\ 2 \\ \vdots \\ i \\ \vdots \\ N \\
\hline
1 \\ \vdots \\ N \\
\hline
\vdots \\
\hline
\end{array}\right]
\begin{array}{c} \\ \\ \\ k=1 \\ \\ \\ k=2 \\ \\ 2n+1 \end{array} . \qquad (12)$$

surfaces this complete system has to be solved simultaneously. The free stream vector is known in special cases (e.g. single propellers working in a nonuniform onflow). Then a harmonic analysis of the free stream velocity can be performed and for each harmonic one system of equations is formulated. A quasi steady solution is obtained, if only the zero coefficients c_{oj} are taken into account and independent sets of equations are built for different times. In the case of steady aerodynamics (single propeller with uniform onflow) only one of these systems needs to be solved.

The influence matrix has the following form (when the differentiations with respect to n_i and n_j are carried out [7])

$$a^{c}_{s\,ijkm} = \frac{1}{8\pi} \cdot W_j^2 \cdot \Delta S_j \int_{-\infty}^{t_k} \begin{matrix} \cos(m\omega t_o) \\ \sin(m\omega t_o) \end{matrix} \left[\frac{\vec{n}_j \cdot \vec{n}_i}{D^3} - 3 \frac{(\vec{D}\vec{n}_j)(\vec{D}\vec{n}_i)}{D^5} \right] dt_o, \tag{13}$$

$i,j=1,2,\ldots N, \qquad k=1,2,\ldots 2n+1, \qquad m=0,1\ldots n.$

Each coefficient contains essentially an integral of a kinematic term (in brackets). This term is evaluated for pure helical motion of the blade element. Figure 2 illustrates the kinematics connected with them. Two cases are shown. First, the upwash point and doublet are moving on the same helix A. That means, the upwash point and doublet are lying on the same propeller blade. The second case shows the doublet moving on a helix B which rotates contrarily to the helix A of the upwash point. This represents two different blades which rotate contrarily about the same axis and move forward. In both cases, $\vec{n}_i$ is fixed at the time t_k, and $\vec{D}$ and $\vec{n}_j$ vary with t_o as the doublet moves along the helix.

The integration is performed numerically with the help of Gauss' quadrature formula over a truncated integration interval. If the distance D is zero anywhere in the interval of integration, the integral becomes infinite and has no meaning in an ordinary sense. In this case the integral is calculated in the sense of principal value formulated by Hadamard and later on extensively described by Mangler. The method of ref. [8] treating such an infinite integral is applied here in the case of propellers. For propellers with two or more blades the periodicity of the pressure distribution is used to reduce the size of the influence matrix.

The derivation of the right hand side in the system of equations is illustrated in Fig. 3. The upwash point of a panel is moving on a helical path. In the moving reference frame $\hat{x},\hat{y},\hat{z}$, attached to the panel, both the free stream velocity, $\hat{W}_i$, and the induced velocity appear. Here only the component of the disturbance velocity $\hat{v}_i$ normal to helix is taken into account. The resultant $\hat{V}_R$ must be tangential to the panel surface and has then to be transformed into the fixed

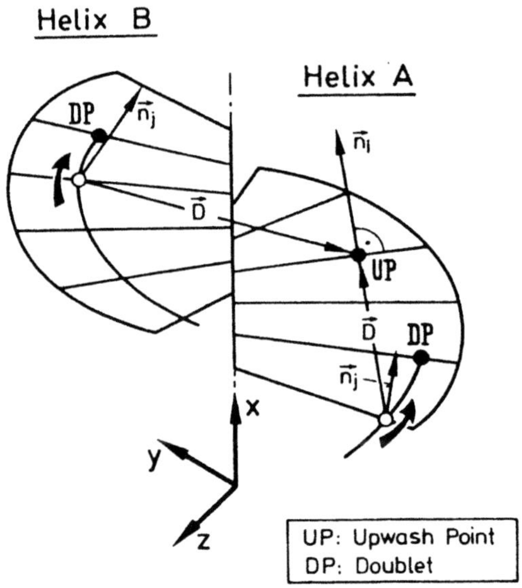

Fig. 2 Kinematics on regular helices

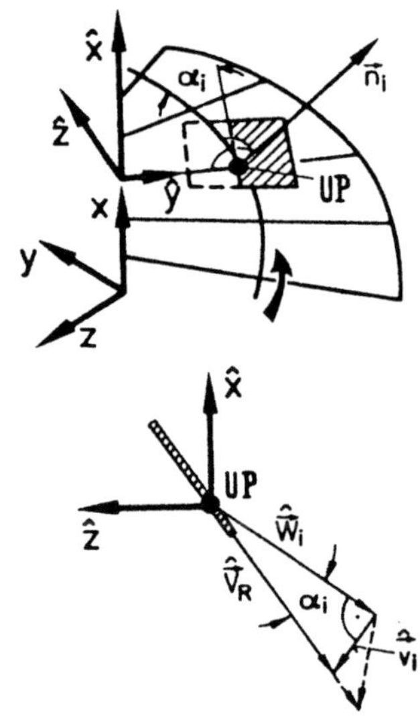

Fig. 3 Tangential flow condition at an upwash point

reference frame x,y,z. This leads to

$$v_i = - \tan \alpha_i \sqrt{U^2 + \Omega_i^2 r_i^2} \, , \qquad (14)$$

where U is the axial velocity of the propeller and Ω the angular velocity. α_i is the angle between the free stream velocity vector and the panel surface. For a single propeller and unsteady conditions α_i is a function of time. In the case of counter rotating propellers it is constant.

The resulting set of linear equations is relatively large - it contains up to 4000 unknowns. The influence matrix is dense and large elements also occur off the diagonal due to three reasons. First, the scheme of additional equations for different times leads to several equation blocks of the same structure. Second, the influence coefficients for fixed i and j and for all m=1,2,...n are of the same order of magnitude. Third the influence of one panel on itself is of the same order of magnitude as its influence on an upwash point which lies behind it. Due to the large off diagonal elements a direct solution method is taken. A Gauss algorithm with partial pivot search is used as an out of core solver. The required computation time of this solver is proportional to the third power of the number of unknowns. Future work should therefore be concerned with a more efficient solution method.

The solution method contains a number of error sources.

- A discretization error due to discrete lifting elements and doublet point scheme.
- An inexact matrix coefficient due to the nature of numerical integration and due to the truncated interval of integration.
- Errors in the solution of the set of linear equations due to a disturbed coefficient matrix and round-off errors.

The influence of the first error source was investigated in a parametric study (see below). It would be possible to determine an upper error boundary for the last two points. However such a calculation requires a large amount of computation time and often overestimates strongly the real error. Therefore, in the code written for the present method, the calculation of the error boundary was not taken up - instead some other error-indicators are calculated. To check the discrepancy in the matrix coefficients the rest of the integral is approximated and the quadrature formula is applied with a different order. To get an indication for the reliability of the matrix solution, the ratio of the maximal and minimal pivot element is determined and the Prager-Oettli bound [9] is determined. The quality of the numerical results is studied with these indicators and new calculations with a more accurate integration is performed if necessary.

4. NUMERICAL RESULTS

Results for steady and unsteady propeller aerodynamics are shown in this chapter. First a parametric study was done

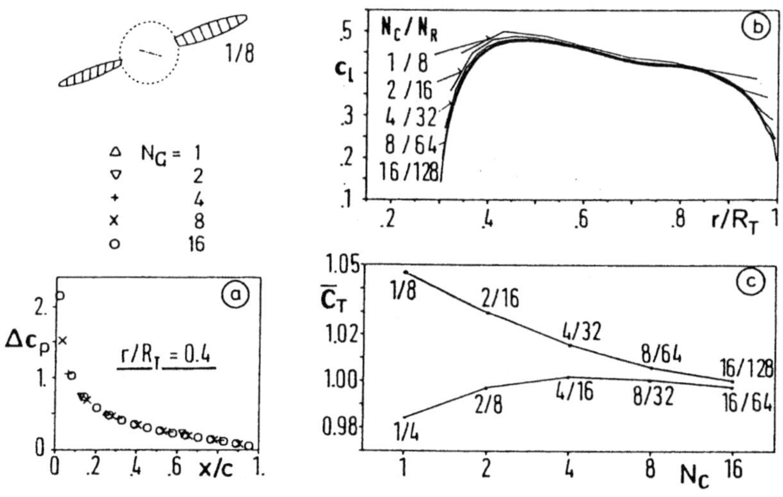

Fig. 4 Influence of discretization in the steady case

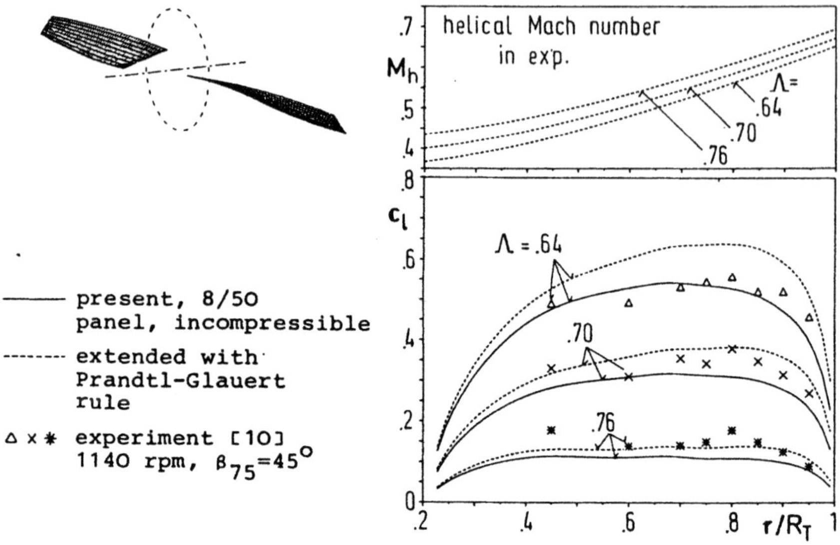

Fig. 5 Comparison between measured and calculated steady lift distributions for a two blade propeller

for a steady case to examine the influence of the discretization. An example propeller was divided in varying numbers of panels. Fig. 4a shows the Δc_p distribution in the chord direction x/c for one section. For increasing panel numbers from 1 to 16 along the chord the values lie nearly on one curve. Fig. 4b shows the lift coefficient versus radius for various refinements of panelling. Already the crude discretization of 8 panels in total shows approximately the behaviour of the finest discretization. For higher numbers of panels only small changes occur at tip and hub regions. In Fig. 4c an related thrust coefficient $\bar{C}_T$ is presented for different numbers of panels along the chord N_C and along the radius N_R. The difference in this integral value between crudest and finest panel distribution is about 2-5%.

Fig. 5 compares calculated and experimental [10] data for a NACA 10-(3)(08)-03 propeller. The lift coefficient versus radius is plotted for three advance ratios Λ. The measurements were done for helical Mach numbers up to 0.65 as shown in the upper figure. To get an idea about the influence of the Mach number, the incompressible results were extended with the Prandtl-Glauert rule. This leads to the dotted line. Two effects tend to act against each other: The compressibility, which increases the lift in comparison with incompressible theory, and viscosity, which decreases it compared with the inviscid results.

Fig. 6 shows the unsteady part of the lift coefficient $\tilde{c}_\ell$ relative to the time average lift (determined from the zero Fourier coefficients) $\bar{c}_\ell$ versus azimuth angle for a counter rotating propeller. Three calculations with different numbers

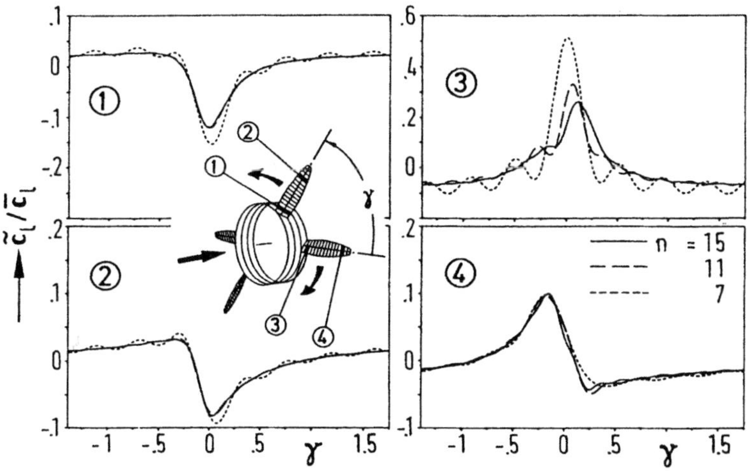

Fig. 6 Relative time dependent part of the lift coefficient for some sections of a counter rotating propeller

of Fourier coefficients n are represented by the three curves. Parts 1 and 2 of the picture show the results for the front propeller for an inner and outer section. The differences between the calculations with 11 and 15 Fourier coefficients are very small. Part 3 and 4 represent this lift coefficient for the same sections for the rear propeller. In the inner section the difference between n = 11 and 15 is relatively large and calculations with more Fourier coefficients seem to be necessary. The lift of the front propeller shows a drop when the two propellers stand behind each other. A peak occurs for the inner section of the rear propeller when it is rotating through the potential wake of the front propeller.

In Fig. 7, calculated and measured [11] efficiency for two different propellers is plotted as function of the advance

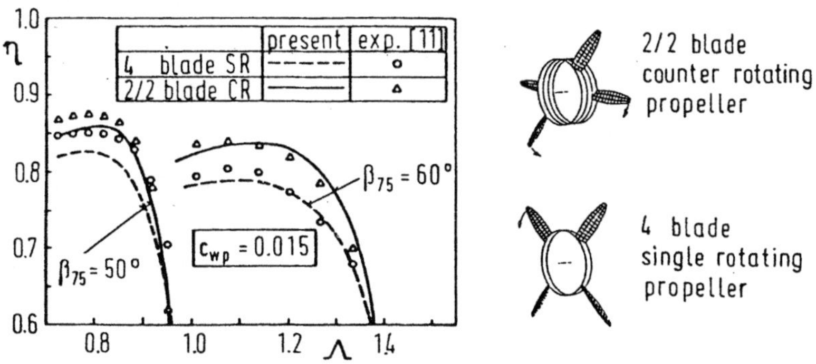

Fig. 7 Comparison between calculated and measured efficiency of single and counter rotating propeller

ratio. An estimated profile drag coefficient c_{wp} is added during the calculation to the induced drag. The improved efficiency of a counter rotating propeller compared with a single propeller is described fairly well with the doublet point method.

Fig. 8 shows the unsteady part of the thrust and power coefficient $\tilde{C}_T$, $\tilde{C}_p$ relative to the design value C_T, C_p and the phase shift ϑ versus advance ratio for a ship propeller in a sinusoidal onflow. Here both results, unsteady and quasi steady doublet point calculations, are included. In comparison with the quasi steady solution, the unsteady doublet point calculation agrees better with the experimental data [12].

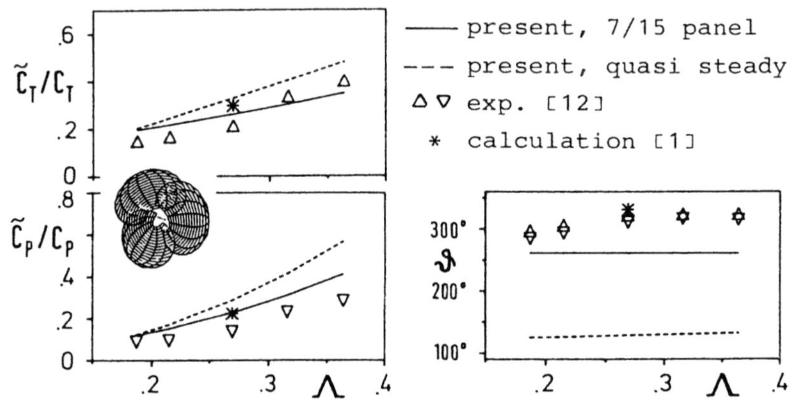

Fig. 8 Results for a ship propeller working in a sinusoidal onflow

5. CONCLUSIONS

This paper presents a lifting surface method for the calculation of unsteady propeller aerodynamics in an incompressible medium. In this model, with single forces distributed over the blades, the unsteady lift is simulated accurately, this method being far superior to the quasi steady formulation. Interfering lifting surfaces, like counter rotating propellers, are treated directly without iterative treatment. Parametric studies, with varying number of panels along chord and radius and varying number of Fourier coefficients, show that reasonable computational effort leads to acceptable results. Comparisons between calculated and experimental results for single and counter rotating propellers show good agreement.

ACKNOWLEDGEMENT

The author would like to thank Prof. Dr. A. DAS, N. KROLL and S. SCHERR for many stimulating and fruitful discussions.

REFERENCES

[1] TSAKONAS, S., JACOBS, W.R., ALI, M.R.:"Propeller Blade Pressure Distribution Due to Loading and Thickness Effects", J. of Ship Research 23 (1979), pp. 89-107.

[2] HANSON, D.B.:"Noise of Counter Rotating Propellers", AIAA-Paper 84-2305 (1984).

[3] HANSON, D.B.:"Compressible Lifting Surface Theory for Propeller Performance Calculation", J. Aircraft 22 (1985), pp. 19-27.

[4] RUNYAN, H.L., TAI, H.:"Lifting Surface Theory for a Helicopter in Forward Flight", NASA-CR-16997 (1983).

[5] LESIEUTRE, D.J., SULLIVAN, J.P.:"The Analysis of Counter-Rotating Propeller Systems", SAE Paper 850869 (1985).

[6] DAS, A.:"A Unified Approach for the Aerodynamics and Acoustics of Propellers in Forward Motion", AGARD-CPP-366(1984), pp. 9.1-9.28.

[7] DAT, R.:"Representation of a Lifting Line in an Arbitrary Motion by a Line of Acceleration Doublets", NASA TT F-12, 952 (1970).

[8] UEDA, T., DOWELL, E.H.:"A New Solution Method for Lifting Surfaces in Subsonic Flow", AIAA Journal 20 (1982), pp. 348-355.

[9] SCHABACK, R.:"Eine rundungsgenaue Formel zur maschinellen Berechnung der Prager-Oettli-Schranke", Computing 20 (1978), pp. 178-182.

[10] EVANS, A.J.:"Propeller Section Aerodynamic Characteristics as Determined by Measuring the Section Surface Pressures on an NACA 10-(3)(08)-03 Propeller under Operating Conditions", NACA RM L50H03(1950).

[11] BIERMANN, D., HARTMANN, E.P.:"Wind-Tunnel Tests of Four- and Six-Blade Single- and Dual-Rotating Tractor Propellers", NACA Rep. No. 747 (1940).

[12] MILLER, M.L.:"Experimental Determination of Unsteady Propeller Forces". Proceedings, Seventh ONR Symposium on Naval Hydrodynamics, Office of Naval Research (1968), pp. 255-290.

CALCULATION OF UNSTEADY TRANSONIC FLOW ABOUT OSCILLATING WINGS BY A FIELD PANEL METHOD

R. Voß
DFVLR Institute of Aeroelasticity
Bunsenstraße 10, D-3400 Göttingen, Germany

SUMMARY

A field panel method is described for calculating unsteady transonic perturbation flows about harmonically oscillating wings. The flow is modelled by dipole distributions on the mean wing and wake surfaces and a source distribution in the flow field to account for the nonhomogeneous compressibility effects in the near field of the wing. The aerodynamic influence coefficients can be calculated largely analytically and the resulting system of equations is solved by a quickly converging block iteration. Coupling with finite difference methods can further increase the computation speed.

INTRODUCTION

Flow problems with strong compressibility effects cannot be treated by surface panel methods, because the governing flow equations are nonlinear, or at least have variable coefficients. Finite difference or volume methods still require much computer time. In some cases efficient alternatives are field panel methods, which use panels also in certain parts of the flow field. Thus the advantages of panel methods, namely the lowered dimension of the problem and the correct treatment of the far field, are retained to some extent. Such a method was developed for unsteady transonic flows about harmonically oscillating wings with freestream Mach numbers $M_\infty \lesssim 1$ (see figure 1). The resulting computer code PTRAN3 is applicable to routine flutter investigations of modern aircraft wings.

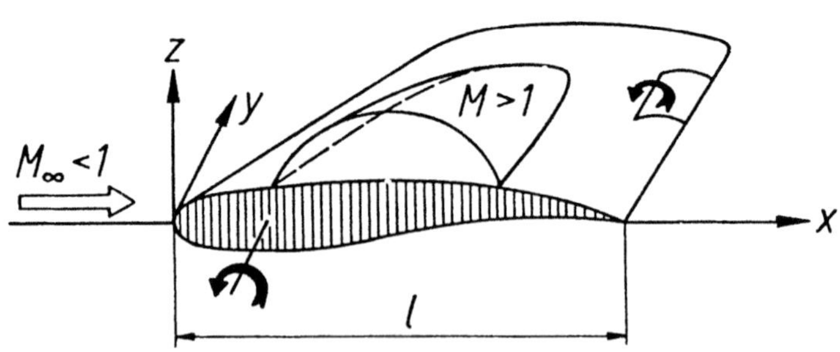

Fig.1 The unsteady flow problem

FORMULATION OF THE INTEGRAL EQUATION PROBLEM

The unsteady transonic flow around oscillating wings of moderate thickness, incidence and sweep is governed by the Transonic Small Disturbance Potential Theory. For flutter investigations it is convenient to assume purely harmonic oscillations of small amplitudes [1]. One may in that case treat the unsteady flow as a superposition of a mean steady flow and a small unsteady perturbation flow and assume for the velocity potential

$$\phi(x,y,z,t) = \phi^0(x,y,z) + \mathrm{Re}\left(\phi^1(x,y,z)\cdot e^{ikt}\right) \qquad (1)$$

with

$$|\phi^0| >> |\phi^1| >> \text{higher harmonics.}$$

All velocities are scaled by the freestream velocity u_∞, space coordinates by ℓ and time by ℓ/u_∞; $k=\omega\ell/u_\infty$ is the reduced frequency.

Neglecting contributions of the second order in the perturbation values the problem is now split into a steady and an unsteady part. The latter is governed by a "time linearised" unsteady potential equation [2],[3] for the unsteady complex potential function ϕ^1:

$$(1-K\phi^0_x)\phi^1_{xx} + \phi^1_{yy} + \phi^1_{zz} - (2i\varepsilon + K\phi^0_{xx})\phi^1_x + k\varepsilon\,\phi^1 = 0 \qquad (2)$$

with

$$K = (\gamma+1)M_\infty^2/\beta^2 , \quad \beta^2 = 1 - M_\infty^2 , \quad \lambda = kM_\infty/\beta^2 , \quad \varepsilon = \lambda M_\infty .$$

The space variables y and z are scaled as follows: $y \rightarrow \beta y$, $z \rightarrow \beta z$. Boundary conditions are to be fulfilled at the wing surface (which may be projected onto the z=0 plane) and for the potential jump across the wake downstream from the wing's trailing edge, in order to account for the unsteady shed vorticity from the wing (h_1 stands for the oscillation amplitude):

$$\phi^1_z = h_{1x} + ik\,h_1 \qquad \text{(at wing)} , \qquad (3a)$$

$$\delta\phi^1_x + ik\,\delta\phi^1 = 0 \qquad \text{(across wake)} . \qquad (3b)$$

Equation (2) is linear but coupled to the steady mean flow ϕ^0 by the spacewise varying coefficients. These may be strongly nonuniform, especially for flows with embedded supersonic regions and shocks. The coefficient of the first term in equation (2) changes sign when the flow conditions change from subsonic to supersonic, so that equation (2) in general has a mixed elliptic/hyperbolic character. The potential equation can be transformed into a nonhomogeneous Helmholtz equation

$$\left.\begin{aligned} L &= \varphi_{xx} + \varphi_{yy} + \varphi_{zz} + \lambda^2\varphi = S \\ \text{with} \quad \varphi &= \phi^1\cdot e^{-i\varepsilon x} , \\ S &= \left[\frac{\partial}{\partial x} + i\varepsilon\right]\underbrace{\left(K\phi^0_x\,(\varphi_x + i\varepsilon\varphi)\right)}_{\sigma} , \quad \lambda = \frac{k\,M_\infty}{\beta^2} . \end{aligned}\right\} \qquad (4)$$

The size of the right-hand side of S is determined by $K\phi^0 \approx \phi_x^0/\phi_x^{0*}$, where ϕ_x^{0*} denotes the sonic value of the local Mach number 1. Equation (4) shows a relation to the problem of acoustic wave propagation in a nonhomogeneous moving fluid. The Helmholtz operator L alone describes disturbance propagations in a compressible uniform subsonic flow with velocity u_∞. The basic influence function for a disturbance source, fulfilling Sommerfeld's radiation condition, reads

$$\psi = \frac{1}{4\pi}\, e^{-i\lambda r}/r \ , \qquad r = \sqrt{(\xi-x)^2 + (\eta-y)^2 + (\zeta-z)^2}\ . \tag{5}$$

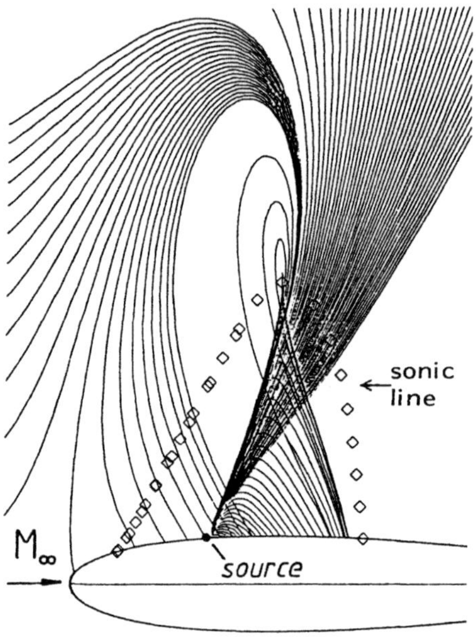

Fig.2 Acoustic ray paths in transonic flow

S accounts for the nonuniform compressibility effects which are determined by the mean flow. The importance of S is restricted to the vicinity of the wing, where local Mach numbers differ significantly from M_∞. The local character of the problem may change from elliptic to hyperbolic, but the global character remains elliptic as long as supersonic regions are finite. Figure 2 shows the paths of disturbance propagation from a source located in a supersonic region embedded in a subsonic flow field. Every part of the flow field is reached by signals on curved rays, which can be calculated for a given flow as shown in [4]. There are no zones of silence as in a uniformly supersonic field. Hence Green's theorem may be applied to equation (4) using equation (5). This, together with our boundary conditions in equation (3), yields a system of two coupled integro-differential equations for a dipole distribution $\delta\varphi$ on the projection area of wing and wake in the z=0 plane and a field source distribution σ in B:

$$w = \varphi_z = -\iint_{\text{wing}} [\delta\varphi\ \psi_{\zeta\zeta}]_{\zeta=0}\ dF - \iint_{\text{wake}} \delta\varphi_{TE}[\psi_{\zeta\zeta}]_{\zeta=0}\cdot e^{-i\nu(X-X_{TE})}\,dF$$

$$- \iiint_B \sigma(\psi_{\xi\zeta} - i\varepsilon\psi_\zeta)\ dV\ , \qquad \text{with} \qquad \nu = k/\beta^2\ ; \tag{6}$$

$$\varphi = \iint_{\text{wing}} [\delta\varphi\ \psi_\zeta]_{\zeta=0}\ dF + \iint_{\text{wake}} \delta\varphi_{TE}[\psi_\zeta]_{\zeta=0}\cdot e^{-i\nu(X-X_{TE})}\ dF$$

$$+ \iiint_B \sigma(\psi_\xi - i\varepsilon\psi)\ dV\ .$$

B is that part of the space where S cannot be approximated by zero. The dipole strength in the wake is related to its trailing edge value $\delta\varphi_{TE}$ by use of equation (3b).

NUMERICAL SOLUTION PROCEDURE

Equations (6) are solved by discretising the projection area of the wing onto the z=0 plane into trapezoidal panels and the wake into semi-infinite parallel strips. The control volume B is divided into volume elements, the top and bottom surfaces of which are located in z=const. planes and are chosen equal to the wing panels just below them. $\delta\varphi$, σ and ϕ_x^0 are assumed to be constant within a panel or volume element, control points are located in the panel centers. Now equations (6) yield a system of linear equations:

$$\begin{aligned} w &= A\delta\varphi + A^* \delta\varphi_{TE} + B\sigma \ , \\ \varphi &= C\delta\varphi + C^* \delta\varphi_{TE} + D\sigma \ . \end{aligned} \qquad (7)$$

The discretisation and the various Aerodynamic Influence Coefficients (AICs) are shown in figure 3. The field source term σ is evaluated by finite differentiation of φ and use of Murman's mixed finite difference operator [3],[5], which switches between central and backward differencing to account for the locally changing elliptic and hyperbolic influence regions. The $\delta\varphi_{TE}$-values are expressed by the two dipole values just upstream of the trailing edge via the Kutta condition of smooth tra -fer of the pressure jump to a trailing edge value of zero [3]. Hence A^* and C^* can be formally included in matrices A and C, yielding a system

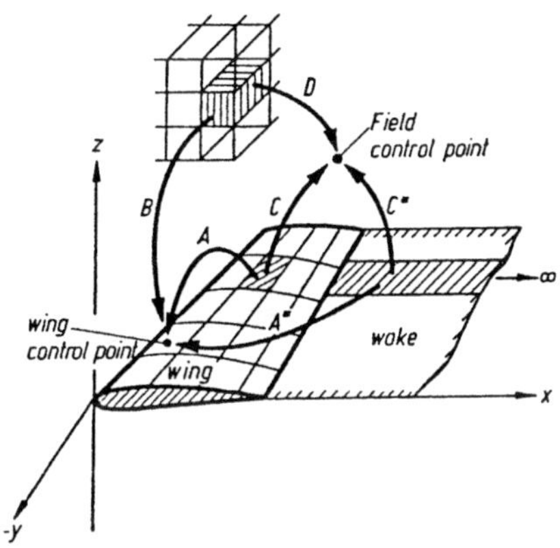

Fig.3 Discretisation and AICs

$$\begin{pmatrix} \widetilde{AB} \\ \widetilde{CD} \end{pmatrix} \begin{pmatrix} \delta\varphi \\ \varphi \end{pmatrix} = \begin{pmatrix} w \\ 0 \end{pmatrix} . \qquad (8)$$

CALCULATION OF THE AERODYNAMIC INFLUENCE COEFFICIENTS

The numerical evaluation of the AIC integrals takes most of the computation time. Fast calculation methods for A and A^* are shown in [6] for the special case of the unsteady subsonic wing theory with $S = 0$. An efficient calculation of the coefficients B, C, C^* and D is achieved by use of cylinder coordinates, expansion of the integrands and partly analytical integration (see figures 4 and 5). In particular one has:

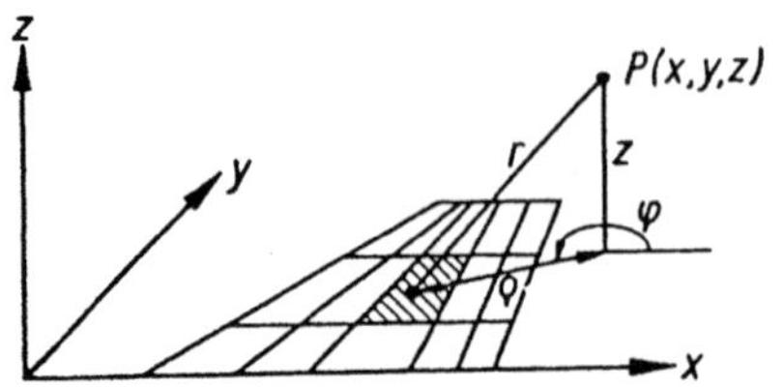

Fig. 4a Aerodynamic C-coefficients

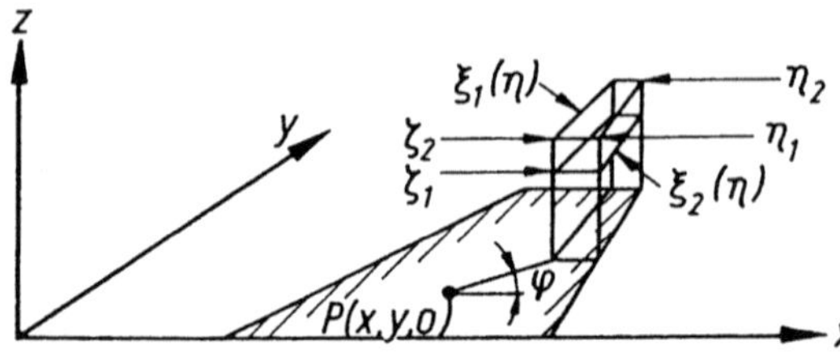

Fig. 4b Aerodynamic B-coeffients

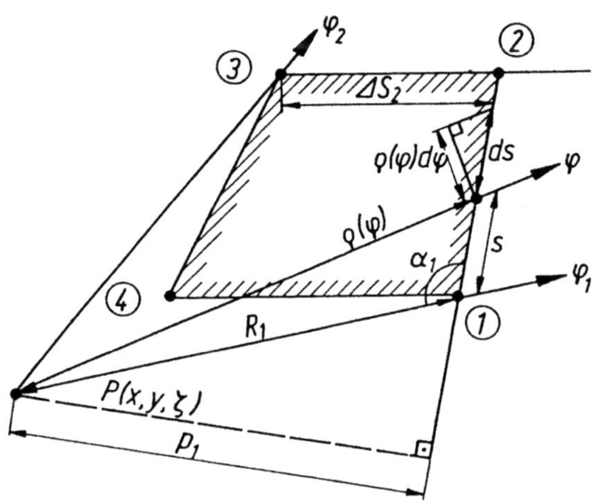

Fig. 5 Line integration of C- and B-coefficients

$$B = \iiint\limits_{\Delta V} (\psi_{\xi\xi} - i\varepsilon\psi_{\zeta})\, dV \tag{9}$$

$$= \int \Big(\psi(\xi_2,\eta,\zeta_2) - \psi(\xi_2,\eta,\zeta_1) - \psi(\xi_1,\eta,\zeta_2) + \psi(\xi_1,\eta,\zeta_1)\Big)\, d\eta$$

$$+ \frac{M_\infty}{4\pi} \sum_{k=1}^{4} p_k \Big(I_B^k(\zeta_2) - I_B^k(\zeta_1)\Big) - \frac{1}{2} M_\infty \delta (e^{-i\lambda\zeta_2} - e^{-i\lambda\zeta_1}) \,,$$

$$C = \iint\limits_{\Delta F} \psi_\zeta \, dF = \frac{\delta}{2} \operatorname{sgn}(z)\, e^{-i\lambda|z|} - \frac{1}{4\pi} \sum_{k=1}^{4} p_k \, I_C^k(z) \tag{10}$$

with

$$p_k = \frac{\vec{R}_k \times \vec{s}_k}{|\vec{s}_k|}\, \vec{n}_z \qquad \text{and} \qquad \delta = 1 \,,$$

if the projection of the control point onto the z=0 plane lies inside the corresponding projection of ΔV or ΔF, otherwise $\delta=0$.

B and C are expressed by sums of line integrals along the 4 edges of the trapezoidal surfaces. These are evaluated in the following manner: (r_{min} = minimum distance between P and the kth edge with length Δs_k). For $\Delta s_k/r_{min} < 0.3$ a 3-point Simpson quadrature is applied. For $\Delta s_k/r_{min} > 0.3$ expansion of the exponential functions in the integrands using the first 20 terms and analytical integration of the resulting integral series yields for instance

$$I_B^k(\zeta) = \sum_{\ell=0}^{20} \frac{(-i\lambda)^\ell}{\ell!} \left(I_{\ell+1}(t_2) - I_{\ell+1}(t_1) \right) ,$$

$$I_C^k(z) = \sum_{\ell=0}^{20} \frac{(-i\lambda)^\ell}{\ell!} \left(I_\ell(t_2) - I_\ell(t_1) \right) , \qquad (11)$$

where

$$t_2 = \Delta s_k - R_k \cos\alpha_k , \qquad t_1 = -R_k \cos\alpha_k .$$

For the above analytical integrals I_ℓ the following recursive relations are valid (different for even and odd indices ℓ):

$$I_1 = \arctan(t/c)/c , \qquad (12)$$

$$I_{2n+1} = \sum_{k=0}^{n-1} \frac{1}{2k+1} \binom{n-1}{k} a^{2(n-1-k)} t^{2k+1} + (a^2-c^2) I_{2n-1} ,$$

$$I_0 = \frac{1}{c\sqrt{a^2-c^2}} \arctan\left(\sqrt{\frac{a^2-c^2}{a^2+t^2}} \frac{t}{c} \right) , \qquad I_{2n+2} = (a^2-c^2) I_{2n} + I^*_{2n+2} ,$$

$$I_2^* = \ln\left(t + \sqrt{a^2+t^2}\right) , \qquad I^*_{2n+2} = \frac{t}{2n} (t^2+a^2)^{n-1/2} + \frac{2n-1}{2n} a^2 I^*_{2n} ,$$

with: $$c^2 = R_k^2 \sin^2\alpha_k , \qquad a^2 = c^2 + \zeta^2 \quad \text{or} \quad c^2 + z^2 .$$

If the series expansion loses accuracy for $\lambda r > 5$ and if $\Delta s_k/r_{min} > 0.3$, numerical quadrature is applied. But this occurs very frequently only for high λ-values and large panel sizes. If the distance between P and the panel exceeds 5 panel diameters, $r/d > 5$, the AICs are approximated by the value of the integrand at the panel center.

The same holds true for D-coefficients. For moderate distances $2 < r/d < 5$, numerical tests have shown that the integrand of D may be approximated by a Taylor series expansion up to the third order, the terms of which can be integrated analytically, which yields (see figure 6):

$$D = \iiint_{\Delta V} g \, dV = \Big(g + \frac{1}{6} \overline{\Delta x}^2 g_{xx} + \frac{1}{6} \Delta y^2 g_{yy} + \frac{1}{6} \Delta z^2 g_{zz}$$
$$+ \frac{1}{6} \Delta y^2 (\tan \alpha_2 + \tan \alpha_1) g_{xy} \Big) \Delta V$$
$$+ (\tan^2 \alpha_2 + \tan^2 \alpha_1) g_{xx} \frac{2}{3} \overline{\Delta x} \Delta y^3 \Delta z$$
$$+ (\tan \alpha_2 - \tan \alpha_1)(2 g_y + \frac{1}{3} \Delta z^3 g_{yzz} + \overline{\Delta x}^2 g_{xxy}$$
$$+ \frac{1}{5} \Delta y^2 g_{yyy}) \frac{2}{3} \Delta y^3 \Delta z + (\tan^2 \alpha_2 - \tan^2 \alpha_1) \times \qquad (13)$$
$$\times (g_x + \frac{1}{6} \Delta z^2 g_{xzz} + \frac{3}{10} \Delta y^2 g_{xyy} + \frac{1}{2} \overline{\Delta x}^2 g_{xxx}) \frac{2}{3} \Delta x^3 \Delta z$$
$$+ (\tan^3 \alpha_2 - \tan^3 \alpha_1) g_{xxy} \frac{2}{15} \Delta y^5 \Delta z$$
$$+ (\tan^4 \alpha_2 - \tan^4 \alpha_1) g_{xxx} \frac{1}{30} \Delta y^5 \Delta z \; .$$

The appearing mixed derivatives (up to the fourth order) of ψ can be easily obtained by an analytical computer program REDUCE2 [7]. In the near field ($r/d < 2$), use of the relation

$$g = \psi_\xi - i\varepsilon\psi = \psi_\xi + i\varepsilon/\lambda^2 (\psi_{\xi\xi} + \psi_{\eta\eta} + \psi_{\zeta\zeta}) \qquad (14)$$

enables one analytical integration. The remaining integrals over the faces of the volume elements are evaluated by numerical quadrature. The semi-infinite wake strip integrals C* are cast by variable transforms and integration in the complex plane into a form which can be easily evaluated by means of a Gauss-Laguerre quadrature.

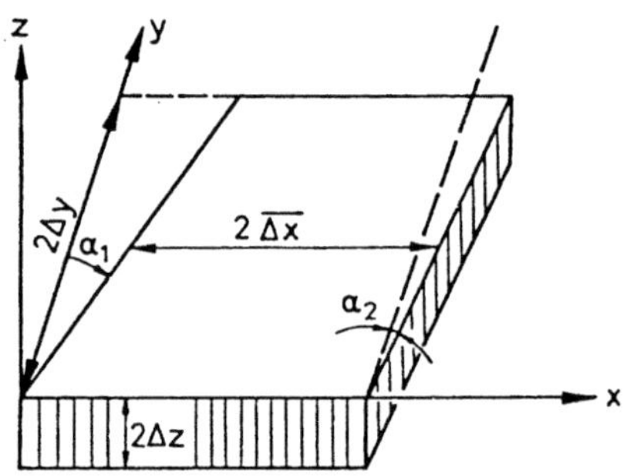

Fig.6 Aerodynamic D-coefficients

SOLUTION OF THE SYSTEM OF LINEAR EQUATIONS

A direct inversion of the AIC matrix in equation (8) is not adequate because of computer time and storage limitations. A reliable solution procedure is obtained by a block iteration strategy. The submatrix including all AICs representing interactions between wing panels and supersonic (index 1) field panels is inverted, while the potential values of subsonic field panels (index 2) are updated in a single step. The iteration scheme reads:

$$\begin{pmatrix} \delta\varphi \\ \varphi_1 \end{pmatrix}^{(n+1)} = \begin{pmatrix} \tilde{A} & \tilde{B}_1 \\ \tilde{C}_1 & \tilde{D}_{11} \end{pmatrix}^{-1} \left\{ \begin{pmatrix} w \\ 0 \end{pmatrix} - \begin{pmatrix} \tilde{B}_2 \\ \tilde{D}_{12} \end{pmatrix} \begin{pmatrix} \varphi_2 \end{pmatrix}^{(n)} \right\} , \qquad (15)$$

$$\varphi_2^{(n+1)} = \tilde{C}_2 \, \delta\varphi^{(n+1)} + \tilde{D}_{21} \, \varphi_1^{(n+1)} + (\tilde{D}_{22} + I)\varphi_2^{(n)} \; .$$

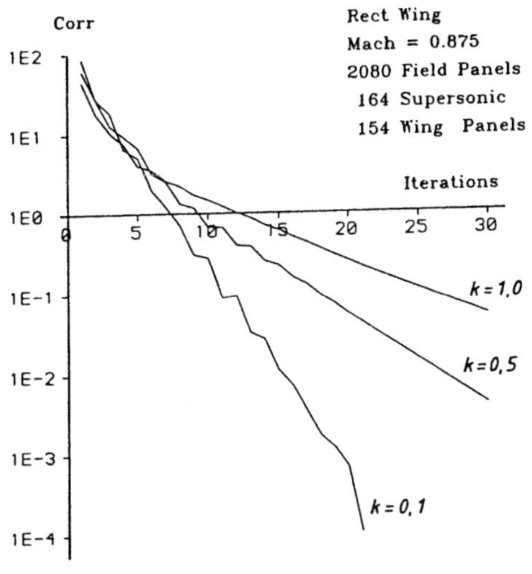

Figure 7 shows the convergence rate achieved by this method for a wing with mixed sub-/supersonic flow conditions. Convergence slows down for increasing frequency, but 30 iteration steps are sufficient for reduced frequencies which are important in flutter investigations to decrease corr, the sum of the absolute changes of all dipole values, by 3 orders of magnitude.

Fig.7 Convergence of the field panel method

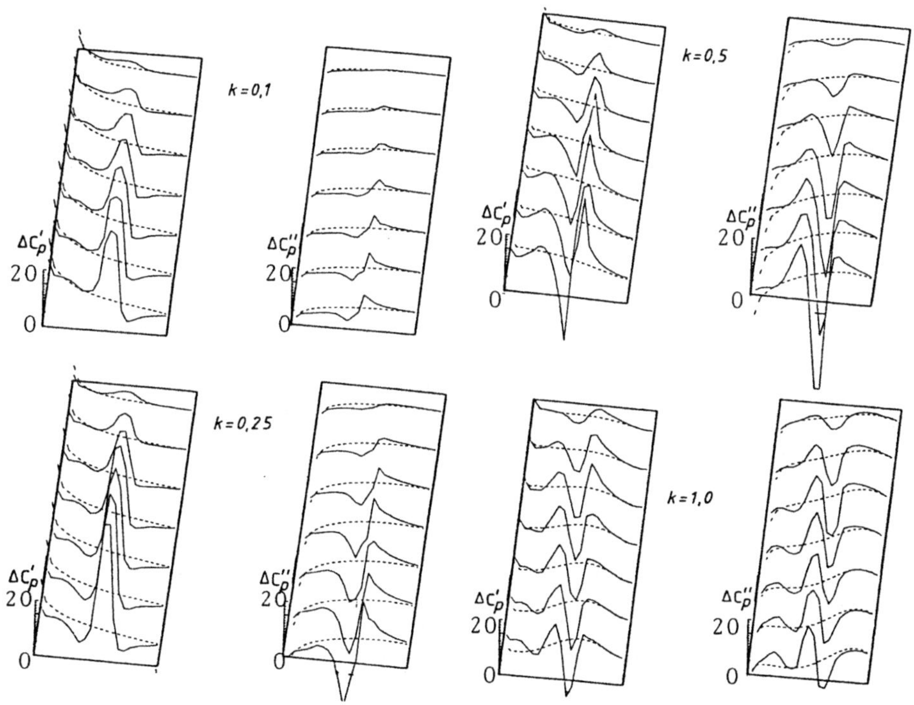

Fig.8 Unsteady transonic pressure distributions
——— PTRAN3, ----- subsonic theory

RESULTS

Figure 8 shows results obtained with the field panel computer code PTRAN3 for a rectangular wing with a halfspan/chord of 2.5 and a symmetric NACA 64A006 profile. The wing is pitching about its 1/4-chord axis at a Mach number of 0.875 and zero angle of attack. The steady flow field was calculated by the finite difference code FL022 [9]. There is a shock between 50 and 60% chord which weakens in wing tip direction (only one halfwing is shown). The unsteady first harmonic components of the pressure jump coefficient, normalised by the amplitude, are calculated by PTRAN3 using the FL022 results as input. Results of the purely subsonic theory (with $S = 0$ in equation (4)) are shown for comparison. The main difference between the two theories is that the transonic theory predicts a large pressure peak representing the contributions of shock oscillation. For low k-values this appears mainly in the in-phase component $\Delta c_p'$; for increasing k the out-of-phase component $\Delta c_p''$ is involved as well. This is due to the finite running times of unsteady perturbations. The time delay at a certain distance from their source, measured in oscillation periods, increases with k.

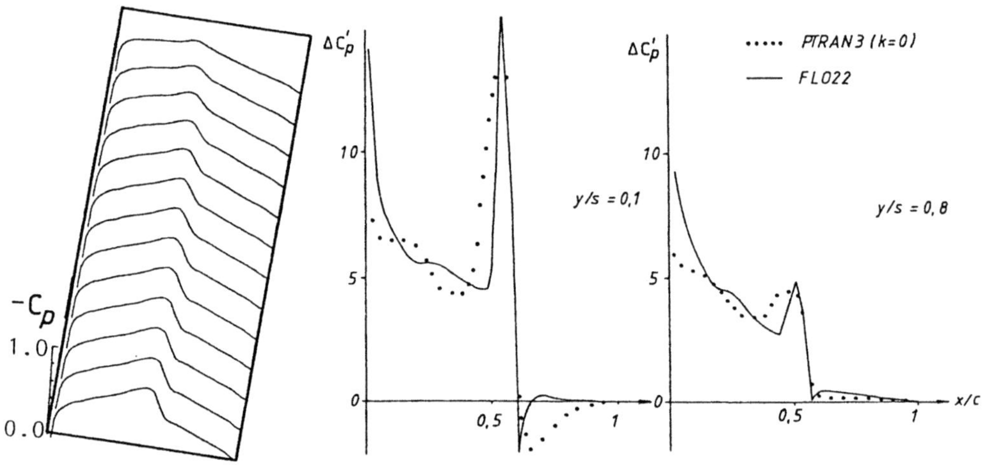

Fig.9 Quasisteady and steady pressure distributions

Figure 9 shows the results for the limiting "quasisteady" case $k = 0$ compared with results obtained from two steady FL022 calculations at ±0.5 degrees angle of attack. These unsteady field panel calculations with 2080 field panels and 154 wing panels took 5 minutes computer time on an IBM 3090. For more complicated wing geometries and flow patterns as in the rectangular wing case more panels (≈ 10,000) are desired to obtain a fine resolution of unsteady pressure details. But such large panel numbers N demand too much computer time and storage, both of which grow quadratically with N. A way to obtain improved results under the same computer limitations is to couple the field panel method with a finite difference method for equation (1), see figure 10. Then

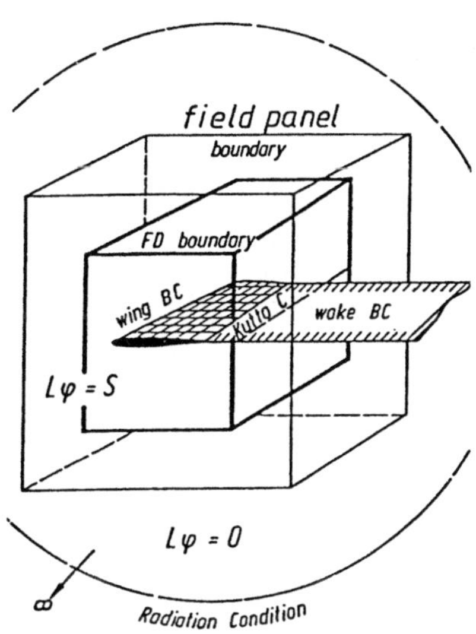

a calculation on a coarse grid (1500 field panels) is followed by a finite difference calculation on a fine grid (15,000 control points) inside the field panel volume. The panel method delivers the outer boundary values and a good starting solution for the finite difference iteration, which better captures the details. For this hybrid method the total computer time and storage do not exceed those of the above-mentioned pure field panel example. For further details see [8]. Unsteady results for the LANN wing with a double lambda-type shock system are shown in figure 11 and compared with experimental results from [10].

Fig 10 Coupling of field panel and finite difference method

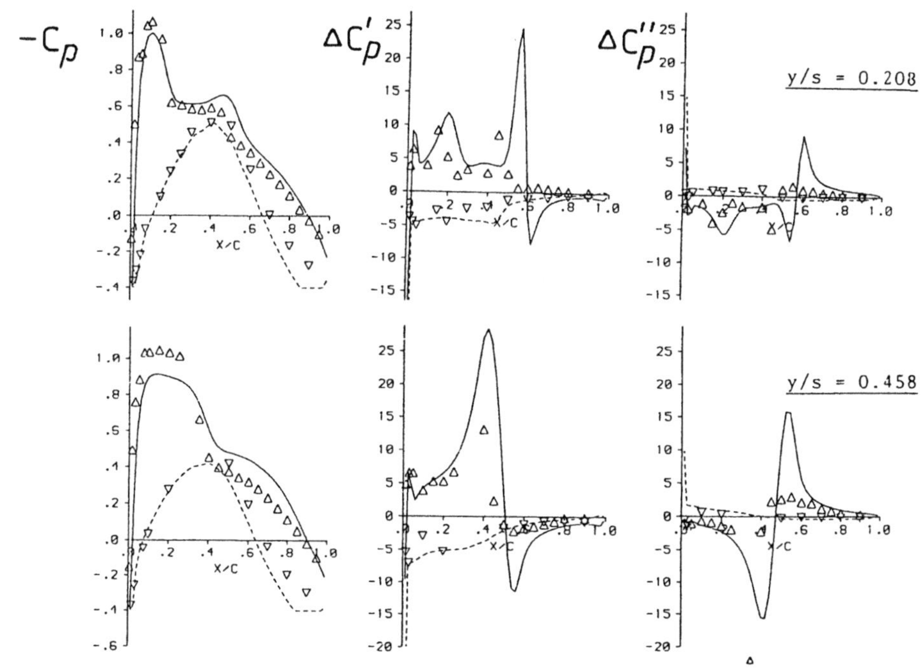

Fig.11 Steady and unsteady pressures at pitching LANN wing, $M_\infty = 0.820$, $k = 0.20$, upper and lower sides, ——— ----- PTRAN3, ΔΔΔ ∇∇∇ Test

REFERENCES

[1] FÖRSCHING, H.: "Grundlagen der Aeroelastik", Springer-Verlag, Berlin/Heidelberg/New York 1974.

[2] LANDAHL, M.T.: "Unsteady Transonic Flow", Pergamon Press 1961.

[3] VOß, R.: "Calculation of 3-D Unsteady Transonic Potential Flows by a Field Panel Method", 2nd Int. Symp. on Aeroelasticity and Structural Dynamics, DGLR Ber.85-02 (1985).

[4] SAUER, R.: "Anfangswertprobleme bei partiellen Differentialgleichungen", Springer-Verlag, Berlin/Göttingen/Heidelberg 1958.

[5] MURMAN, E.M.: "Analysis of Embedded Shock Waves Calculated by Relaxation Methods", AIAA Journal, Vol.12, May 1974, pp.626-633.

[6] GEIßLER, W.: "Ein numerisches Verfahren zur Berechnung der instationären aerodynamischen Druckverteilung der harmonisch schwingenden Tragfläche mit Ruder in Unterschallströmung", DLR FB 77-15 (1977).

[7] HEARN, A.C.: "REDUCE2 User's Manual, Computational Physics Group, University of Utah, Report No. UCP-19 (1973).

[8] VOß, R.: "Unsteady Time-Linearized Potential Theoretical Calculations of Transonic Flows about Oscillating Wings", DFVLR FB (to appear).

[9] JAMESON, A., CAUGHEY, D.A.: "Numerical Calculation of the Transonic Flow Past a Swept Wing", ERDA Report. C00-3077-140, New York Univ. (1977).

[10] "Compendium of Unsteady Aerodynamic Measurements", Addendum No.1, Data Set 9, AGARD-R-702 (1985).

A VORTEX-LATTICE METHOD FOR THE CALCULATION OF WING-VORTEX INTERACTION IN SUBSONIC FLOW

S. Wagner, Ch. Urban, R.Behr
University of the Armed Forces
of the Federal Republic of Germany,
Munich, Werner-Heisenberg-Weg 39, D-8014 Neubiberg

SUMMARY

Flow interferences between wings and free vortex sheets (e.g. canard-wing with vortex separation) can influence the aerodynamic characteristics of a configuration to a considerable extent. If viscous and turbulent effects are negligible, potential flow methods represent a suitable way to calculate wing-vortex interactions and in many cases lead to simpler mathematical structures than Euler methods.
The present vortex lattice method [1] is based on a time-dependent, discrete procedure with zero order doublet distributions. Except of the kinematic flow condition at the wing, all other boundary conditions are implicitely fulfilled by the mathematical model itself. Subsequently, iterations or the prescription of a starting solution is not necessary. At the shear layers, a continuous velocity distribution is indispensable for a successful analysis of interference effects. This problem, which all low order panel methods have to face, compare [2], is solved by spline functions and a differencing scheme.
The method is designed for subsonic flow and single or coupled plane, thin wings with or without vortex separation at the sharp wing edges. Potential flow effects (e.g. primary vortex) and their influence on the wing loads and the velocity field are reproduced in good agreement with experiments. In some cases, hints on a beginning vortex break down are possible. Variations of the free stream $\vec{U}_\infty$ (t) can be simulated, low Strouhal numbers presumed.
With these attributes, the vortex lattice method offers a reliable and versatile concept to gain insight in basic wing-vortex interactions at low computational effort.

THEORETICAL APPROACH

The discontinuity surfaces are represented by a set of discretely distributed doublet elements of stepwise constant strength μ, which is equivalent to a network of vortex rings with $\Gamma \equiv \mu$, Fig. 1 . During a time dependent, quasisteady procedure, starting with an impulsive motion of the wing W, the developement of the free vortex sheets V is observed until steady wing loads are attained, Fig. 2 .

The paper is based on research work funded by the Deutsche Forschungsgemeinschaft (DFG), contract No. Wa 424/3.

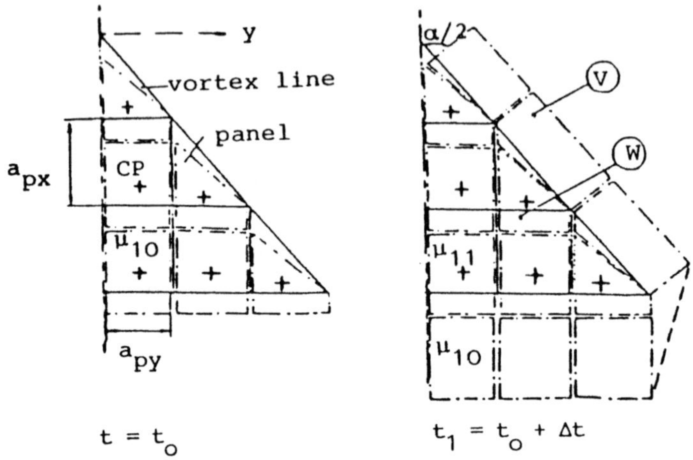

Fig. 1 Discretization of wing W and free vortex sheet V

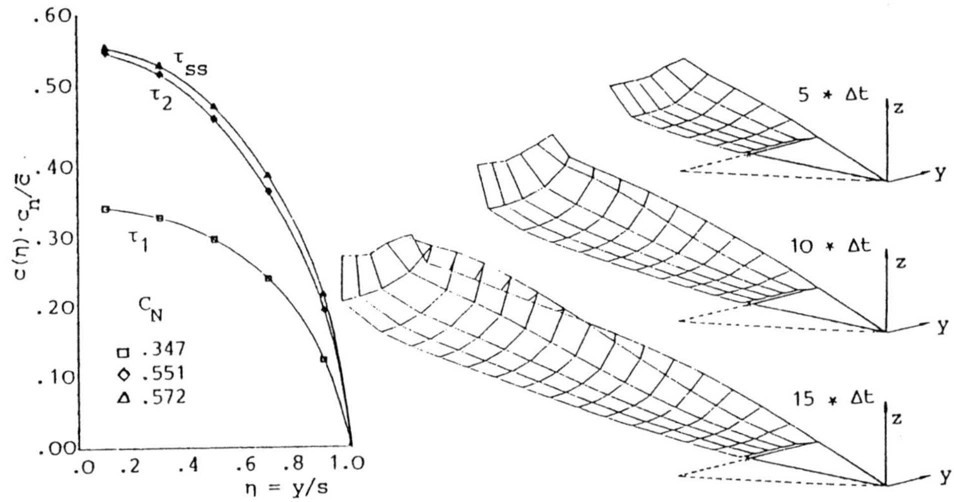

Fig. 2 Time-dependent developement of normal force coefficient and free vortex sheet of a delta wing (AR = 2)

Subsonic compressible flow is integrated in the method by the Prandtl-Glauert-transformation of the linearized potential equation. This simplification has proved to be admissible up to medium Mach numbers, although the preconditions of a linearization are not fulfilled in the case of wings with vortex separation. At present for higher Mach numbers a more accurate analysis of compressible effects is under developement. Computation time differs from 5 ÷ 1000 CPU-sec (Cyber 175) depending on the grade of discretization and the flow problem. On an average, 500 ÷ 600 CPU-sec are sufficient to interfering wings with vortex separation.

BOUNDARY CONDITIONS

The velocity potential Φ of the doublet distributions produces irrotational flow except at the location of the shear layers W and V. Simultaneously the function Φ fulfills the continuity equation. The doublet elements emanating from the wing edges at a certain moment retain their actual singularity strengths for all further time steps ($d\Gamma/dt = 0$). Evidently, the vortex conservation laws are observed by the application of closed vortex rings. After each time step Δt, the new positon $\vec{r}_V$ $(t+\Delta t)$ of the free vortex sheet elements is determined by the stream flow vector $\vec{w}$ $(\vec{r}_V,t)$ at the element corners $\vec{r}_V(t)$ (kinematic flow condition).

At wing edges with separation (these are defined as separation lines), the Kutta condition is implemented by $d\Gamma/dt = 0$ (see above) and by the chosen time increment

$$\Delta t = a_{pW}/U_\infty \qquad (1)$$

with a_{pW} as a characteristic size of the wing panels.

INDUCED VELOCITIES

In the far-field the Biot-Savart-law is used to calculate the velocities induced by vortex lines. Yet, applying this law to rolled-up vortex sheets or free shear layers located closely to wing surfaces leads to unrealistic values of both amount and direction of the vortex-induced velocities. In order to solve this problem of the near field, the velocity of near field points is determined by a linear interpolation between the velocity at the far-field boundary (Biot-Savart-law) and the limiting velocity at the discontinuity surfaces ($\pm \frac{1}{2}$ $\nabla.\Gamma$ plus far-field inductions). The mean values of $\nabla.\Gamma$ are derived by spline function describing the shear layer geometry and by interpolation of the discrete circulation strengths. The far-field boundary, i.e. at a distance a_n normal to the shear layers, is set to

$$a_n = \pm a_p/2 \qquad (2)$$

depending on the medium local size a_p of the vortex rings.

As an example, Fig.3 shows the velocity distribution for a point approaching the wing surface $z \to 0$. The velocities calculated by the pure Biot-Savart-law do not satisfy the flow conditions at the wing surface. On the contrary, the procedure described above leads to an adequate velocity distribution. The tangential fluid tansport is guaranteed. Leackage flow through the the wing does not appear. In addition, the kinematic boundary condition is artificially enforced on the complete wing surface

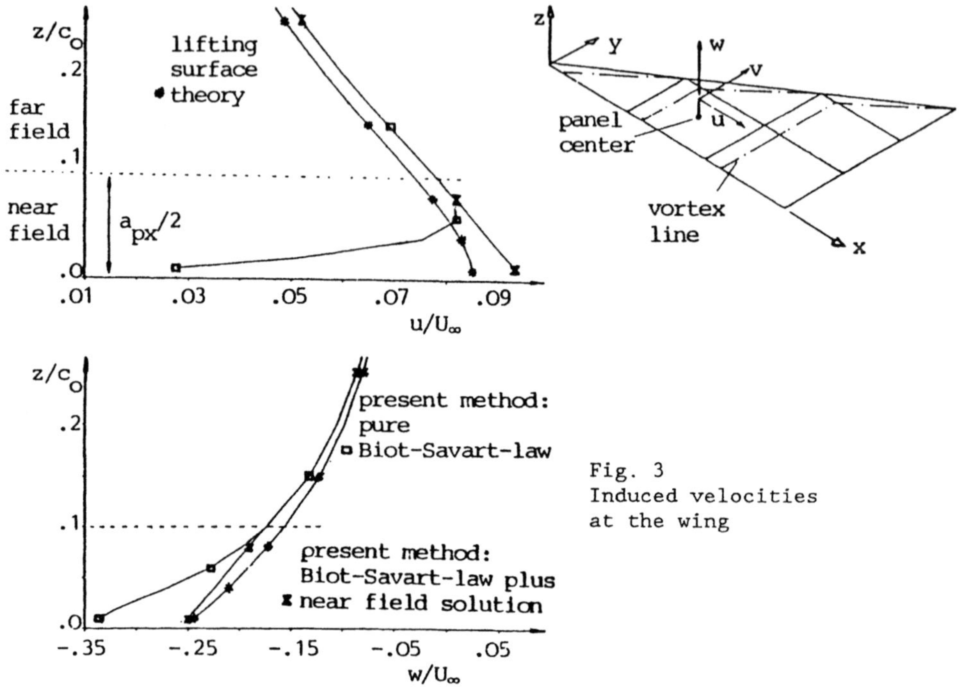

Fig. 3
Induced velocities at the wing

SOLUTION

After each time step a system of linear equations for the wing-bound doublet stengths μ_{Wi} (t) is derived by demanding the kinematic flow condition at the wing control points i:

$$[\vec{C}_{Wij} \cdot \vec{n}_{Wi}]\{\mu_{Wi}(t)\} = -\vec{U}_\infty \, \vec{n}_{Wi} - [\vec{C}_{Vik}(t) \cdot \vec{n}_{Wi}]\{\mu_{Wk}\}. \quad (3)$$

With i,j = $1 \ldots N_W$, k = $1 \ldots M_V(t)$ and the unit influence coefficients $\vec{C}$. At the time t = 0 free shear layers do not exist; therefore the starting solution corresponds to a displacement flow at the wing.

WING LOADS

The normal force coefficients of the wing are calculated by

$$\Delta \, c_{pW} \quad \frac{2}{U_\infty^2} \left[\frac{\partial \, \Gamma_W}{\partial \, t} + \nabla . \Gamma_W \left(\vec{U}_\infty + \nabla . \Phi_m \right) \right] ; \quad (4)$$

$\nabla.\Phi_m$ contains the induced velocities of the free vortex sheets and an additional wing (if present). In attached flow the thrust coefficient - and thereby the induced drag coefficient - can be approximated by

$$c_t\ (\eta) = \frac{\sqrt{1+\beta^2\cot^2\phi_{LE}}}{2\pi\ \cot\phi_{LE}} \quad [\ c_n\ (\eta) + 4\ c_m\ (\eta)\ ]^2 \tag{5}$$

with $\beta = \sqrt{1-M_\infty^2}$.

RESULTS

Wings without vortex separation

Compared with a plane wake the induction of a free wake leads to some variations in the flow field of the wing ($\Delta\ c_{p_{SE}} \rightarrow \infty$, $\partial\Gamma/\partial x_{TE} \neq 0$). The differences are insignificant concerning the overall aerodynamic wing loads, Fig. 4 . However, this is only valid for single wings in symmetric flow; in all other cases the free vortex sheets have to be considered.

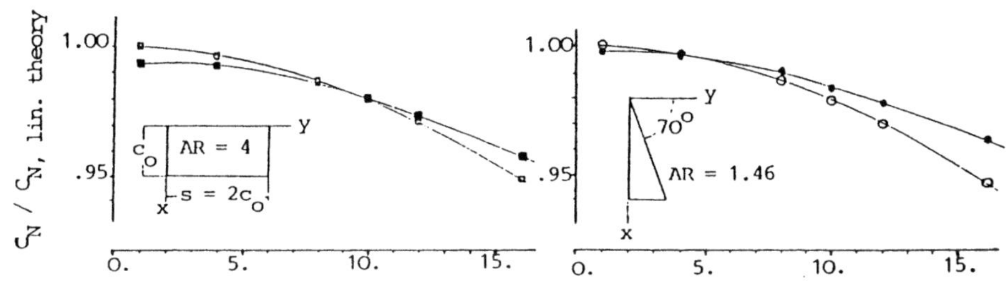

Fig. 4 Influence of plane wake and free wake on total normal force coefficient
□ ○ c_N plane wake (exact trigonometric functions)
■ ● c_N free wake (exact trigonometric functions)

An example for unsteady and steady interference effects at a delta-delta-configuration is shown in Fig. 5: The lift of the leading wing remains approximately constant for all times. When the starting vortex (or any similar shedded vorticity) of the leading wing is carried through the vicinity of the trailing wing, its lift is diminished simultaneously. In steady state, the lift values calculated with a free wake do not coincide with the results of the simple plane-wake model.
The second example refers to a wing in yawed flight, Fig. 6: If a free developement of the vortex sheets is admitted (model II), the pressure difference $\Delta\ c_p\ (\eta)$ is, compared with the plane-wake model I, significantly increased at the lagging side of the wing. At higher angles of attack, vortex separation at the leading side edge of the wing can be assumed; the resulting $\Delta\ c_p\ (\eta)$ is also added in Fig. 6 .

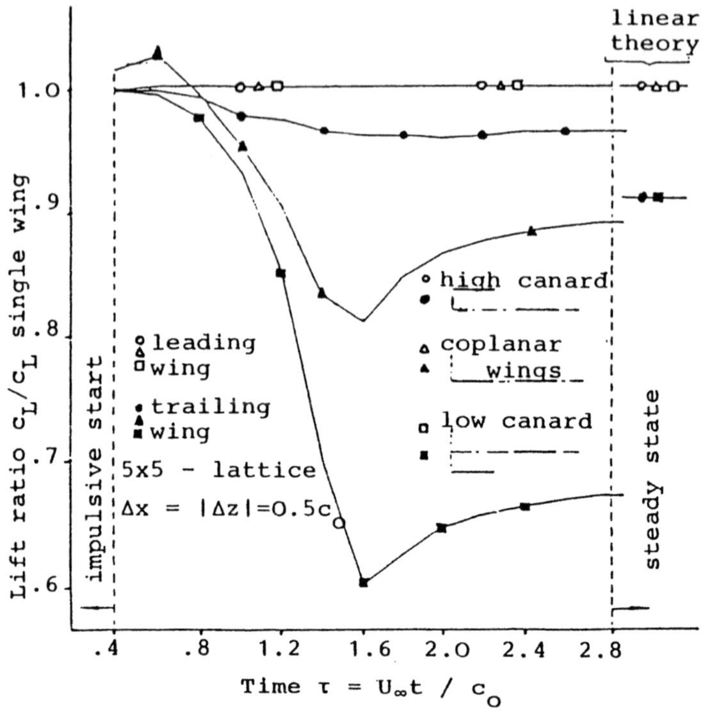

Fig. 5 Time-dependent developement of lift coefficients of interfering delta wings without leading-edge separation, AR = 1, α = 15°

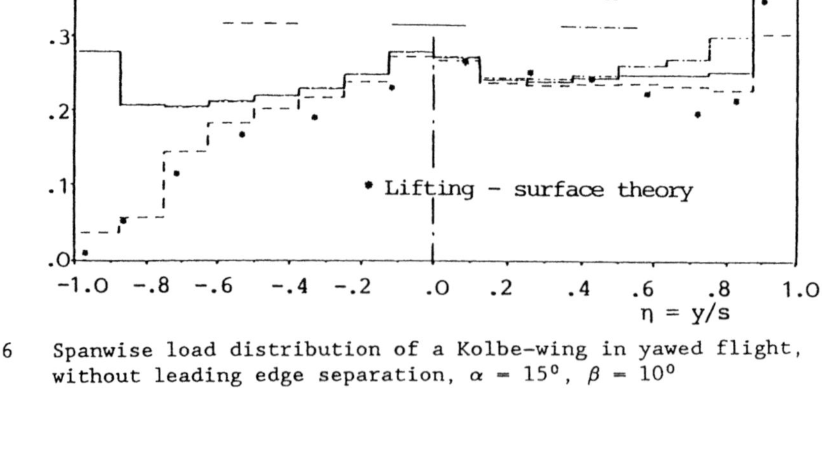

Fig. 6 Spanwise load distribution of a Kolbe-wing in yawed flight, without leading edge separation, α = 15°, β = 10°

Wings with vortex separation

Comparison with experimental results [3] of a 76°- delta wing proves that the present method reproduces the velocity field of the leading edge vortex with acceptable accuracy, Fig. 7 . At present, some deficiences in the near field of strongly bended element structures have still to be tolerated. Referring to the same wing, a compilation of results $\Delta c_p (\eta)$ gained by different panel methods is shown in Fig. 8 .

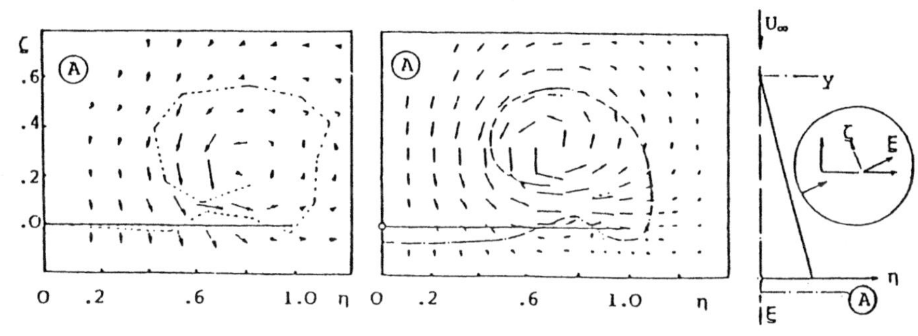

Fig. 7 Velocity-field and position of vortex-sheet behind a delta wing with leading-edge separation, $\alpha = 20{,}5°$

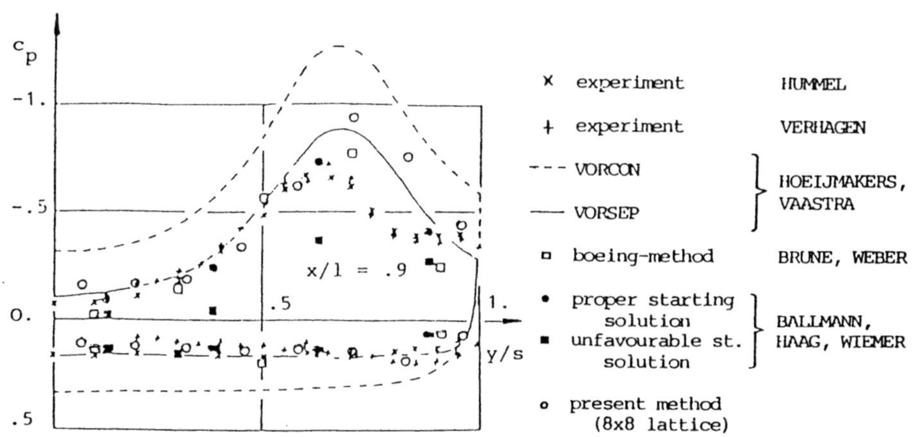

Fig. 8 Comparison of theoretical and experimental pressure distributions of a 76° - delta-wing, with leading edge vortex separation, $\alpha = 20{,}5°$

Vortex break down as a viscous and turbulent flow phenomenon limits the range of application of potential flow methods. In some cases, hints on a beginning vortex break down can be derived by the analysis of the inviscid pressure field inside the rolled up vortex sheets. The limiting suction pressure coefficient of the real flow is given by an empirical formula [4]:

$$c_{p_{lim}} = - \frac{2}{\varkappa M_\infty^2} \left[\frac{Re_\infty \cdot 10^{-6}}{Re_\infty \cdot 10^{-6} + 10^{(4-3M_\infty)}} \right]^{0,05 + 0,35(1-M_\infty)^2} . \qquad [6]$$

A comparison with experimental results [5], Fig. 9, shows, that vortex break down can be assumed, if the inviscid value of $c_p \leq c_{p_{lim}}$ in the vortex core. (Deviations of the experimental from the theoretical velocity distribution in Fig.9 result from a slightly different position of the vortex axis)

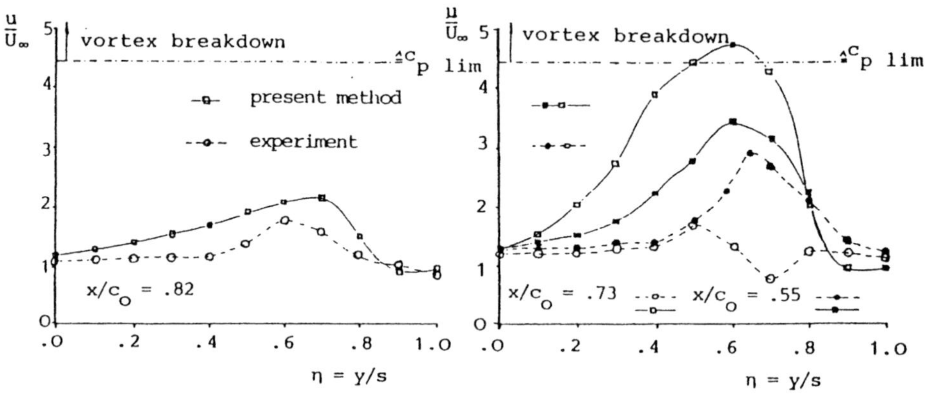

Fig. 9 Calculated and experimental [5] velocity distributions in the vortex core over a 70°-delta wing

Finally some results for delta-canard configurations are presented in Fig. 10 . In every case, the downwash of the canard diminishes the circulation of the main wing, also compare Fig. 5 . Note, that the developement of the leading edge vortex of the main wing is strongly disturbed in the case of the 60° - canard , Fig. 10 a . If the aspect ratio of the canard is reduced or if the distance between canard and main wing is increased as expected the characteristics of c_n (η) of the main wing are approching the values of a single delta wing.

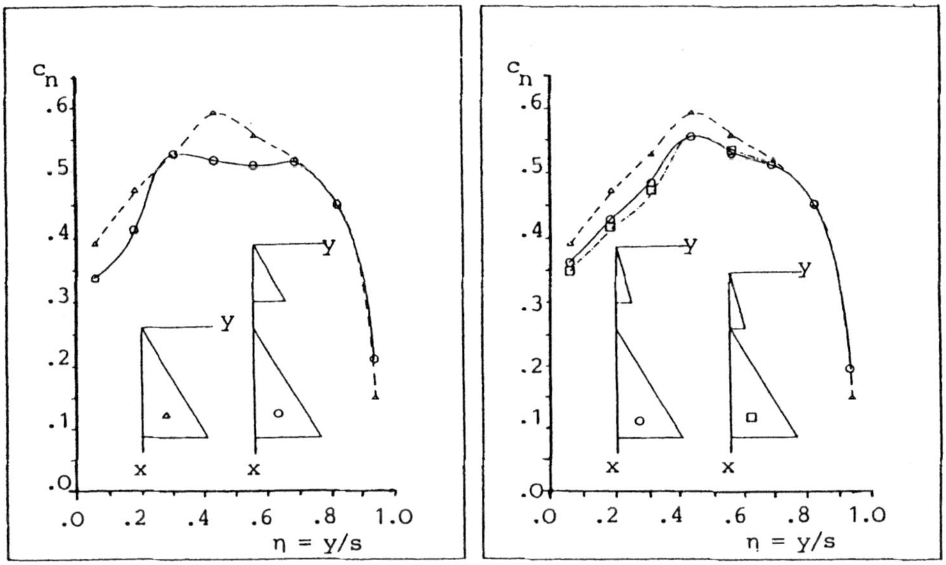

Fig. 10 Spanwise distribution of normal force coefficient at main wing depending on position and shape of the canard ($\alpha = 10^{o}$)

REFERENCES

[1] URBAN, C.; WAGNER, S.: "Ein Wirbelgitterverfahren zur Berechnung nichtlinearer, aerodynamischer Charakteristika von interferierenden Tragflächen". DGLR-Nr. 85-122, DGLR-Jahrestagung, Bonn-Bad Godesberg 1985.

[2] WAGNER, S.; URBAN, C.: "Current Activities in Basic Research Work on Panel Methods in Germany", in: Finite Approximations in Fluid Mechanics (E.H.Hirschel, ed.), Vol. 14 of Notes on Numerical Fluid Mechanics, Vieweg, Braunschweig/Wiesbaden, 1986, pp. 273-294.

[3] HUMMEL, D.: "On the Vortex Formation over a Slender Wing at Large Angles of Incidence", AGARD-CP-247, Oct. 1978.

[4] CARLSON, M.W.; MACK, R.J.: "Studies of Leading-Edge Thrust Phenomena", Journal of Aircraft, Vol. 17, No. 12, 1980.

[5] ANDERS, K.; WEDEMEYER, E.: "Messungen von Geschwindigkeitsverteilungen im Vorderkantenwirbel eines Deltaflügels mit dem Laser-Doppler-Verfahren", DGLR-Symposium, "Strömung mit Ablösung", Stuttgart 1981.

APPROXIMATION OF FREE AND BOUNDED VORTEX SHEETS AT DELTA-WINGS

P.Wiemer, K.Haag, J.Ballmann
RWTH Aachen, Lehr- und Forschungsgebiet Mechanik,
Templergraben 64, D-5100 Aachen, Germany

SUMMARY

The influence of the free vortex sheets on the pressure distribution of delta-wings is analysed with a higher order panel-method. The bounded vortex sheets are represented with isoparametric elements in order to get a continuous doublet distribution without singular vortex-lines for the wing itself. The free vortex layers, which detach already at small angles of attack from the sharp edges of the delta-wings, are paneled along the vortex-lines specified by the wing doublet distribution. With two independent geometry parameters for each panel as used in this "vortex-line adjusted panel-method" a successful modeling of the free vortex sheets is possible.

INTRODUCTION

The fluid flow detaches even at small angles of attack at the sharp edges of a delta wing and rolls up over the suction side of the wing. These free vortex layers in the vicinity of the wing have a great influence on the wing pressure distribution, generating a nonlinear increase of the normal-force with the angle of attack [1,2].

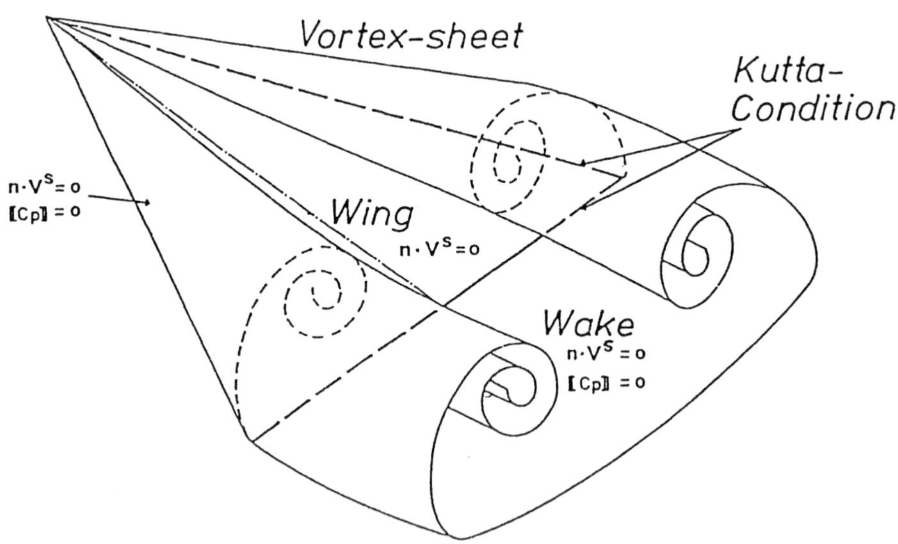

Figure 1: Delta wing with vortex-separation

SOLUTION OF THE PROBLEM

The study is based on the assumptions that the flow is stationary and the fluid is inviscid. The potential theory with bounded and free sheets of thin discontinuity is applied to the problem. Therefore the LAPLACE-equation
$\Delta\Phi = 0$ must be solved with the kinematic boundary condition
$nV_s = 0$ (no normal velocity at all vortex layers),
$[\![C_p]\!] = 0$ (no pressure jump at the free vortex sheets) and
the KUTTA-condition at the sharp edges of the wing.
We use the scalar potential $\Phi(P) = \iint\limits_{S_Q} \mu(Q)\, \nabla_Q \left(\frac{1}{4\pi R}\right) n(Q)\, dS$

with $V(P) = \text{grad}\ \Phi$, $V_s = 0.5*(V_u + V_l)$.

THE FREE VARIABLES

Unknowns in this model are the doublet distribution $\mu(Q)$ on the wing and the geometry of the free vortex sheets S_Q.
In order to prevent a priori approximation errors as far as possible for a suitable modeling of the physics, the following free variables are introduced:
a) The doublet distribution on the wing surface.
a1) For each panel (wing and free sheets) a quadratic doublet distribution in both local directions $e_{x'}, e_{y'}$ is used. A weighted least square fit is applied to determine the six free coefficients μ_1 up to μ_6 with 8 (9) discrete doublet values μ^*. The location of the discrete doublet values μ^* is shown for the wing in figure 2 and for the free vortex sheets in figure 3a).

$$\mu(x',y') = \mu_1 + \mu_2 x' + \mu_3 y' + 0.5*\mu_4 x'^2 + \mu_5 x'y' + 0.5*\mu_6 y'^2$$

$$\begin{pmatrix} \mu_1 \\ \mu_2 \\ \cdot \\ \cdot \\ \mu_6 \end{pmatrix} = C_{(6,M)} * \mu^*_{(M)}$$

$$M = 8\ (9)$$

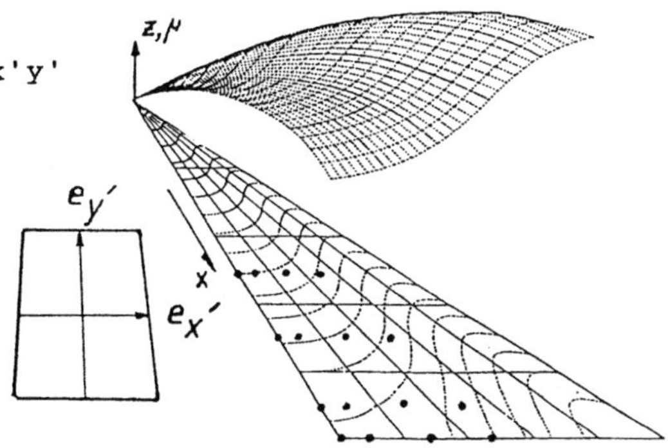

Figure 2: Doublet distribution (wing, half configuration)

a2) For the method with isoparametric elements the distribution is formed in the directions g_s, g_t with the parameter s and t [-1,+1], $g_s = \frac{\partial Q}{\partial s}$, $g_t = \frac{\partial Q}{\partial t}$, $Q = Q_0 + Q_s s + Q_t t + Q_{st} st$.

The 4 vectors Q_0 to Q_{st} follow from the 4 corner-points of the panels [3].

b) The geometry-variables of the "vortex-line adjusted panel model"

The circulation along the sharp edges of the wing is specified with the doublet distribution on the wing. This causes the strength of the circulation to be fixed for the complete vortex sheets, but the geometrical position of the free sheets, generated by lines of constant μ, is unknown. In order to describe the separated vortex-layers in a suitable manner, we paneled them along the lines of constant μ which coincide with vortex lines, see figure 3. As the independent geometry-variables for every panel two edge parameters are chosen: angles Θ_i and coordinates X_i or Y_i (for variable wakes), i panel index.

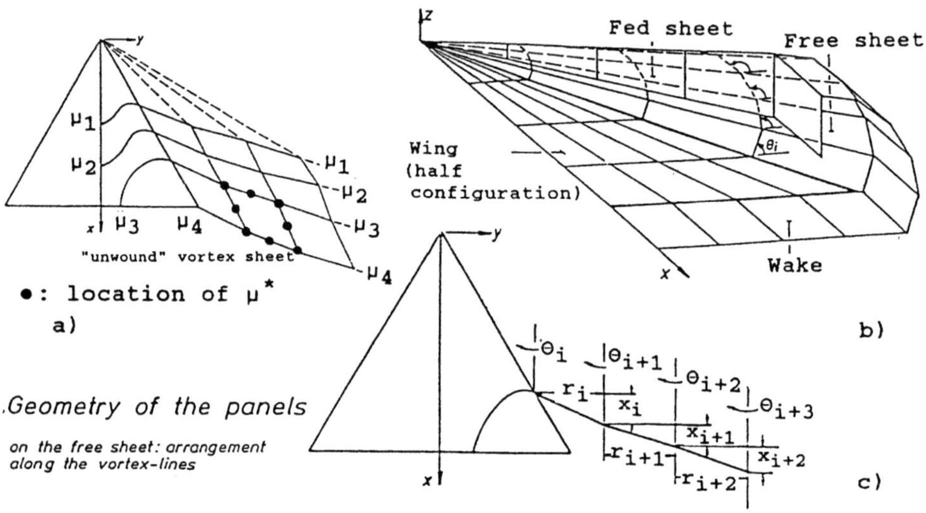

Figure 3: Vortex-line adjusted panel model

Because of the high number of independent geometry-variables in this "vortex-line adjusted panel model" a successful arrangement of the free vortex sheets is possible.

THE NUMERICAL PROCEDURE

It is not assumed to find an analytical solution of the integral equation for all field points on the surfaces, where the boundary conditions are to be satisfied. Therefore a collocation method is used.

All surfaces, the wing and the free vortex sheets, divided into panels have a control point placed in the panel-center, where the boundary condition shall be fulfilled. The equations for the boundary conditions are nonlinear, because of the unknown geometry of the free vortex sheets. This nonlinearity brings up a system of nonlinear equations for the discretized problem. This system is solved with a modified NEWTON algorithm.

A TAYLOR expansion of the nonlinear system of equations yields the correction vector $q^T=(\Theta_i, X_i, \mu_i)$:

$(\Delta q)_k=(-\sigma J^{-1} f)_k$; $(0<\sigma\leq 1)$ for the k-th iteration-step.

In order to realize a good convergence rate, great efforts were done to calculate the boundary conditions f analytically as well as the complete JACOBIAN. Exact calculations were possible for the flat panel approximation.
When the boundary conditions are solved within a certain computer time, one needs additionally only about twice this time to calculate the integrals for the JACOBIAN too, using a recurrence formula to calculate the integrals. On the other hand: with the exact calculated JACOBIAN the samplesize of converging calculations and the rate of convergence were improved [4].

RESULTS

Figure 5 shows a qualitative result of the iteration process. The sequence of figures shows the variation of the pressure distribution and of the geometry from the starting solution up to the 5-th iteration step. After 5 iteration steps the stable solution is reached. The residuals are smaller than 10^{-3} of the free streem velocity and the free variables of the model change not any more significantly in further iteration steps.
Some quantitative results of the rate of convergence are shown in figure 4 for examples with a homogeneous discretization.

Delta-wing

$$\bar{R} = \left[\frac{\sum_{i=1}^{N+M} (\underline{n}\underline{V}^S)_i^2 + \sum_{j=1}^{M} (\Delta c_p)_j^2}{N+2M} \right]^{1/2}$$

N: Number of wing-panels
M: Number of free sheet panels

—·—·— $N=9,\ M=9$
— — — $N=16,\ M=16$
——— $N=25,\ M=20$

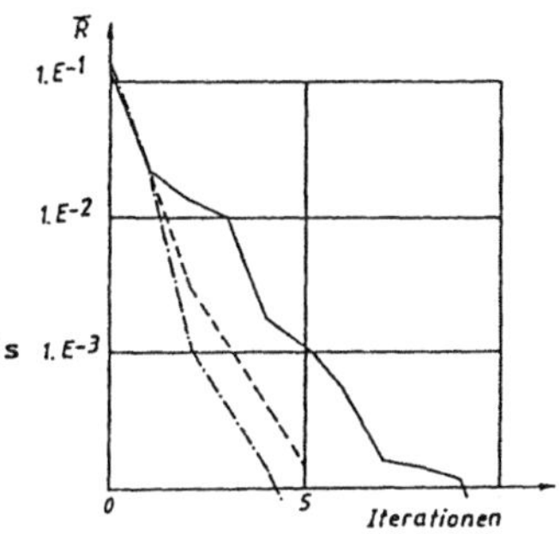

Figure 4: Mean residual error is plotted with respect to the number of iterations for three different numbers of control-points

It seems that the rate of convergence is better for a coarsed panelling than for a fine one. Analysing this problem we assumed, that the reduction of the convergence rate may be a local effect. The region around the downstreem corner of the wing's trailing edge is the starting point of the reduced convergence, accompanied with strong geometrical variations of the free sheets in many cases.
There are physical and mathematical reasons which support this behavior. The physical reason: secondary-separation takes place near the corner of the trailing edge and the vortex sheets of the free sheet and of the wake become closer to each other and are rolled up very strongly. The mathematical reason: in this region the gradient of the doublet distribution becomes zero and the existence of a coherent vortex sheet is no longer guaranteed, since the vorticity vector changes its direction. This is consistent with the physical observations.

The JACOBIAN becomes singular. In some cases, with the use of a NEWTON-factor $\sigma \leq 1$, the "overshooting" of the correction-vector could be prevented, but the rate of convergence became very small. A simple method to reach an acceptable convergence-rate in an acceptable computation time for fine meshed geometry may be the use an extended free-sheet-wake from the corner of the wing's trailing edge. This model is shown with an example for the "vortex-flow-model" in figure 7. The last strip of the free vortex sheets is part of the fixed wake, without free geometry-parameters.
Quantitative results for the delta wing with an aspect ratio AR=1 and an angle of attack $\alpha=20°$ are shown in figure 6 and for the "vortex-flow-model" in figure 8. The agreement with the experiments [5,6] and with the comparable expensive calculation "VORSEP" [2] is very good. These results were found with an excellent convergence behaviour of the described method.

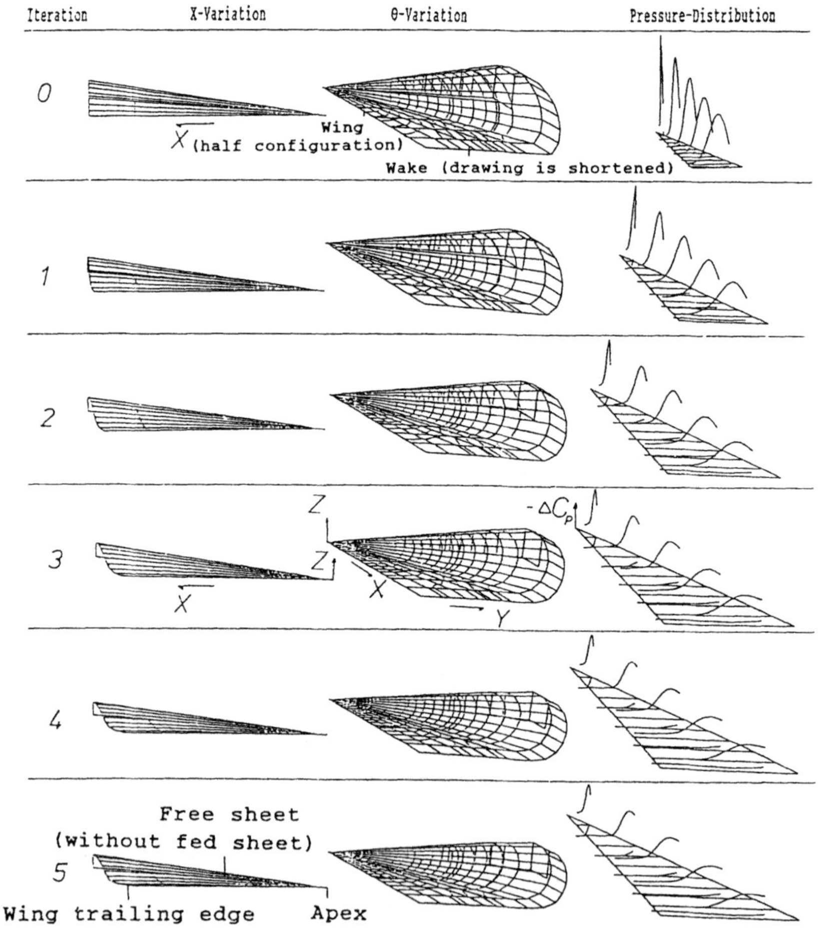

Fig. 5 : Iteration-study (Drawing of the wake is shortened)

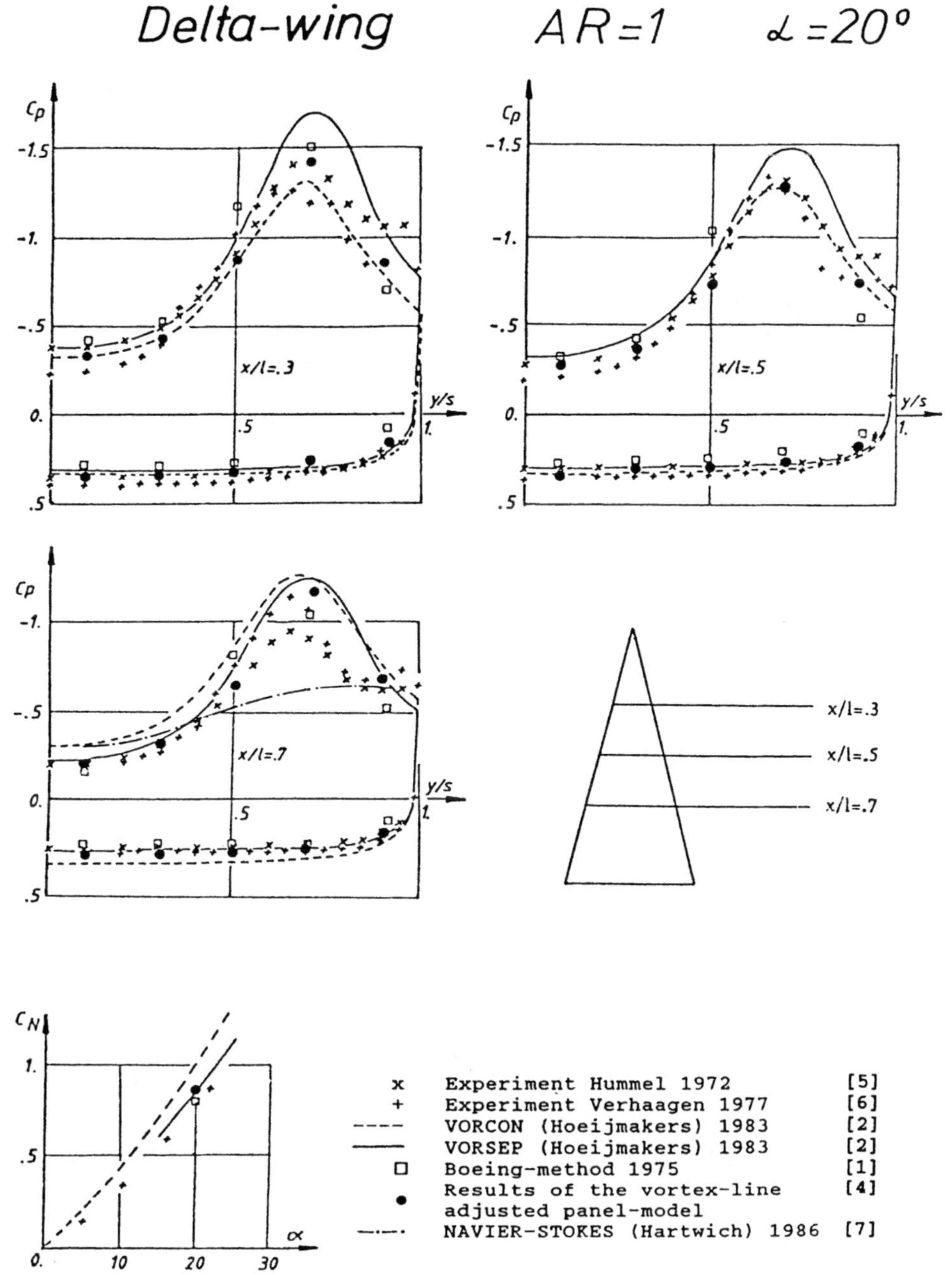

Fig. 6 : Vortex-line adjusted panel-model ;
Comparison of the pressure distribution and the normal force coefficients with other numerical methods and experiments.

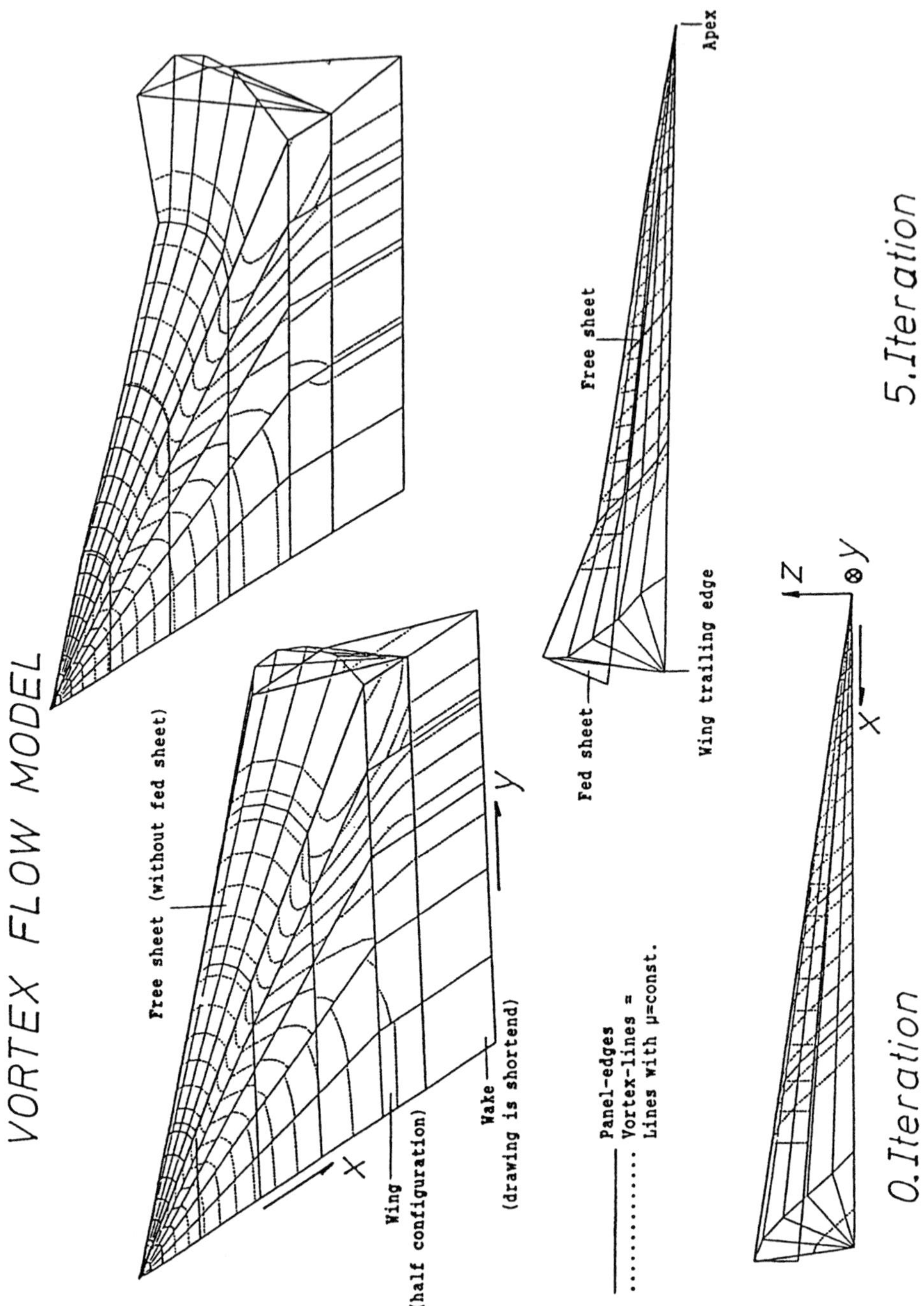

Fig. 7 : Variation of the free sheet ("Vortex Flow Model")

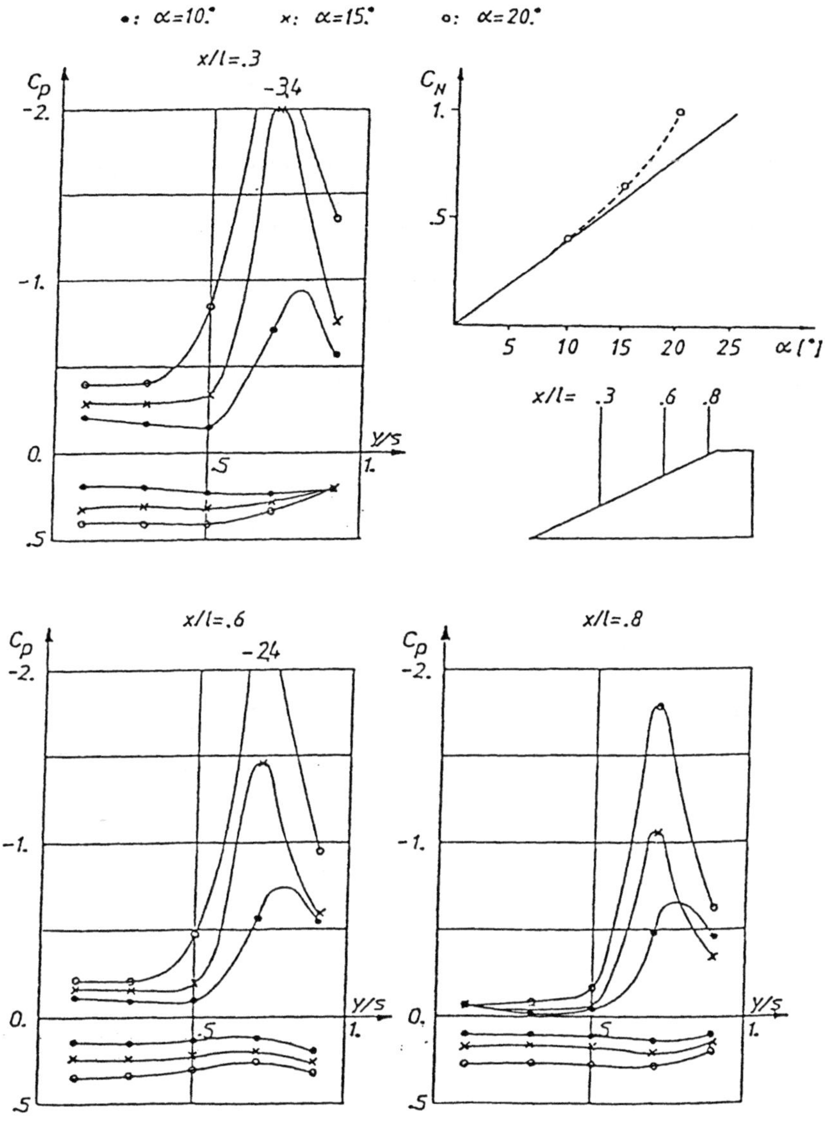

Fig. 8 : Calculations of the pressure distribution and the normal force coefficient for the "Vortex Flow Model"

REFERENCES

/1/ G.W.Brune, J.A.Weber, F.T.Johnson, P.Lu, P.E.Rubbert: A Three-Dimensional Solution of Flows over Wings with Leading-Edge Vortex Separation, NASA Cr 132709 Sep. 1975.

/2/ H.W.M. Hoeijmakers: Numerical Computation of Vortical Flow about Wings, National Aerospace Laboratory NLR, The Netherlands, NLR MP 83073 U (15.12.1983), Presented at: Lecture Series "Computational Fluid Dynamics", Rhode-St-Genese, Belgium, March 12-16, 1984.

/3/ K.Haag,P.Wiemer: Vorderkantenwirbel an Deltaflügeln mit gekrümmten Anströmkanten, SFB 25 "Wirbelströmungen in der Flugtechnik" RWTH Aachen, Forschungsbericht 1985.

/4/ P.Wiemer: Modellierung freier Wirbelschichten an Deltaflügeln mit einer wirbellinienangepaßten Panelmethode, Dissertation RWTH Aachen ,1987, submitted.

/5/ D.Hummel, G.Redeker: Experimentelle Bestimmung der gebundenen Wirbellinien sowie des Strömungsverlaufs in der Umgebung der Hinterkante eines schlanken Deltaflügels, Abhandlungen der Braunschweigischen Wissenschaflichen Gesellschaft, Vol.22, 1972.

/6/ N.G.Verhaagen: Measurment of the Pressure Distribution on a Biconvex Delta Wing of Aspekt Ratio 1, Dept. of Aeronautical Engeneering, Delft University of Technology, Delft, The Netherlands, unpublished report, 1977.

/7/ P.M.Hartwich, C.H.Hsu: An Implicit Flux-Difference Splitting Scheme for Three-Dimensional, Incompressible Navier-Stokes Solutions to Leading Edge Vortex Flows, AIAA 4th Applied Aerodynamics Conference June 9-11, 1986 San Diego, California (AIAA 86-1834-CP).

List of Participants

H. Antes, Prof. Dr., Mechanik II, Ruhr-Universität Bochum, Universitätsstr. 150, 463 Bochum 1
F. Arnold, Hermann-Foettinger-Institut, Techn. Universität, Str. des 17. Juni 135, 100 Berlin 12
J. Ballmann, Prof. Dr.-Ing., Mechanik, RWTH Aachen, Templergraben 64, 5100 Aachen
R. Behr, Inst. für Luftfahrttechnik, Hochschule der BW München, Werner-Heisenberg-Weg 39, 8014 Neubiberg
V. Bertram, Inst. für Schiffbau, Universität Hamburg, Lämmerrsieth 90, 2000 Hamburg 60
D. Braess, Prof. Dr., Mathematisches Institut, Ruhr-Universität Bochum, 4630 Bochum 1
E. Bruch, Ir., Institut du Génie Civil, Université, 6, Quai Banning, B-4000 Liège
A. Das, Prof. Dr., Inst. SM-EA, DFVLR Braunschweig, Flughafen, 3300 Braunschweig
Ehrmann, Dipl.-Math., MBB, Abt. TE 1, Hünefeldstr. 1- 5, 2800 Bremen
R. Eppler, Prof. Dr., Institut A für Mechanik, Universität Stuttgart, Pfaffenwaldring 9, 7000 Stuttgart 80
L. Fornasier, MBB Flugzeuge, 8000 München 80
J. Geissler, Dr.-Ing., Inst. für Aeroelastik, DFVLR-AVA, Bunsenstr. 10, 3400 Göttingen
S. Grilli, Dr. Ir., Institut du Génie Civil, Université, 6, Quai Banning, B-4000 Liège
K.-H. Haag, Mechanik, RWTH Aachen, Templergraben 64, 5100 Aachen
W. Hackbusch, Prof. Dr., Informatik, Christian-Albrechts-Universität, Olshausenstr. 40, 2300 Kiel 1
H.W.M. Hoeijmakers, Dr., NLR, POB 90502, NL-1006 BM Amsterdam
G. Hofmann, Dr., ITK Informationstechnologie Kiel GmbH, Holzkoppelweg 5, 2300 Kiel
D. Hummel, Prof. Dr.-Ing., Strömungsmechanik, Techn. Universität, Bienroder Weg 3, 3300 Braunschweig
Jakob, Dipl.-Math., MBB, Abt. TE 1, Hünefeldstr. 1-5, 2800 Bremen
G. Jensen, Inst. für Schiffbau, Universität Hamburg, Lämmersieth 90, 2000 Hamburg 60
O.A. Kandil, Prof. Dr., Dept of Mech. Engineering, Old Dominion University, Norfolk, VA 23508 - USA
E. Katzer, Dr.-Ing., Inst. für Aeroelastik, DFVLR-AVA, Bunsenstr. 10, 3400 Göttingen
J.A. Kolodziej, Dr.-inz., Techn. University, Inst. of Applied Mechanics, POB 5, PL-60965 Poznan

P. Kreuzer, Fachbereich Aerodynamic, Techn. Universität Darmstadt, Petersenstr. 30, 6100 Darmstadt

K. Kubrynski, ul. Boya 2 m. 28, PL 00-621 Warszawa

S. Lämmlein, Fachbereich Aerodynamic, Techn. Universität Darmstadt, Petersenstr. 30, 6100 Darmstadt

P. Loetstedt, TKLUB 2, SAAB Aircraft Div, SAAB-SCANIA AB. S - 58188 Linkoeping

D. Lohmann, Dr., DFVLR Braunschweig, Flughafen, 3300 Braunschweig

E. Martensen, Prof. Dr., Mathematisches Inst. II, Universität Karlsruhe, Englerstr. 2, 7500 Karlsruhe 1

B. Maskew, Dr. Analyt. Methods, Inc, 2133 152nd Ave.N.E., Redmond, WA 98052, USA

J. Meyer, FORD-Werke Köln, Computer Anwendungen MD/PN-50, John-Andrews-Entwicklungzentrum, 5000 Köln 71

Z.P. Nowak, Dr., 850 Bidwell, Room 203, Vancouver B.C., Canada V6G 2J8

H. C. Oelker, Strömungsmechanik, Techn. Universität, Bienroder Weg 3, 3300 Braunschweig

E. Ortiz, Prof. Dr., Dept. of Mathematics, Imperial College, Queen's Gate, London SW7 2BZ

R. Rannacher, Prof. Dr., Fachbereich Angewandte Mathematik und Informatik (10), Universität des Saarlandes, 6600 Saarbrücken

J. Ryan, ONERA, B.P. 72, F-92322 Chatillon Cedex

J. Saranen, Prof. Dr., Dept of Mathematics, Faculty of Techn. University of Oulu, Linnanmaa, SF-90570 Oulu

H. Schippers, Dr., National Lucht- en Ruimtevaart Laboratorium, Postbus 153, NL-8300 AD Emmeloord

Schmid-Goeller, Institut A für Mechanik, Universität Stuttgart, Pfaffenwaldring 9, 7000 Stuttgart 80

J. Schoene, Entwurfsaerodynamik, DFVLR Braunschweig, Flughafen, 3300 Braunschweig

E. Stein, Prof. Dr.-Ing, Institut für Baumechanik und Numerische Mechanik, Universität Hannover, Callinstr. 32, 3000 Hannover

B. Thomas, Dr., SUPRENUM, Hohe Str. 73, 5300 Bonn 1

R. Verführth, Prof. Dr., Inst. f. Angew. Mathematik, Universität Heidelberg, Im Neuenheimer Feld 294, 6900 Heidelberg

R. Voss, Inst. für Aeroelastik, DFVLR-AVA, Bunsenstr. 10, 3400 Goettingen

S. Wagner, Prof. Dr.-Ing., Inst. f. Luftfahrttechnik, Hochschule der BW München, Werner-Heisenberg-Weg 39, 8014 Neubiberg

W. Wendland, Prof. Dr., Mathematisches Inst. A, Universität Stuttgart, Pfaffenwaldring 57, 7000 Stuttgart 80

P. Wiemer, Mechanik, RWTH Aachen, Templergraben 64, 5100 Aachen

Addresses of the editors of the series „Notes on Numerical Fluid Mechanics":

Prof. Dr. *Ernst Heinrich Hirschel*
Herzog-Heinrich-Weg 6
D-8011 Zorneding
FRG

Prof. Dr. *Keith William Morton*
Oxford University
Numerical Analysis Group
8–11 Keble Road
Oxford OX1 3QD
Great Britain

Prof. Dr. *Earll M. Murman*
Department of Aeronautics and Astronautics
Massachusetts Institut of Technology (M.I.T.)
Cambridge, MA 02139
U.S.A.

Prof. Dr. *Maurizio Pandolfi*
Dipartimento di Ingegneria Aeronautica e Spaziale
Polytechnico di Torino
Corsa Duca Degli Abruzzi, 24
I-10129 Torino
Italy

Prof. Dr. *Arthur Rizzi*
FFA Stockholm
Box 11021
S-16111 Bromma 11
Sweden

Dr. *Bernard Roux*
Institut de Mécanique des Fluides
Laboratoire Associé au C.R.N.S. LA 03
1, Rue Honnorat
F-13003 Marseille
France

GPSR Compliance

The European Union's (EU) General Product Safety Regulation (GPSR) is a set of rules that requires consumer products to be safe and our obligations to ensure this.

If you have any concerns about our products, you can contact us on ProductSafety@springernature.com

In case Publisher is established outside the EU, the EU authorized representative is:

Springer Nature Customer Service Center GmbH
Europaplatz 3
69115 Heidelberg, Germany

Batch number: 10577876

Printed by Printforce, the Netherlands